Birkhäuser

Trends in the History of Science

Trends in the History of Science is a series devoted to the publication of volumes arising from workshops and conferences in all areas of current research in the history of science, primarily with a focus on the history of mathematics, physics, and their applications. Its aim is to make current developments available to the community as rapidly as possible without compromising quality, and to archive those developments for reference purposes. Proposals for volumes can be submitted using the online book project submission form at our website www.birkhauser-science.com.

Andrea Verdugo Rohrer · Joerg Zender
Editors

History of Mathematics and Its Contexts

Essays in Honor of Gert Schubring

Birkhäuser

Editors
Andrea Verdugo Rohrer
Departamento de Modelagem Computacional
Rio de Janeiro State University
Rio de Janeiro, Brazil

Joerg Zender
Department of Education and Digital Learning
RheinMain University of Applied Sciences
Wiesbaden, Germany

ISSN 2297-2951 ISSN 2297-296X (electronic)
Trends in the History of Science
ISBN 978-3-031-60827-8 ISBN 978-3-031-60828-5 (eBook)
https://doi.org/10.1007/978-3-031-60828-5

This book is published under the imprint Birkhäuser, www.birkhauser-science.com by the registered company Springer Nature Switzerland AG
The registered company address is: Gewerbestrasse 11, 6330 Cham, Switzerland

Prof. Dr. Gert Schubring

Universität Koblenz/Philipp Stieffenhofer

Preface

It always seems impossible until it's done.

Nelson Mandela

Editing a book is always a challenging task, but this one I took with joy since it collected essays in honour of Gert Schubring. It all started with a video call between Andrea and me around January in 2023, a year before Gert's 80th birthday. His birthday is on the 22nd of February, a date that I easily remember, since I was a boy scout for many years and it is the birthday of Baden-Powell, founder of the Scout Movement. Happily, this is something Gert and I share. I was with the boy scouts, and Gert was with the *Wandervogel*, a German youth movement that considerably influenced the later German Scouts.

In fact, we share much more than that. For instance, I always remember our seminar on the presentation of mathematical topics, held in 2009 at Bielefeld, in collaboration with the Faculty of Design and Art of the University of Applied Sciences Bielefeld. The students produced beautiful posters on various subjects that were later presented in the Faculty of Mathematics. Gert was our consultant and gave us a lot of advice for all topics. In the end, he also created a poster series on Sona sand drawings from Angola.

Furthermore, we share a passion for mathematics, its history, its education, and even the history of mathematics education. This book is all about the passion for the history of mathematics and mathematics education. It was an honour for me to work with so many people that authored articles and books I have read. Everyone contributed with, not only a piece of their work, but also their time to write, review, and, finally the last stage, help with the final edits of their articles.

I am thankful to have had the opportunity to edit this book for Gert and to have contacted so many fine people. I specially thank Frida Trotter from Birkhäuser for supporting this project. And last, but not least, I am grateful to have worked with Andrea, whom I haven't seen for 13 years since she finished her doctorate in Bielefeld, this book would not have been possible without her.

Wiesbaden, Germany Joerg Zender

Contents

Introduction

Andrea Verdugo Rohrer

Gert Schubring received in 2019 the Hans Freudenthal Award in recognition of his outstanding contributions, which have strongly influenced mathematics education research internationally. His research programme on the history and epistemology of mathematics has proven to be profound, novel and sustainable.

This book, published on the occasion of Gert Schubring's 80th birthday, brings together a collection of articles by scholars who have contributed to the development of the history and epistemology of mathematics. A *Festschrift*, a European tradition originating in Germany, honours special achievements in science and culture. We are more than flattered to be able to honour Gert Schubring in his lifetime.

This *Festschrift* compiles 16 essays from Gert's lifelong colleagues, students and fellows. It starts with Luis Carlos Arboleda and Jhon Bello analysing the mathematical practice with diagrams developed by van Schooten, when he edited and commented on the 2nd edition of *Descartes' Geometry*, translated from Latin in 1637.

Considering a gender approach within the history of mathematics education, Ausejo studies the impact of the Spanish Enlightenment and, later on, the Spanish Civil War on the opportunities women had to study higher education mathematics and to research mathematics.

Claudio Bernardi and Marta Menghini discuss and compare the intensive teaching of advanced courses for teacher training with the extensive teaching of elementary mathematics courses, based on Felix Klein's book *Elementary Mathematics from a higher standpoint*, translated to English by Gert Schubring.

A. V. Rohrer (✉)
State University of Rio de Janeiro, Nova Friburgo, Brazil
e-mail: ursula@iprj.uerj.br

A. Verdugo Rohrer and J. Zender (eds.), *History of Mathematics and Its Contexts*, Trends in the History of Science, https://doi.org/10.1007/978-3-031-60828-5_1

Mónica Blanco discusses the difficulty of teaching Leibnizian calculus through Guillaume de L'Hôpital's *Analyse des infiniment petits* and how Aimé-Henri Paulian's commentaries were appended to the third edition of *Analyse*.

Since 1998, Brazil's national government assesses elementary, middle and high school textbooks via a specific programme abbreviated to PNLD. João Bosco Pitombeira studies how this programme has affected the academic production of dissertations and theses.

Joseph W. Dauben's chapter brings to attention how mathematics and pedagogy have featured significantly in Chinese culture from the most ancient times.

Sergey S. Demidov presents in his paper how two mathematical communities, namely the Moscow and the Petersburg school, were confrontational, until Stalin reformed Soviet Science as a pyramid controlled by its headquarters. The Soviet mathematical school was conceived after the two communities had to search for a common language to unify their ideas.

The article of Menso Folkerts presents the prehistory and history of the foundation of the mathematics–physics seminars at the University of Göttingen in 1850, considering the influence of Carl Friedrich Gauss' radical remarks.

Christian Gilain examines d'Alembert's general conception of mathematics, disagreeing with the utilitarian view usually presented in historiography, and arguing that mathematics could play an important role in the emancipation of society, as he wrote in the *Encyclopédie*.

In forgetting past science—reconstructing past science, Jens Høyrup pointed out that mathematicians have traditionally a strong interest in its history, which is also what has occupied Gert Schubring's attention for half a century.

Vatslav Mrocheck is the main subject of Alexander Karp's paper, acknowledging him as a promoter of radical methodological ideas during the pre-revolutionary Russian mathematics education.

Vinicius Couto Pereira highlights the importance of the criticisms Gert Schubring states on the traditional concept of transmission since, on the one hand, it neglects that the transmitted concept is no longer identical to the original because it enters a different culture. On the other hand, this transmission conceives historical developments of a certain science in terms of a "centrism." Gert Schubring has made several contributions on the effects of colonization and decolonization on the development of mathematics education.

Reinhard Siegmund-Schultze uses the personal examples of 16 German mathematicians who stayed in the country during the Nazi rule 1933–1945 to discuss various questions relevant to the development of mathematics and its applications under extreme political circumstances. Questions relating to anti-semitism among mathematicians, collaboration with, and resistance to the regime and international relations were raised. Post-war apologia and "coping with the past" after 1945 are also discussed.

Circe Mary Silva da Silva discusses how textbooks analysis shows us contents, teaching methods, textual documents or imagery and ideologies in a given geographic, cultural, social and historical context, considering Gert Schubring's programme social history of international mathematics education.

Peter Ullrich attempts to study the beginnings of Karl Weierstraß' academic career, considering Gert Schubring's findings that he almost failed his teacher exam. Ullrich also suggests that Weierstraß received a rejection on his attempt to publish in Berlin, due to vanity.

Finally, Gregg de Young analyses specific issues of *I-VI Elements* by Euclides, from 1824, edited by Thomas T. Thomason for schools from the East India Company. These particular issues can be understood within the context of the Orientalist-Anglicist controversy.

Due to one of Gert Schubring's research programmes, the historian of mathematics education attempts to determine how the teaching of mathematics was perceived at any given time period, what role it played, and, most importantly, why certain changes took place and how these were connected to the larger social and economic processes. The changes in the understanding of the tasks to be undertaken in this field of research echo the changes that have taken place in mathematics education as a whole, where the social aspect now receives far greater attention than ever before. In the history of mathematics education, these changes are unquestionably linked to the name Schubring.

In particular, Schubring advocated a comprehensive examination of ongoing processes, whereby the biographies of prominent figures in mathematics education were studied not merely as an amusing addendum to the discussion of the subjects comprised in their textbooks, for example, but as part of a single subject, which aids us understand the thought process behind the composition.

Within the framework and along the lines of the methodological principles he had formulated, Schubring conducted important research in the history of mathematics education in Germany. We highlight how his approach and programme are of great relevance, since regional and national processes are treated as part of a larger international one. The history of mathematics education is, hence, indebted to Schubring.

We invite you to read this book, that collects some of the many impacts Gert Schubring's research programmes have in our scientific community.

The Dissemination of Cartesian Ideas. An Analysis of Practice with Diagrams in Van Schooten's Reading of Descartes

Luis Carlos Arboleda and Jhon Helver Bello

Introduction

The first edition of *Geometry* in French was published as part of the *Discourse on Method* (Descartes 1637). Its diffusion was limited, among other reasons because the scientific documents of the time were then written in Latin. A few years later Frank van Schooten Jr., who had been the illustrator of the work, became interested in its mathematical content and conceived the idea of making a second edition in Latin. In addition to the translation, van Schooten included *Commentaries* with his own reflections and the conclusions of some of the disputes that Descartes' work had aroused among the mathematicians of the time, among them, Huygens, Roberval, and van Schooten himself. A study of these controversies is found in Dopper (2014). He also included some of the earliest productions in Cartesian geometry by his most prominent students: Florimondi de Beaune, Johannis Hudden, Henrici van Heuraet, and Johannis de Witt. Van Schooten collected and edited these materials with the firm purpose of promoting the study and theoretical development of Cartesian geometry.

In the present work, the first dimension of the scheme that Schubring (1987, 2023) proposes for the study of mathematical texts, the analysis of editions, is applied, specifically, the diagrams that van Schooten used in the commentaries of the work. We argue that the network of signs of the diagram mobilizes a whole way of thinking that allows us to move from geometric generalization to

L. C. Arboleda (✉)
Universidad del Valle, Cali, Colombia
e-mail: luis.carlos.arboleda@gmail.com

J. H. Bello
Universidad Distrital Francisco José de Caldas, Bogotá, Colombia
e-mail: jhbelloc@udistrital.edu.co

A. Verdugo Rohrer and J. Zender (eds.), *History of Mathematics and Its Contexts*, Trends in the History of Science, https://doi.org/10.1007/978-3-031-60828-5_2

the thematization of algebraic curves, as noted in van Schooten's *Commentaries*. In the mathematical practice of Descartes, diagrams are used from the beginning of *Geometry*, both to define operations represented in segments and to solve problems by means of the analytical method, study mechanical instruments, and solve equations. Diagrams are essential to understanding Cartesian practice. In van Schooten's explanations, on the other hand, diagrams appear with other functions. They are also a means of justifying properties of curves and identifying an object of a class designated by a symbolic expression. These new uses of diagrams involve new ways of reasoning and proceeding in mathematical practice that van Schooten recognized in Descartes' work.

Practice with Diagrams in Descartes

According to Schubring (1987, 2023), one of the challenges of the history of mathematics in terms of the study of its production and reproduction through texts is to recognize the interaction of the teaching and invention of theories in their respective sociocultural context (Schubring 2001). The period of production and reproduction of the new field of Cartesian geometry begins in 1637 and stabilizes with the introduction of calculus by Newton and Leibniz (O'Halloran 2005). *Geometry* began an era of profound transformations in mathematical practice. New relationships were established between arithmetic, algebra, and geometry based on principles of representation, justification, and logical rigor that distanced themselves from those established in antiquity. New approaches to geometry include the use of curves to solve problems, the classification of geometric curves from their symbolic representation, the study of the representation, and geometric character of curves by means of the construction of instruments other than the ruler and the compass, a form of representation of curves by means of Cartesian axes, and the constitution of a theory of curves for the study of their fundamental properties. Studies on these topics can be found in Bos (2001), Clarke (2005), Dopper (2014), Liu (2017), Macbeth (2014), Maronne (2007), Rabouin (2016, 2017).

In the analysis of the diagrams of the *Geometry,* it is observed that the geometric constructions are based on a new model of representation of the magnitude. Indeed, Descartes defines at the beginning of his work the algebraic operations between abstract magnitudes, which are interpreted as geometric constructions of segments whose relationship is supported in the Eudoxo-Euclidean theory of ratios and proportions. With the operations thus defined Descartes constructs methods to solve the equations obtained above, designs mechanical instruments for the drawing of curves, and introduces the tool of the Cartesian axes to represent curves. The first sentence of Book I is very significant in this regard:

> Any problem in geometry can easily be reduced to such terms that a knowledge of the lengths of straight line is sufficient for its construction. (Descartes 1954, p. 2)

A fundamental resource of Cartesian practice is the use of the analytical method of solving problems and geometric situations, which consists of six stages: enunciate what is given; enunciate with the same practice what is sought; from the same operative treatment to what is given and sought, an equation is deduced that functions as a principle of the given, analysis; the resolution of the equations; and the construction of the solution (Hintikka and Remes 1974, cited by Arboleda 2012, p. 2).

With the introduction of a practice of algebrization of geometry with the above characteristics, Descartes initiates a change in the semiotic nature of the problems of science.

The change that diagrams allow from the geometric mode of representation to the symbolic mode using algebraic calculations provides different ways to express a curve. However, the description of the generality of the curve is expressed in an equation that represents all curves that satisfy proportional relationships. The diagram plays a synthetic role in the presentation of the properties of the curve, and the principle of identity of Descartes' algebraic geometry allows the diagram to initiate a two-way relationship between the geometric and algebraic structure, which entails some invariance between the reasoning product of the elaboration of the diagram and the arguments that support the equations.

In this sense, we agree with Rabouin (2017) that it is not so much the symbolic language of algebra that marks the emergence of modern mathematics. What allows a new way of reasoning in mathematics is the reorientation of the relationships between segments of lines in natural language, to their symbolic representation as arbitrary quantities and the algebraic treatment of such relations.

In the development of the analytical method of problem-solving, Descartes resorts to the diagram at least in three moments: in assuming a solution of the problem, in the solution of equations by geometric means, and in the identification of the geometric place that represents an algebraic expression. In the first case, the diagram directs, organizes, and systematizes reasoning (Rabouin 2016). It allows the imagination of the solver to bring into play the connections and operations necessary to show points, segments, and relationships that solve the situation. In this sense, the guiding principle of Descartes' mathematization is mainly the practice and systematization of the imagination (Sepper 2016). Let us pause to comment on this idea.

Descartes affirms that in the understanding of those questions that do not deal with the corporeal, it is not necessary to appeal to the resource of the senses, memory, and imagination. This is what happens with the objects of geometry whose study mobilizes the imagination through their properties of extension, figure, and movement. It is true that pure understanding can assign an abstract meaning to extension, figure, number, surface, line, or unity. But, in Cartesian geometry this abstract sense does not work, since you always must appeal to the imagination to reflect on its objects:

> The lines of Cartesian geometry cannot be separated from material objects. This does not imply in any way that when reflecting on such lines, the whole of the determinations of the

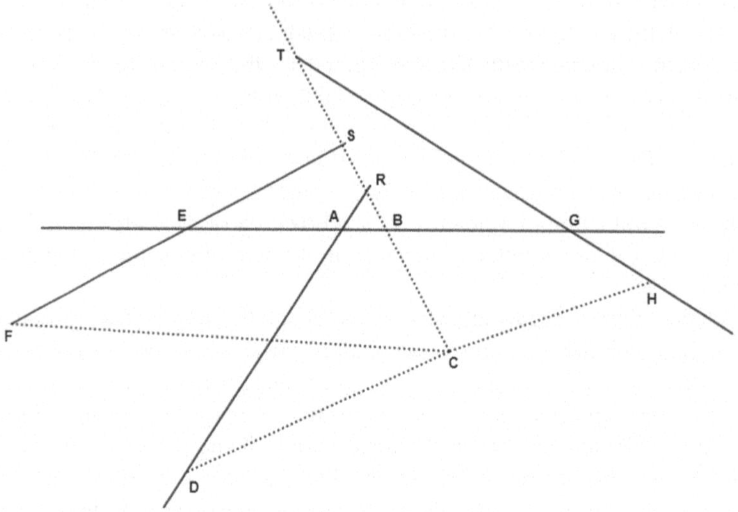

Fig. 1 Diagram solution of Pappus' problem (Descartes 1954, p. 26). Drawn by Bello

> bodies is covered; one can – or even must – focus attention on a particular mode of the thing, on one (or two) dimensions, making abstraction of the rest of its determinations. (Jullien 1996, p. 13)

For example, in the approach to the solution of Pappus' problem for four lines, the following diagram is posed (Fig. 1):

Where AB, AD, EF y GH are four lines given in one position, C is a point that belongs to the geometric place being searched, and the CB, CD, CF y CH are the segments that meet the proportionality condition $CB \times CF = CD \times CH$. The construction of the equation depends on the considerations made in the diagram.

In the case of Pappus' problem, as Maronne (2007) shows in detail, the situation of the point C in a given region of the drawing, the position of the auxiliary segments CB, CD, CF and CH, and the choice of a known line $AB = x$ and another unknown $CB = y$ to represent the points that are part of the geometric place, they condition the geometric relationships from which the different forms of the equations solution of the problem are constructed by mediating the algebraic operability. This problem is raised in Descartes' treatment of the signs of symbolic representation. That is, each line is represented symbolically depending on its position, for example: If R is between C y B, the segment $CR = y - \frac{bx}{z}$, and when C is between B y R, then $CR = -y + \frac{bx}{z}$ (Descartes 1954, p. 29).

The second case of recourse to the diagram occurs when Descartes analyzes the curve product of a collection of points generated by the interception of two curves of lower order. In these cases, the diagram configures a network of concepts that allows to study the object and its properties. Figure 2 shows the curve drawn by

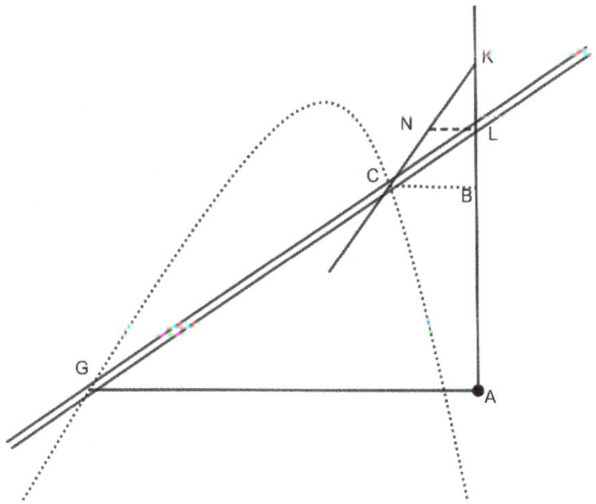

Fig. 2 Hyperbolograph (Descartes 1954, p. 51). Drawn by Bello

a hyperbolograph from the interception of two lines, according to the procedure original:

> Suppose the curve *EC*, to be described by the intersection the ruler *GL* and the rectilinear plane figure *CNKL* whose side *KN* is produced indefinitely in the direction of *C* and which, being moved in the same plane in such a way that its side *KL* always coincides with some part of the line *BA*, (produced in both directions), imparts to the ruler *GL* a rotary motion about *G* (the ruler being hinged to the figure *CNKL* at *L*). If I wish to find out to what class this curve belongs, I choose a straight line, as *AB* to which to refer all its points, and in *AB* I choose a point *A* at which to begin the investigation. (Descartes 1954, p. 51)

In this innovation of the artifacts arranged for geometry, the diagrams synthesize the properties of the curves, all the representations are hidden in the relationships that make the image possible, and from there, the symbolic representation of each element that composes it is structured. The diagram represents the curve and outlines the relationship that allows to move from the figurative register to the algebraic register, and two objects that are in different ontologies, the curve as a geometric object and as an algebraic equation, are described. Each point on the curve is described as the root of an equation when one of the two variables is fixed, allowing the points of the curve to be displayed without having to do the geometric construction for each point (Maronne 2007).

In the third case of diagrammatic practice, after obtaining the analytic principle represented in the equation, Descartes focuses on showing the type of curves that are solution. In the case of Pappus' problem for four lines, perform a construction to determine the relationship between the points that make up the solution and the lines taken as axes (Fig. 3).

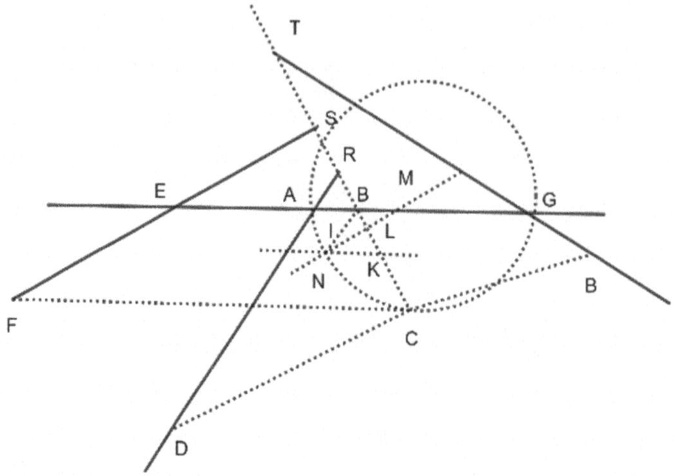

Fig. 3 Construction of a conic in the solution of the Pappus problem (Descartes 1954, p. 69). Drawn by Bello

Note that all dotted lines, including the points of the conic, are part of a strategy to show the relationship that each point of the circumference should have with the reference axes, but it does not allow to validate that all points are constructible. The position of the line depends on the conic that you want to analyze and its symbolic representation of the position with respect to the line. Descartes interprets BC as $y = m - \frac{n}{z}x + \sqrt{m^2 + ox - \frac{p}{m}x^2}$, reduced form of the general equation of second degree, where m, $\frac{n}{z}$, o, $\frac{p}{m}$ are given in terms of segments defined by the positions of the straight lines of the Pappus problem. Again Descartes warns the relationship between the position of the line segments in the diagram and the sign of the algebraic representation:

> I take the point K between L and C, since the equation contains $-\frac{n}{z}x$; if this were $+\frac{n}{z}x$, I should take L between K and C; while if $\frac{n}{z}x$ were equal to zero, I should not draw IL. (Descartes 1954, p. 310)

Descartes performs this same sign analysis for the term $\frac{p}{m}x^2$ in LC of the equation, where $LC = \sqrt{m^2 + ox + \frac{p}{m}x^2}$, which solves Pappus' problem for four lines, as conveniently studied by Dopper (2014, pp. 93–94). However, Descartes does not perform the process shown in the diagram for each case; the explanation of the solution curves is submerged in the analysis of the sign of the algebraic expression.

After advancing the process of generalization that starts from the diagram leads to the construction of the equation that represents the curve solution, Descartes goes on to deduce the curves determined by the algebraic expression. In this last part, the diagram is not essential as it was for obtaining the algebraic expression. This second analysis is therefore purely algebraic. On the other hand, Descartes leaves the procedure to determine and construct the conic solution from the equation of the geometric place in Pappus' problem, since he only makes this procedure explicit for the case of a conic, leaving in the hands of the reader the work of completing the analysis. As we will see in the next part of the article, it will be van Schooten who will perform the task of analyzing the different cases of conic solutions (Maronne 2007).

Van Schooten's Explanations in the *Commentaries*

In the preface to the second Latin edition, van Schooten states that the *Commentaries* and the writings accompanying the translation are intended to make the reader understand the content of *Geometry* and the importance of Descartes' method of analysis. This edition commented and expanded with other documents can be considered as part of a communicative strategy of van Schooten in the dissemination and teaching of Cartesian ideas, in particular, in terms of helping a keen reader to understand and complete the tasks that Descartes leaves posed in his work. In that sense, van Schooten's edition incorporates a clearly pedagogical function for the real understanding of Descartes' new algebraic geometry. This undoubtedly contributed to the wide dissemination and reception of this edition and its rapid appropriation in other domains of mathematical practice.

This is a historical example of how an explanatory text of a paradigmatic work can contribute to the consolidation and reproduction of a new mathematical practice and its extension in other directions. This issue is considered by Schubring (2019) when he criticizes the notion of didactic transposition of a text because he considers that it does not consider the eventual importance of the preparation of materials for teaching in the practice of mathematical production. Schubring (2001, 2019, 2023) indicates that, for its part, historical studies of mathematics must reveal the importance of teaching not only in the reproduction of mathematical culture, but also in the practice of mathematical production.

In this section, we review some moments of van Schooten'*s Commentaries* in which we consider that significant aspects of Cartesian practice are transformed or enhanced with diagrams. One such moment is Descartes' use of diagrams in solving Pappus' problem for four lines, specifically in the identification of the conics that represents the general equation of second degree. The problem facing van Schooten is to clarify the method followed by Descartes to analyze the equation and deduce the relationship between the conic, its symbolic expression, and the geometric construction of the diagram.

In the interval between the appearance of the 1637 edition and the elaboration of the *Commentaries* of the second edition from 1659 of the Latin translation, the mathematicians of the time, mainly Roberval and Huygens, study and question the position of each of the conics and the fact that Descartes presents only a part of the hyperbola. The controversies to which this gave rise are reconciled with the introduction of commentaries B and BB of the Latin version. The history of these controversies has been studied in detail in (Dopper 2014).

Van Schooten's BB commentary refers to the relationship that Descartes establishes between the position of the point C in the diagram of Pappus' problem for four lines (Fig. 1) and the angles at which the given lines meet. That is, in modern terms, it comprises the loss of generality in the determination of the equation of the geometric place solution of the problem, by restricting the position of C in a certain region of the plane. This restriction of the geometric space in which the solution occurs was one of the reasons for controversy of the time. Descartes argued that he had implicitly indicated how to extend his procedure to other situations, so that any thoughtful reader could do it on his own. In his letter to Mersenne of March 31, 1638, he states:

> I have the construction as architects build structures, giving the specifications and leaving the actual manual labor to carpenters and masons. (Descartes 1954, p. 63)

If examined in detail, Descartes' procedure consists of interpreting and studying the position of the figure in terms of the algebraic equation, with which the emergence of the analysis of the equation is separated from its geometric interpretation, leaving the reader the need to construct for each case the solution curve.

Van Schooten understands the importance of this intention of Descartes, and in his commentary, BB tries to clarify the procedure. To do this, he appeals to the problem posed by Huygens in his letter to Descartes of December 6, 1656, where the problem of the geometric situation is explained in terms of the analysis of signs of the equation.

The problem concerns the construction of a space d from the position of C about points A y B of a line. Van Schooten makes explicit the analogy of the solution with the reasoning of Pappus's problem in each of the cases of identification of the signs of the equation (Fig. 4).

The construction of d from the position of C relative to A y B models the condition of the signs of the algebraic expression in Pappus' problem with respect to the solution point. If C is to the left of AB and $CA = x$, then $xx + ax = d$. If C is between A y B, the equation is $ax - xx = d$; and finally, if C is to the right of

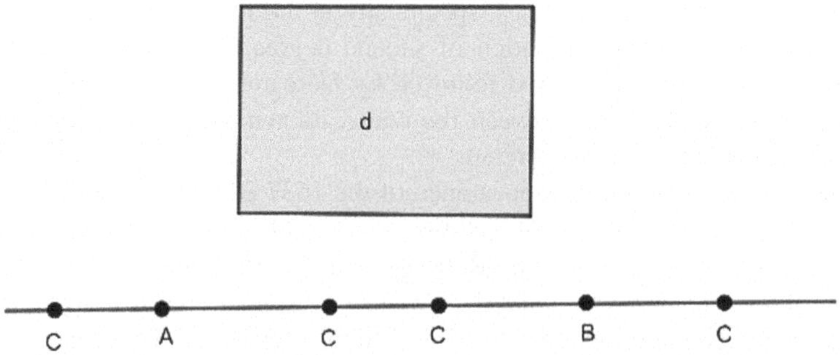

Fig. 4 Diagram associated with the BB comment example (Schooten 1659, p. 180). Drawn by Bello

B, then, the equation is $xx - ax = d$. Van Schooten uses a less complex problem in this commentary to explain the problem of signs in the symbolic representation of the equation.

This problem and its solution aim to recognize two significant moments of the Cartesian procedure. Understand the prefiguration of the use of oriented magnitudes that retain a specific value independent of the change of orientation and the adoption of the symbolic convention of alternation of signs to consider all possible cases in which the magnitudes are related. At the same time, this commentary BB by van Schooten underlines the importance of algebraic treatment in the clarification of a geometric problem.

The commentary CC of Book II takes up in part the discussions between van Schooten and Huygens on the algebraic analysis of the equation and the deduction from it of the solution curves of Pappus' problem. In total, 4 cases are studied for the parabola, 5 cases for the circle, 5 cases for the ellipse, and 10 cases for the hyperbola. In the 1637 edition, Descartes explains the procedure of construction and presents the conclusions regarding the relationship between the conic solution and the respective changes in the form of the equation given by, $y = m - \frac{n}{z}x + \sqrt{m^2 + ox - \frac{p}{m}x^2}$. However, it does not clarify the relationship between the points on each curve and the general equation. This is the purpose of the CC comment.

Van Schooten starts from Descartes' conclusion about the type of conic solution that follows from the equation when considering the term $LC = \sqrt{m^2 + ox - \frac{p}{m}x^2}$:

In particular, if the term $\frac{p}{m}x^2$ is zero, the conic section is a parabola; if it is preceded by a plus sign, it is a hyperbola; and, finally, if it is preceded by a minus sing, it is an ellipse. An exception occurs when a^2m is equal to pz^2 and the angle ILC is a right angle, in which case we get a circle instead of an ellipse. If the conic section is a parabola, its *latus rectum* is equal to $\frac{oz}{a}$ and its axis always lies along the line IL. To find its vertex, N, make IN equal to $\frac{am^2}{oz}$, so that the point I lies between L and N if m^2 is positive and ox is positive; and L lies between I and N if m^2 is positive and ox negative. It is impossible that m^2 should be negative when the terms are arranged as above. Finally, if m^2 is equal to zero, the points N and I must coincide (Descartes 1954, p. 68).

Van Schooten constructs for each case the position for the lines IL, IN y LK and identifies the position of each curve, then the *latus rectum,* the vertex, and the diameter of each curve as described by Descartes and displays its corresponding equation. These diagrams clarify the problem of change of sign that Descartes had raised and the nullity of some of the terms of the general equation. Figure 5 shows the diagram of the first case for the parabola used in the CC commentary.

In the text accompanying this diagram van Schooten (1659) states: the section is a parabola in which the LC line exists; and the ordinates applied to the diameter are always on the IL line. The vertex N on the other side of the point L should be taken with respect to the point I, $LC = \sqrt{mm + ox}$, to find, the *latus rectum.*

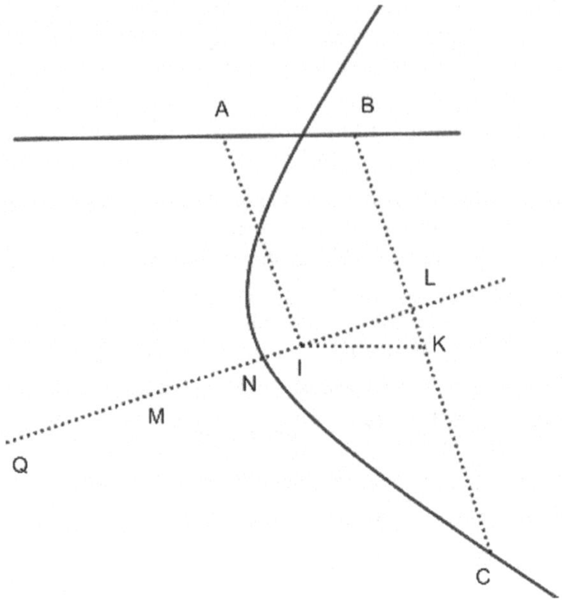

Fig. 5 First case of the conic solution of Pappus' problem (van Schooten 1659–1661, p. 182). Drawn by Bello

Van Schooten makes a definitive change with respect to the treatment of Descartes, in that his commentary aims to clarify the idea of curve and the relationship with its symbolic representation. At this point in the Cartesian practice, we move from constructing the curve to establishing the relationship between its known properties through an algebraic treatment. As indicated by Maronne (2007), this analysis is purely algebraic. The diagram is a reference for the situation, but it is not decisive in obtaining its symbolic expression.

For all cases, van Schooten shows the relative positions of the points N, L, I, C and line segments IK, IN, LC, LM, LN, explaining the statements Descartes made in *Geometry* about the relationship between the equation representing the problem and the segments being constructed in the diagram in Fig. 5.

In van Schooten's explanation of conics solution of Pappus' problem for four lines, the diagram is used as a conceptual network that allows to move from the point analysis of the curve to an algebraic expression that represents all the points of the curve and that allows to analyze its relative position from the use of the lines that are taken as axes. The 24 diagrams in this commentary are illustrative of the known characteristics of the curve, when taken as given.

These explanations of the conic solution of the Pappus problem continue in the CCC commentary, which is related to the following phrase of *Geometry*: it is therefore easy to determine this parabola, according to the first problem of the first book of Apollonius (Descartes 1954, p. 68). Van Schooten understood the methodological approach proposed by Descartes: to find the analytic principle of the curves that represent the solution to the problem and then show that every conic follows from the equation expressing this principle. The usual notion of conic at the time was that introduced by Apollonius, as a cut of a cone. In this commentary, van Schooten shows that the symbolic representation of the class of curves of Apollonius is equivalent to that deduced from the equation solution of Pappus' problem and, therefore, is identified with the geometric place of the solution points of such a problem.

In the introduction to this commentary, van Schooten indicates that his intention is for readers of *Geometry* to understand the solution and properties of conics, even if they do not have access to the book of Apollonius. Van Schooten took proposition 11 from Book I of *Conics* and interpreted the definition for each curve in the conceptual terms of Cartesian geometry. On the same diagram that Apollonius uses, he designates quantities of magnitude with a lowercase letter (AB becoming *a*, BC becoming *b*, etc.) and operates with them algebraically. The Euclidean diagram interpreted in the Cartesian double script is the reference support to show the equivalence between the figural representation of the curve and its algebraic representation (Bello 2021) (Fig. 6).

In the case of the parabola van Schooten observes that if $AB = BC$, the base of the cone is a circle of diameter AC, and the section KFL is also a circle parallel to the base. Names $AB = a, BC = b, AC = c, EB = d, EI = x, FI = y$, then as the triangle ABC is similar to KEI and to the triangle EBS, it deduces that $KIL = FI$ and, therefore, the relation that identifies the conic can be represented by $\frac{ccd}{ab}x = yy$ which is a second-degree curve.

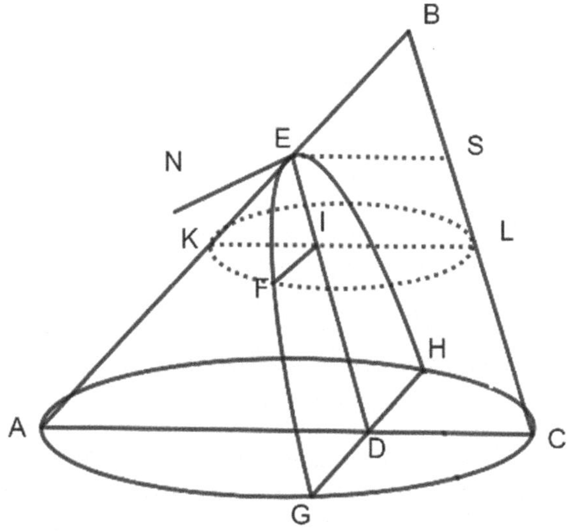

Fig. 6 Diagram used by van Schooten from Proposition 11 of Book I of Apollonius (Schooten 1659, p. 207). Drawn by Bello

Van Schooten's algebraic treatment of geometric construction is based on the Cartesian procedure of extending Euclidean magnitudes to the operative field of abstract magnitudes and using the Euclidean theory of ratios and proportions to determine the symbolic expression of each conic. This implies a whole understanding of the nature of the generalization of the notion of algebraic curve product of Cartesian analysis. Van Schooten's identification of the background of the thematization that Descartes introduced in Euclidean geometry turns in commitment in its dissemination and public explanation through the annotated edition. Another proof of this is the inclusion that he makes in the second edition (1659–1661) of two writings devoted to discussing the fundamental principles of Cartesian mathematics and the Cartesian method of solving problems based on the algebraic treatment of its interpretation by geometric constructions, *Principia Matheseos Universalis Seu Introductio Ad Geometriae Methodum Renati Des Cartes* (Schooten 1661a) in which he emphasizes the treatment of algebraic expressions and their operations. This text does not contain problems, nor diagrams, and the universal principles focus on the construction of a theory of symbolic expressions for the application of the Cartesian method. The second paper *Tractatus de Concinnandis Demostrationibus Geometricis ex Calculo Algebraico* (Schooten 1661b) presents the Cartesian method of problem-solving based on the parallelism between algebraic treatment and geometric construction.

Conclusions

In the present work, we have tried to show the interest of an analysis of historical texts, in this case the first editions of *Geometry*, to understand the complex processes of communication and formation of mathematical culture around a paradigmatic work, the theory of algebraic curves. We have seen that the study of this communicative function and mathematical education appears closely related in the beginnings of Cartesian practice with the main function of knowledge production. Van Schooten's communicative strategy regarding the dissemination and teaching of Cartesian ideas was aimed at satisfying the needs of a proactive reader to know the new geometry, but also to carry out the tasks of perfecting the method that Descartes had left in his work. The interrelation between the pedagogical function and the epistemological function in van Schooten's edition undoubtedly contributed to the wide dissemination and reception of this edition and its rapid appropriation in other domains of mathematical practice.

We have also shown that the study of diagrams in these editions provides significant elements to understand mathematical practice in a specific historical context. About the role of diagrams in the establishment of Cartesian geometry, the analysis of the *Commentaries* highlights the important role that van Schooten played in clarifying in his time the function of the diagram in the constitutive process of the algebraic curve from the geometric configuration of Pappus' problem. In modern words, we would say that the *Commentaries* explain with new elements the use of the diagram in *Geometry* and its function as a conceptual network that allows to move from the point analysis of the curve to an algebraic expression that represents all the points of the curve, through the characterization of its relative position from the use of the lines that are taken as axes.

Van Schooten as editor and commentator understood the background of the novelty of Descartes' method and the importance of directing readers of the Latin edition in that direction. We corroborate this in his careful presentation of the extension of Euclidean magnitudes to the operative field of abstract magnitudes, as a necessary condition for using the Euclidean theory of ratios and proportions in terms of a convenient symbolic representation and for constituting the algebraic curve as a radically new mathematical object. We have used the term *thematization* on several occasions to refer to this procedure.

This notion was introduced by Jean Cavaillès to distinguish the logical construction of theories from a simple generalization. As clarified in (Arboleda 2011): "thematization is the process by which an operation that has previously been performed on a field of objects, is the object of a second operation, which in turn becomes the object of a third operation, and so on". Paraphrasing the article, we can say that the specific art of Cartesian geometry is thematization. What we wanted to expose through this study of the editions of *Geometry* is that Descartes did not arrive at the object algebraic curve by abstractions of real objects through the description of their main characteristics, but through a complex practice of chaining logical operations on preconstituted mathematical domains.

References

Arboleda, Luis Carlos (2011). Objetividad matemática, historia y educación matemática. En Recalde, L. C. y G. I. Arbeláez (2011). *Los números reales como objeto matemático. Una perspectiva histórico-epistemológica.* Cali: Editorial Universidad del Valle; pp. 19–38.

Arboleda, Luis Carlos (2012). El análisis cartesiano en la solución del problema de Pappus y la introducción de las curvas algebraicas. *13° Encuentro Colombiano de Matemática Educativa*, 764–777. Medellín: Universidad de Medellin. http://asocolme.org/images/eventos/13/MATEMATICA_EDUCATIVA_13_Encuentro_Colombiano%20ECME.pdf

Bello, Jhon (2021). Diagramas y práctica matemática en la Geometría Cartesiana (1637–1750). Contribución de la historia de la matemática a la formación de profesores. Cali: Universidad del Valle. https://bibliotecadigital.univalle.edu.co/browse?type=author&value=Bello%20Chavez,%20Jhon%20Helver

Bos, Henk (2001). Redefining Geometrical Exactness: Descartes' Transformation of the Early Modern Concept of Construction. New York: Springer.

Clarke, Desmond (2005). Descartes' philosophy of science and the scientific revolution. Dans *The Cambridge Companion to Descartes* (éd. Décima, 258–285). Cambridge : Cambridge University Press.

Descartes, René (1637). La geometrie. Dans R. Descartes, Discours de la methode pour bien conduire sa raison et chercher la verite dans les sciences; plus La dioptrique Les meteores et La geometrie qui sont des essais de cete methode (pp. 297–413). Leiden.

Descartes, René (1659 - 1661). *Geometria a Renato Des Cartes anno 1637 gallice edita, postea autem...* (F. Schooten, Éd.) Apud Ludovicum & Danielem Elzevirios, Amstelædami: Apud Ludovicum & Danielem Elzevirios.

Descartes, René (1954). *The Geometry.* (D. Smith, & M. Latham, Trads.) New York: Dover Publication.

Dopper, Jantien (2014). A life of learning in Leyden. The Mathematician Frans van Schooten (1615–1660). Doctoral Dissertation, Utrecht University. https://dspace.library.uu.nl/handle/1874/288935

Hintikka, Jaakko & Remes, Unto (1974). *The Method of Analysis. Its Geometrical Origin an Its General Significance.* (Boston Studies in the Philosophy of Science, 25). Dordrecht & Boston: Reidel.

Jullien, Vincent (1996). *Descartes. La "Géométrie" de 1637.* Paris: Presses Universitaires de France.

Liu, Cathay (2017). Re-examining Descartes' Algebra and Geometry. *Analytic Philosophy, 58*(1), 29–57. https://doi.org/10.1111/phib.12093

Macbeth, Danielle (2014). Viète, Descartes, and the Emergence of Modern Mathematics. *Graduate Faculty Philosophy Journal, 25*(2), 87–117. https://doi.org/10.5840/gfpj200425212

Maronne, Sébastien (2007). La théorie des courbes et des équations dans la Géométrie cartésienne: 1637–1661. Université Paris - Diderot , Paris VII. Paris : *Histoire et perspectives sur les mathématiques [math.HO].* https://tel.archives-ouvertes.fr/tel-00204125v2

Rabouin, David *(2016).* Mathesis universalis et algèbre générale dans les Regulae ad directionem ingenii de Descartes. *Revue d'histoire des sciences, 69, 259–309.* https://doi.org/10.3917/rhs.692.0259

Rabouin, David (2017). Logic of imagination. Echoes of Cartesian epistemology in contemporary philosophy of mathematics and beyond. Synthese, 195:4751–4783. https://doi.org/10.1007/s11229-017-1562-1

Schooten, Frans van (1659). Commentarii. En F. Schooten (Ed.), Geometria à Renato Des Cartes (Vol. 1, págs. 143–400). Amsterdam: Apud Ludovicum & Danielem Elzevirios. Vol. 1 (1659), vol. 2 (1661). https://gallica.bnf.fr/ark:/12148/btv1b8601510z/f11.item

Schooten, Frans van (1661). Principia Matheseos Universalis seu Introductio Ad Geometriae Methodum Renati Des Cartes. En F. Schooten (Ed.), *Geometria à Renato Des Cartes* (Vol. 2, pp. 1–48). Amsterdam: Apud Ludovicum & Danielem Elzevirios.

Schooten, Frans van (1661a). Tractatus de Concinnandis Demostrationibus Geometricis ex Calculo Algebraico. En F. Schooten (Ed.), *Geometria à Renato Des Cartes.* (Vol. 2, pp. 341–420). Amsterdam: Apud Ludovicum & Danielem Elzevirios.

Schubring, Gert (1987). On the methodology of Analysing Historical Textbook: Lacroix as Textbook Autor. *For the Learning of Mathematics*, Vol. 7, (3), 41–51. https://www.jstor.org/stable/40247906

Schubring, Gert (2001). Production mathématique, enseignement et communication. Remarques sur la note de Bruno Belhoste, "Pour une réévaluation du rôle de l'enseignement dans l'histoire des mathématiques" parue dans la RHM 4 (1998), p. 289–304." *Revue d'histoire des mathématiques* 7.2 (2001): 297–308. http://eudml.org/doc/252108.

Schubring, Gert (2019). The Impact of Teaching Mathematics Upon the Development of Mathematical Practices. Dans G. Schubring (Éd.), *Interfaces between Mathematical Practices and Mathematical Education.* Springer. https://doi.org/10.1007/978-3-030-01617-3_5

Schubring, Gert (2023). *Analysing Historical Mathematics Textbooks.* Switzerland: Springer. https://doi.org/10.1007/978-3-031-17670-8

Sepper, Denis (2016). Descartes. Dans A. Kind (Éd.), *The Routledge handbook of philosophy of imagination* (pp. 27–39). New York: Routledge

Women's Access to Mathematical Research in Spain (1868–1936): The State of the Art

Elena Ausejo

Introduction

This paper starts in the eighteenth century, because education was one of the main interests in the time of the Spanish Enlightenment. Women's education was developed, although only up to a basic elementary level, and not considering women's independent professionalization. However, the mathematical skills of two single daughters of different bourgeois families were appreciated, either by the Catholic clergy or by military men, as the second section shows.

The third section deals with the long and difficult access of women to higher mathematics education in Spain, the organization of Sciences Faculties, and the rate of women‘s incorporation from 1910 onward, when their full access to the universities was regulated on equal terms with men.

Finally, the fourth and last sections explain the modest women's access to mathematical research in Spain, a project that the Spanish Civil War curtailed.

Two Women's Mathematics Textbooks in Eighteenth-Century Spain

María Andresa Casamayor de La Coma (1720–1780), mathematician and girls' teacher, was born into a wealthy family of textile merchants, when the French colony dominated the trade of Zaragoza and a large part of the kingdom of Aragon, forming a large population group with strong commercial and family relationships. In 1738, at the age of eighteen, she published the book *Tyrocinio Arithmetico* (Casamayor y de La Coma 1738) under the name of Casandro Mamés de la

E. Ausejo (✉)
University of Zaragoza, Zaragoza, Spain
e-mail: ichs@unizar.es

A. Verdugo Rohrer and J. Zender (eds.), *History of Mathematics and Its Contexts*, Trends in the History of Science, https://doi.org/10.1007/978-3-031-60828-5_3

Marca y Araioa, the anagram of her full name and surnames. This is the first preserved mathematics textbook written by a Spanish woman—one single copy in the Spanish National Library.

The *Tyrocinio* is a book on Arithmetic to teach the four rules, deliberately short to make it affordable, and with an eminently practical content aimed at trade. She handled with ease the units of weight, currency, and length that existed in Aragon and Castilla, as well as the intricate exchange operations between them, showing her great arithmetic ability. The censors of the book were her friend and collaborator Pedro Martínez (1664–1739)—the rector of the San Vicente Ferrer's School—and Juan Francisco de Jesús—a Piarist and professor of mathematics.

The main supporters of the young María Andresa gradually disappeared in a few years. Her father died in 1738 and Pedro Martínez in 1739. Her maternal family went into debt in 1740 and lost all their real estate. The rest of her life she had to work for a living. She worked as a teacher for girls and, for a long part of her life, as a teacher of first letters in the public schools of the city, which provided her with a house in which to live as part of her salary as a teacher (Bernués and Miana 2019).

The female author of the second textbook in eighteenth-century Spain was the Italian mathematician María Gaetana Agnesi (1718–1799), the author of *Instituzioni Analitiche* (Agnesi 1748), a book in two volumes. The second dealt with differential and integral calculus, a most innovative aspect in the field of mathematics in that century. The book was welcomed in Italy, translated into French in 1775, and into English in 1801. As for Spain, it is well known that Agnesi's book was purchased for the Military Mathematical Society, founded in 1755, for it was recorded in the inventory of the Society library, dated 1761 (Cuesta Dutari 1985, pp. 218–226). However, its use and influence in Spain was unknown.

In 2021, the study of an unpublished, until then unexplored manuscript dated in 1775 was published. It was the course on Integral and Differential Calculus at the Royal College of Artillery of Segovia—dictated by Professor Isidoro Gómez between December 5, 1775, and April 12, 1776, and written down by Cadet Martín García Loygorri. The author was Cipriano Vimercati, mathematician and First Professor at the College, and Agnesi's influence, among other authors, was proved (Velamazán 2021).

Women's Access to Higher Mathematics Education in Spain

The Royal Decree of the Ministry of the Government of June 8, 1843, established a Faculty of Philosophy in Madrid with the same rank as the other major faculties and job expectations in teaching (Ausejo 2007, p. 510). This law defined the degrees of bachelor's, master's, and doctor in Sciences, which were included in September 17, 1845, in the Royal Decree of the Ministry of the Government on the General Study (Ausejo 2007, p. 514).

In September 9, 1857, the Law of Public Instruction founded the Faculties of Sciences, first organized in three sections—physics-mathematics, chemistry, and natural sciences. One year later, the sections became Exact, Physical, and Natural Sciences by the regulations of November 7, 1858. The sciences faculties awarded the degrees of bachelor's, master's, and doctor in Science with a common syllabus for the bachelor's degree and specialized sections syllabi for the master's and doctor degrees. However, by 1860, the only comprehensive university capable of teaching all three degrees was that of Madrid, while another six (Barcelona, Granada, Santiago, Sevilla, Valencia, and Valladolid) could only teach up to a bachelor degree.

Women's exclusion from Spanish universities was implicit, but not forbidden by law—a proof of their social invisibility. However, during the Revolutionary Sexennium (1868–1874), under Minister Ruiz Zorrilla decrees of October 21 and 25, 1868, which proclaimed broad freedom of education and reorganized secondary and higher education, women proposed a modern reading of legal regulations, considering that if their access to the university was not forbidden, it was allowed. But this did not last long, for since 1882–83 the restrictive sense of the Orders of the General Directorate of Public Instruction was stabilized with the Royal Order of June 11, 1888, which required the express permission of the academic authorities for women's entry.

Under these conditions, until 1900, twenty-five women graduated and seven received doctorates, none in science. Three women enrolled in Faculties of Sciences were studying the first preparatory course for the Faculty of Pharmacy: M. Dolores Martínez Rodríguez (Valencia, 1888–89), Mª Blanca de Lucía Ortiz (Seville, 1889–90), and Mª Dolores Pujalte Martínez (Valencia, 1897–98 and 1898–99).

Dolores Closas Morera was the first woman enrolled in a Faculty of Sciences in Spain (Barcelona, 1886–87), but in 1887, once she passed the exams, her registration was canceled according to the above-mentioned Orders. In this context, the next application for enrollment at the University of Barcelona did not take place until 1895, and it was first denied.

The situation did not improve in the first decade of the twentieth century, for only two women enrolled in Science Faculties in Spain. Pilar López Barea (Valencia, 1906–07) did not finish her studies. As for María Sordé Xipell, she enrolled in Barcelona from 1906–07 to 1911–12, and after passing her degree exam on February 11, 1914, she became the first Spanish graduate in science, the only one among the twenty-eight university graduates admitted between 1900 and 1910 (Flecha 1996).

At this point, it is worth considering that in 1900, 71% of women and 55.8% of men were illiterate in Spain.[1] However, from mid-twentieth century on the main goal for Spanish Women was the access to all educational levels rather than the right to vote.

[1] Percentages are obtained from the total population. Discounting inhabitants under 10 years, figures show 66% female and 45.3% male illiterate (Capel 1982, pp. 362–365).

In 1900, a Ministry of Public Instruction and Fine Arts was instituted in Spain. Minister García Álix' Royal Decree of August 3, 1900, established a new syllabus and organized Faculties of Sciences in four sections: Exact Sciences, Physics, Chemistry, and Natural Sciences. As usual, the only complete Faculty was still Madrid, the only one with a Natural Sciences Section. The University of Salamanca had no Faculty of Sciences. Among the remaining eight universities, only three had complete sections: Exact Sciences, Physics, and Chemistry in Barcelona; Exact Sciences and Chemistry in Zaragoza; and Chemistry in Valencia.

The professional opportunities associated with the immediate applicability of chemistry made the chemistry section stand out among all the others, for chemistry was available in four faculties at the Spanish northeastern quadrant. As for the exact sciences section, it was available in three faculties at the Madrid-Barcelona axis. Professional opportunities in mathematics were due to the growing presence of mathematics in teaching, which required qualified teachers.

Regarding the physics section, it should be mentioned that the new syllabus shared nine of the twelve subjects with the exact sciences section. The first two years were common for physics and mathematics; the third year shared two of the three subjects; and the fourth year only one of the three subjects. Moreover, not only did the exact sciences section ensure three quarters of the physics studies, but also the chemistry section ensured the first two years of mathematics and physics—plus two of the three subjects of the respective third years, and the first year of natural sciences.

As for contents, García Álix' syllabus was insufficient at all levels, but especially erroneous in its mathematical approach and particularly useless for future chemists, since it was mainly devoted to geometry—especially projective geometry—in a rather obsolete way of development. Zoel García de Galdeano (1846–1924)—the founder of the first Spanish mathematical periodical (*El Progreso Matemático*), the most internationally connected Spanish mathematician of his time, and the main importer of modern mathematics into Spain during the end of the nineteenth century and the beginning of the 20th—publicly criticized Álix' syllabus unsuccessfully (Ausejo and Hormigón 2002, pp. 52–53).

The first women mathematicians in Spain were trained according to García Álix' new syllabus from 1910 on, when their full access to the university was regulated on equal terms with men. The rate of incorporation was slow, but it should be noted that it was faster in the Faculties of Sciences than in the university as a whole. In 1915–1916, 1.8% students were women at the universities and only 1.5% at the Faculties of Sciences; but in 1932–1933, 6.3% students were women at the universities, while women were 11.1% at the Faculties of Sciences (Ausejo and Magallón 1994, pp. 544–545).

As regards mathematics, we have data on classes in one of the three Faculties of Sciences with an Exact Sciences Section, the one in Zaragoza (Tomeo 1962).

At the University of Zaragoza, the percentage of women in the 1932–33 class (7.6%) was higher than the national average (6.3%). In its Faculty of Sciences, the rate of incorporation (15.7%) also exceeded the national average (11.1%). In all, 739 students graduated between 1882 and 1936 in Zaragoza.

Table 1 Ten women who graduated in mathematics at the University of Zaragoza between 1919 and 1936

Class	Name	Origin	Professional destiny
1926–30	Carolina Jiménez Butigieg	Madrid	Private education, Teresian sister
1927–31	Mercedes Soler García	Bañolas(Girona)	Private education, sister
1928–32	Mª de los Remedios Ruiz Feixas	Córdoba	Sister
1930–34	Rosa Obrador Parpal	Mahón(Baleares)	Private education, sister?
1931–35	Vicenta Esponera Andrés	Zaragoza	Private education Sister of the Sacred Hearts of Jesus and Mary
1931–35	Carmen López de Carrión de Tobaruela	Madrid	Private education Sister of the Sacred Hearts of Jesus and Mary
1932–36	Teresa Coll Ortega	Barcelona	Private education, sister?
1932–36	María Ganuza del Riego	Santander	Private education Sister of the Sacred Hearts of Jesus and Mary
1932–36	Isabel Gurrea Ortiz de Zárate	Logroño	Private education, sister?
1932–36	Josefa Mingo Ruiz de Gordejuela	Medinaceli (Soria)	Private education, sister?

However, female graduates' presence—always less than 10—was not constant until the 1925–29 class. Number 9 was the maximum achieved in the republican period (1931–35 and 1932–36 classes).

Of the thirty-nine women who graduated between 1919 (1915–19 class) and 1936 (1932–36 class), 10 did so in mathematics (25.9%). These data place women's preference for mathematics in an intermediate place between chemistry (66.6%) and physics (7.7%). These are the data we have (Table 1).

These data show the majority presence of Catholic sisters, possibly due to the Catholic Church interest in the education of the new female teachers.

Women's Access to Mathematical Research in Spain

The aforementioned regulations of 1858 initialized research in faculties of sciences on the basis of the doctorate, which was not conceived as an original research work, but as a historical synthesis understood as a compilation of knowledge on a specific subject that extended up to the current state.

In 1900, the doctoral thesis was defined as research work "on a doctrinal or practical research point". Later on, in 1907, the *Institución Libre de Enseñanza* (Free Institution of Education)—a private institution founded in 1876—created a new institution to support scientific research, the *Junta para Ampliación de Estudios e Investigaciones Científicas* (Board for Advanced Studies and Scientific Research). Throughout its existence (1907–1939) the Board awarded a total of 121 scholarships to women, increasing the percentage of these scholarships from 4% in the period 1908–1919 up to 12% in the period 1931–1934 (Ausejo and Magallón 1994, pp. 549–550).

With regard to mathematics between 1910 and 1915, the Board—a well-financed institution independent from the underfinanced universities—awarded scholarships to six mathematicians in order to study teaching methods and syllabi abroad (Berlin, Leipzig, Munich, Geneva, Paris, London, Cambridge). It also gave financial support to García de Galdeano's participation in the International Commission on Mathematical Instruction (Ausejo and Millán 1989, pp. 263–264; Ausejo 2014, p. 289).

In 1915, the Mathematical Laboratory and Seminar was founded as part of the Board. Its first director was Julio Rey Pastor (1888–1962), the most outstanding disciple of García de Galdeano and most famous Spanish mathematician at that time. He developed a line of work on preparation for teaching in secondary education—parallel to research in the field of mathematics. Thus, the laboratory adequately combined research with teaching, for secondary teaching was a highly attractive professional opportunity that enjoyed positive social consideration, both for men and women.

For the modest Spanish mathematical community, the laboratory showed the institutional recognition of research as a necessary activity for the social justification of mathematicians. In addition, the development of a line of work parallel to research in the field of secondary school teacher's preparation was a very peculiar realization of the double teaching and research function that characterized the modern mathematical profession.

Initially, Italian mathematics were established as a main point of reference for the modernization process of Spanish mathematics, which was organized by sending research students abroad and engaging guest professors from different European universities.

From 1915 to 1920, Spanish mathematical research focused on projective geometry, which still had an international network of researchers; mathematical analysis, with Adolf Hurwitz, Karl Fueter, and George Polya in Zurich were among the international leaders; and nomography and numerical analysis, which reflected a timid introduction of the work of Carl Runge in Spain. From 1920 to 1930, the research wheel turned mainly around relativist mechanics and mathematical physics (with Levi–Civita and Hermann Weyl as the primary foreign references) (Ausejo and Hormigón 2002, p. 56).

In the analysis of the Mathematical Laboratory and Seminar, some seventy-five individuals were related in one way or another to the mathematical activity that developed throughout the twenty years from 1915 to 1935. Of these, thirty-six were candidates for positions in secondary teaching, that is, they were preparing to take the competitive examinations prefatory to work in secondary education; thirty-three were researchers; and ten, of whom four came from the group of researchers, were mentioned as research directors. Thirty-five of these 75 names appear as authors in Spanish mathematical journals. It can thus be said that the Laboratory researchers were producing Kuhn's sense of "normal mathematics" by international standards. Nothing can be said against their average yield or the modernity of their mathematical production. However, thirty-five is a modest number in terms of human capital, especially when compared with the number of students, graduates, and doctors in science belonging to the Spanish Mathematical Society in 1912—that will be shown below.

The only woman in this group was Carmen Martínez Sancho (1901–1995), the first female doctor in mathematics in Spain, whose scientific biography is well known (Millán 1990). She was also the first woman to receive a scholarship from the Board to carry out postdoctoral studies abroad. She worked with Schur and Bieberbach in Berlin for eighteen months, starting in January 1931. According to her own testimony, on her return to Spain she did not hesitate to interrupt her research activity to join the recently inaugurated Institute-School at Seville, perhaps because in the environment of the forces of progress associated with the Free Institution of Education, education was the first and main field of social and political action.

On a personal level, it is interesting to point out that Carmen Martínez Sancho shared a family environment close to the Free Institution of Education, which promoted women's education—for which the figure of the father was fundamental. She also relied—unlike others—on the support of her husband, Alberto Meléndez, which allowed her to continue in her chair after their marriage and after the advent of Francoism, once the corresponding purge file had been passed (Negrín 2005, p. 524). Already in the fifties, her condition as an early widow and mother protected her from the policy of deaf obstruction to the professional practice of women, so common at the time, coming to join the University of Seville as an assistant in mathematics for chemists.

Carmen Martínez Sancho was not the first female mathematician to hold one of these university positions, since in the 1931–1932 academic year María Capdevila (1905–1993), the second mathematician to receive a scholarship from the Board, was an assistant in General Astronomy and Physics of the Globe at the University of Barcelona.

María Capdevila graduated in mathematics at the University of Barcelona in 1928. In this year, she was appointed interim professor of mathematics at the National Institute of Zafra. In 1930, she obtained the chair of French language and literature at the Institute of Alcoy and, in 1933, that of mathematics at the Institute of Figueras. In 1934, a Board scholarship enabled her to spend nine months at the Sorbonne studying function theory with Gaston Julia. Although she studied the doctoral subjects at the Central University of Madrid, there is no record that she received her doctorate prior to the Civil War (Millán 1990, p. 515; Esbribano et al. 2007). After the advent of Francoism, she also managed to overcome her purge files: by resolution of August 3, 1939, she was confirmed as a teacher in Logroño, and later, by resolution of February 24, 1940, she was confirmed as a professor in Figueras (Negrín 2005, pp. 525 and 528).

As far as mathematics is concerned, the twentieth century began in Spain with the creation of two mathematical institutions of capital importance in the process of homologation of Spain to European standards: the aforementioned Laboratory and Mathematical Seminar and the Spanish Mathematical Society—an interesting source as regards this paper.

The Spanish Mathematical Society (SME) was founded in 1911. It brought both the stabilization of a centralizing forum, channeling the professional relationships among Spanish mathematicians abroad and the consolidation of a specialized periodical—the *Revista Matemática Hispano-Americana*—as the most advanced line of Spanish mathematical expression in the international context. The Society's List of Members published in the sixth number of the first volume (February 1912) of the *Revista de la Sociedad Matemática Española*[2] enables a quantitative approach to the level of professionalization of this community, as it included a reference to the professional situation of its members. The then recently created Society had at that time 423 members. As regards the 358 founders, the following professional profile emerges for the 89.7% of this block. The panorama in round numbers would be as follows: 38% teachers, 25% undecided, 18% technicians, and 9% miscellaneous ("other"). This classification shows that the Spanish mathematical community had only reached a 56% level of professionalization in the most optimistic case (combining teachers and technicians) or a 38% level in a more restricted interpretation.[3]

Of the 38% teachers (136 individuals) 65 were in secondary education, 29 taught at the university level, 27 were employed in high technical schools, 2 were school teachers, and 3 were mercantile teachers. Of the 25% undecided (89 individuals), 57 were students, 26 science graduates, and 6 Ph.D. The 18% technicians (65 individuals) broke down this way: 34 engineers, 11 technicians, 8 officials, 6 astronomers, 4 statisticians, 1 seaman, and 1 employee of the Bank of Spain. The 9% in the category of "other" (31 individuals) were 24 military men, 2 presbyters, 2 pharmacists, 1 chemist, 1 mechanic, and 1 clock-maker.

The Society's List of Members published between 1912 and 1932 and the papers published in the above-mentioned Society's journals show eleven women active at the Society, among them three members of the Board of Directors, which only lacked female representation initially (until June 1912) and between 1921 and 1926.

The first two were Josefa Barrera and María de la Rigada (1863–1930), both professors at the High Normal School of female teachers in Madrid and vocals at the Board of Directors until December 1920. María de la Rigada was, most certainly, María Encarnación de la Rigada y Ramón, an important personality in the

[2] See Revista de la Sociedad matemática española **1**(1912): 223–233.

[3] Tables and graphs can be seen in (Ausejo and Hormigón, 2002. pp. 54–57; Ausejo 1998, p. 218-222).

field of special and social education and the co-author of *Aritmética Elemental*—together with Mercedes Tella y Comas—a book published in Madrid in 1896 and 1924 (Colmenar 2009).

Subsequently, in October 1914, María Oña joined the society, presented by the mathematicians José Mingot Shelly and Vicente García de Robles Vegas.

In 1924, the *Revista Matemática Hispano-Americana* published the first exercises solved by a woman, Mª del Carmen Martínez Sancho, who repeated in 1925 together with Águeda Gimeno—from Priego, Córdoba—and Mª del Pilar Martínez E.; in 1926 also with Águeda Gimeno and in 1927 again alone.

In 1925 Felisa Martín, Josefa V. García—from Logroño and Martínez Sancho joined the society, and so did Águeda Gimeno in 1927.

In 1933 more women joined the Society: Mª Dolores Pardo Gayoso, Professor of Mathematics at the Institute of Ceuta; Irene Roig Mota, Professor of Mathematics at the Institute of Orihuela; in 1935, Mª Silvia Colino appears as a subscriber to the *Revista Matemática Hispano-Americana.*

It should be noted that Mª Dolores Pardo Gayoso appears as surplus at the time of the resolution of her purge file (August 14, 1943). Irene Roig Mota is in Alcoy when she is sanctioned with transfer outside the province, disqualification from the exercise of managerial positions, and disqualification from trust in cultural and educational institutions (Negrín 2005, pp. 539–540; BOE 27August 1942, p. 6522).

Particularly noteworthy is the appointment of Mª del Carmen Martínez Sancho as member of the Board of Directors on January 8, 1927. On February 27 of that same year, she became the first female Ph.D. in mathematics in Spain. Her thesis dealt with differential geometry—on Bianchi spaces, a specialty that was arriving to Spain in parallel with relativistic mechanics and was developed in the following years. Subsequently, in May 1928, she was appointed a member of the Editorial Committee of the *Revista Matemática Hispano-Americana*, which shows the recognition and consideration that she enjoyed at the Spanish Mathematical Society, not only for her doctorate, but also for becoming the first woman Professor of Mathematics in Spain—at the Institute of El Ferrol—(Millán 1990, pp. 509–510).

Final Remarks

The sexist discourse about the incapacity of women for scientific activity disappeared from Spain as soon as the legal framework for admission to universities was regulated on equal terms with men.

The interest of women in their scientific training was not initially quick, but it did increase constantly, even above the growth rate of their universities as a whole. However, the interest in mathematics never approached or exceeded the interest in Chemistry, although it did in physics.

Nevertheless, the Spanish Mathematical Society shows that 9% of their undecided members (32 individuals) were 26 science graduates and 6 Ph.D., possibly unemployed or employed in a nonscientific job, but still interested in mathematics.

These were the kind of social and economic problems that a peripheral country like Spain found on trying to ensure mathematics institutionalization, start professionalization, and get involved in the international mainstream for twenty years.

Unfortunately, the Spanish Civil War (1936–1939) and the following forty years long Franco's dictatorship purged the Laboratory, the universities, the institutes, and non-exiled mathematicians and cut the civil rights of married adult women.

References

Agnesi, M.ª Gaetana. 1748. *Instituzioni Analitiche*, Vol. II. Milano, Nella Regia-Ducal Corte.

Ausejo, Elena. 1998. El oficio de matemático en la Edad Contemporánea (1808–1936). In *Matemática y región: La Rioja, Actas Simposio sobre Julio Rey Pastor 3. (1996).* Instituto de Estudios Riojanos, Logroño, 211–226. https://dialnet.unirioja.es/descarga/articulo/2012772.pdf

Ausejo, Elena. 2007. La enseñanza de las ciencias exactas, físicas y naturales y la emergencia del científico. In *El Ochocientos: Profesiones e Instituciones civiles*, Vol. 5. *Técnica e ingeniería en España,* ed. Manuel Silva. Real Academia de Ingeniería, Institución Fernando el Católico, Prensas Universitarias de Zaragoza, Zaragoza, 507–550.

Ausejo, Elena. 2014. Mathematics Education in Spain and Portugal: Spain. In *Handbook on the History of Mathematics Education,* ed. Alexander Karp and Gert Schubring. Springer Science+Business Media, New York, Springer, 283–291. https://doi.org/10.1007/978-1-4614-9155-2_14.

Ausejo, Elena and Hormigón, Mariano. 2002. Spanish Initiatives to Bring Mathematics in Spain into the International Mainstream. In *Mathematics Unbound: The Evolution of an International Mathematical Research Community 1800–1945,* ed. Karen Hunger Parshall, and Adrian C. Rice. "History of Mathematics", 23. ISBN 0-8218-2124-5. Providence, RI (USA) / London, American Mathematical Society / London Mathematical Society, 45–60.

Ausejo, Elena and Magallón, Carmen. 1994. Women's Participation in Spanish Scientific Institutions (1868–1936). *Physis, Rivista Internazionale di Storia della Scienza* 31(2): Nuova Serie, 537–551.

Ausejo, Elena and Millán, Ana. 1989. La organización de la investigación matemática en España en el primer tercio del siglo XX: El Laboratorio y Seminario Matemático de la Junta para Ampliación de Estudios e Investigaciones Científicas (1915–1938). *LLULL, Revista de la Sociedad Española de Historia de las Ciencias y de las Técnicas* 12(23): 261–308. https://dialnet.unirioja.es/descarga/articulo/62064.pdf

Bernués Pardo, Julio and Miana Sanz, Pedro. 2019. Soñando con números, María Andresa Casamayor (1720–1780). *Suma: Revista sobre Enseñanza y Aprendizaje de las Matemáticas,* 91, 81–86. https://dialnet.unirioja.es/servlet/articulo?codigo=7071219

Capel, Rosa María. 1982. *El trabajo y la educación de la mujer en España (1900–1930).* Ministerio de Cultura, Madrid.

Casamayor y de La Coma, María Andresa. 1738. *Tyrocinio arithmético:instruccion de lás quatro reglas llanas / que se saca a la luz Casandro Mamès de la Marca y Araioa.* Zaragoza, Joseph Fort. http://bdh-rd.bne.es/viewer.vm?id=0000119089&page=1

Colmenar, María del Carmen. 2009. Contribución de Mª Encarnación de la Rigada a la Educación Especial y a la Educación Social. In *El largo camino hacia una educación inclusiva: la educación especial y social del siglo XIX a nuestros días: XV Coloquio de Historia de la Educación, Pamplona-Iruñea, 29, 30 de junio y 1 de julio de 2009, ed.* M.R. Reyes Berruezo y Susana Conejero, Universidad Pública de Navarra, Pamplona, Vol. 1, 593–604. https://dialnet.unirioja.es/descarga/articulo/2962791.pdf

Cuesta Dutari, Norberto. 1985. *Historia de la invención del análisis infinitesimal y de su introducción en España*. Salamanca, Ediciones Universidad de Salamanca.

Escribano, José Javier and Español, Luis, and Martínez García, María Ángeles. 2007. El doctorado en matemáticas durante la II República española. *LLULL, Revista de la Sociedad Española de Historia de las Ciencias y de las Técnicas* 30(65): 51–64. https://dialnet.unirioja.es/descarga/articulo/2958118.pdf

Flecha, Consuelo. 1996. *Las primeras universitarias en España 1872–1910*. Narcea, Madrid.

Millán, Ana. 1990. Sobre la incorporación de la mujer a la actividad científica en España: la primera doctora en Matemáticas. In *Història, Ciència i Ensenyament*, ed. Roser Codina and Rosa Llobera. EU del Professorat d'EGB/Sociedad Española de Historia de las Ciencias y de las Técnicas, Zaragoza, 505–515.

Negrín, Olegario. 2005. La depuración del profesorado de los Institutos de Segunda Enseñanza. Relación de los expedientes resueltos por el Ministerio de Educación Nacional (1937–1943). *Historia educación* 24, 503–542. https://dialnet.unirioja.es/servlet/articulo?codigo=2222461&orden=279325&info=link

Tomeo, Mariano, 1962. *Biografía científica de la Universidad de Zaragoza*. Universidad de Zaragoza, Zaragoza, 1962.

Velamazán Gimeno, M. Ángeles. 2021. Un curso de cálculo diferencial e integral en la Academia de Artillería de Segovia: el cuaderno escrito por el cadete Martín García Loygorri al dictado del profesor Isidoro Gómez (1775–1776). *Cuadernos Dieciochistas*, *22*, 89–131. https://doi.org/10.14201/cuadieci20212289131

The Training of Mathematics Teachers: A Higher Standpoint

Claudio Bernardi and Marta Menghini

Klein's Volumes on "Elementary Mathematics" and Their Translations

In 2016 a new revised and completed English edition of the first two volumes of Felix Klein's "Elementarmathematik vom höheren Standpunkte aus" appeared, together with the first English translation of the third volume of the series. The 2016 translation of volumes I and II (Klein 2016a; b) by Gert Schubring (Fig. 1) corrects several errors, omissions, and terminological inaccuracies in previous translations.

The translation of volume III by Marta Menghini (Klein 2016c) in collaboration with Anna Baccaglini Frank and Schubring as advisor makes up for a long and inexplicable lack.

The recent editions also remedy a mistranslation of the term "höher", now translated as "higher" and no longer as "advanced". This mistranslation was pointed out by Jeremy Kilpatrick (2008). It is not a trivial matter; it is a question of correctly interpreting the aim of the three books.

The new translations play a very important role in the diffusion of this seminal work of Klein's, still relevant for the training of teachers and for all those who are interested in the foundations of mathematics. As is well known, the volumes are lecture notes of courses that Klein offered to future secondary school mathematics teachers at Göttingen university, published between 1902 and 1908, proposing a teacher training that became a model for many mathematicians and remained valid and effective until today. According to Kilpatrick (2019), the volumes of Klein's

C. Bernardi · M. Menghini (✉)
Sapienza, Università di Roma, Rome, Italy
e-mail: marta.menghini@uniroma1.it

A. Verdugo Rohrer and J. Zender (eds.), *History of Mathematics and Its Contexts*, Trends in the History of Science, https://doi.org/10.1007/978-3-031-60828-5_4

Fig. 1 Gert Schubring with Marta Menghini at the presentation of the new editions of the "Elementary Mathematics from a higher standpoint" at ICME 13 in Hamburg, 2016

textbook have been used in countless courses for prospective and practicing teachers and provide excellent early examples of what today is termed *mathematical knowledge for teaching.*

As Klein explains in the introduction to volume I, one of the aims of his courses is to abolish the "double discontinuity", that is, the lack of a link between what is studied in school and university courses, and then between university courses and the work as a teacher. This double discontinuity means that the period of studies at the university becomes only a "parenthesis" in the teachers' work, who will teach to their students exactly in the way they were taught when they were students.

Elementary Mathematics from a Higher Standpoint

Which is the meaning of *elementary mathematics*? And what does it mean to look at them from a *higher standpoint*? The three volumes of the series seem to give different answers to these questions, being very different from each other.

The first of the three volumes is devoted to "*Arithmetic, algebra and analysis*" and takes a closer look to contents of school mathematics and to difficulties that a teacher could face. The second volume is devoted to "*Geometry*" and gives a comprehensive overview of the field, describing only in the last part how the teaching of geometry is implemented in various countries. The third volume is devoted to "*Precision and approximation mathematics*" and aims at underlining the relation between the exactness of the idealized concepts and the approximations to be considered in applications, showing how mathematics arises from observation.

As to the contents of volume I, they deal with natural numbers and their extension to other numerical systems, from integers to complex numbers. In the section about algebra, the main topics concern the use of graphical and geometric methods in the theory of equations. The section about analysis concerns elementary transcendental functions and calculus. Well known is the proposal to introduce the logarithm of a positive number a as the oriented area between the hyperbola $xy = 1$, the x-axis, the ordinate $x = 1$, and the ordinate $x = a$, first approximating the area as a sum of rectangles and then taking the integral (see Kilpatrick 2019). Everywhere in the book the mutual connections between the various fields are underlined.

Volume II about geometry also has three parts. The first concerns elementary geometric entities and configurations; the second, geometric transformations; the third part is a systematic discussion of geometry and its foundations, which includes non-Euclidean geometries. Klein considers such an overview important for the prospective teacher. As you can imagine, geometries are characterized on the base of the invariants of their symmetry groups, in line with Klein's famous Erlangen program (see Bass 2005; Schubring 1989; Bernardi and Menghini 2003). Here too the links between the synthetic and the analytic point of view are underlined.

In volume III concerning the applications of mathematics, Klein illustrates how, starting from an intuitive and sometimes practical approach, we can develop abstract concepts working in rich "mathematical environments". There are many examples, e.g., the concept of an empirical function and the comparison with an idealized curve; point sets obtained through circular inversion that lead to recognize the behavior of rational numbers among the real numbers; and the "continuous" transformation of curves with the help of a point moving in space (Menghini 2019).

The three volumes were re-edited in the years 1925–28. Klein participated, together with Fritz Seyfarth, in the whole project, but he died in 1925, after the first two volumes had appeared. The third volume was, therefore, edited only by Seyfarth (Klein 1928); changes and insertions had been nevertheless discussed with Klein. Probably this is the reason why the third volume contains no particular indications of its importance for the training of future mathematics teachers.

Such an indication came instead from the Italian geometer Guido Castelnuovo. In 1913/14 Castelnuovo gave a series of lectures at the University of Rome entitled "Matematica di precisione e matematica di approssimazione" (Castelnuovo 1913/14). The course was explicitly inspired by the one delivered by Felix Klein in 1901/02, which at that time was still entitled "Anwendung der Differential- und Integralrechnung auf die Geometrie: eine Revision der Prinzipien".

Castelnuovo states clearly at the beginning of his lecture course that the teaching and learning of mathematics would be more successful if it included, besides the logical procedures that lead to the theorems, also the way in which concepts arise starting from observations, and how they can be verified in practice.

In what sense do the three volumes deal with *elementary mathematics*? The meaning of elementary is not that of a "simple" mathematics as opposed to an "advanced" one. Klein underlines that by no means he addresses himself to beginners, but he takes for granted that his students are acquainted with the main features of the most important disciplines of mathematics (Klein 2016a, p. 1).

Schubring clears that the "basic epistemological issue implied by the concept of *elementarization* is the relation between school mathematics and academic mathematics" (Schubring 2019, p. 171). There is no separation between the two aspects because "the *elements* are understood as the fundamental concepts of mathematics, as related to the whole of mathematics and according to its restructured architecture" (ibid., p. 172). The epistemological aspect of Klein's work consists in explaining connections between different areas and connections between subdisciplines, "pointing out the links of particular mathematical issues and questions with a synthetic view on mathematics" (ibid., p. 174).

The historical point of view that Klein adopts is an aid in this sense, because the historical evolution of mathematics "entails a process of restructuration of mathematics where new theories [...] become well connected to other branches of mathematics and effect a new architecture of mathematics, based on re-conceived elements and thus on a new set of elementarized concepts" (ibid. p. 175).

In the three volumes, the emphasis is put on the elementarization of mathematics, looking for common elements, applications, space perception, and the historical development of concepts (Kilpatrick 2019). That allows prospective teachers to "stand above" their subject.

In the preface to Volume I, Klein writes "I shall allow myself free excursions as the changing stimulus of surroundings may lead me to do in the course of the actual lectures" (Klein 2016a, xiii). These "free excursions" remind of the "extensive teaching"—as opposed to "intensive teaching"—that Castelnuovo wants to give to prospective teachers in his university courses: Extensive teaching means giving priorities to the connections (also with the other sciences), the general ideas, and the examples, rather than to the deep knowledge of the most advanced techniques and the search of the maximum generality (De Marchis, Menghini, and Rogora 2020). This going "around" a topic in various directions allows for the "higher" point of view.

University Courses and the Training of Mathematics Teachers

What about the double discontinuity today? Let us refer to the Italian situation. However, many of the following considerations apply to other countries as well.

In Italian Universities, the prospective mathematics teachers do not have a dedicated degree but are first of all mathematicians: After high school, they usually follow a bachelor's degree in Mathematics (*laurea triennale*, 3 years) and then a master's degree in Mathematics (*laurea magistrale*, 2 years). In general, students who take a master's degree in Mathematics choose their curriculum among some options; in most universities, one of the options is "Mathematics Education" or "Education and History". Students who choose this curriculum have to attend—among others—dedicated courses, such as "Elementary Mathematics from a higher standpoint", "Complementary Mathematics", "Mathematics education" (see Sect. "Extensive Teaching in the Curriculum "Mathematics Education""). At the end, a student must write a dissertation (*tesi*).

The First University Courses in the Bachelor's Degree

A first question: Is there still a school-to-university discontinuity in Italy today? Compared to a few decades ago, the situation has partly improved, because there is greater awareness of the problems. To encourage an informed choice by students enrolling at the university, many forms of guidance have been created. In particular, the *Piano Lauree Scientifiche* (Science Degrees Plan) organizes summer schools and workshops for high school students; summer schools and workshops are often run by high school and university teachers working together. Just the presence of high school teachers in these activities constitutes an "active training" of teachers: The teachers who plan and implement these workshops will oppose the idea of mathematics as a computation-based subject in their students. Moreover, tutoring has been introduced for first-year university students.

In addition, most university teachers know (or should know) that if the percentage of students who pass the examination in a course is too low, part of the responsibility lies with the teacher.

However, thinking about the contents, a clear impression remains in the students who move from high school to university: The mathematics studied at university is completely different from that studied in high school. Such a belief is erroneous on an epistemological level and harmful on an educational level. On the contrary, our opinion is that almost all university courses could usefully be linked to contents addressed in secondary school. This is not only important for those who want to become a teacher later, but for all students in the bachelor's degree. Nevertheless, if we look at the usual courses taken by first-year mathematics students, there is in fact a discontinuity in the contents. Let's see some examples.

Let's start with *Linear Algebra*. It is a course on geometric transformations where everything is done to hide the geometric aspect, which is the very origin of the vector space structure. In the secondary school, the student met vectors in the study of physics. However, vectors in linear algebra are different from vectors in physics, both on a conceptual level and, at least in part, on an operational level. Many *Linear Algebra* textbooks are almost entirely figureless; in any case, the number of geometrical figures is much less than the number of matrices. We are obviously not discussing the contents of *Linear Algebra* courses, but the presentation of the concepts. On the one hand, we have algebraic and matrix calculations, while on the other, we have the reference to the geometric interpretation, a space in which symbols acquire meaning; in logical terms, there is a *syntax* as opposed to a *semantics*.

The usual university teaching favors calculation at the expense of geometric intuition, which instead has a fundamental cognitive value. In the study of linear algebra, the geometric model made up of straight lines and planes in space, a model well known to the student, constitutes an excellent heuristic tool, allows an effective control of the results, and provides a valid mnemonic support. Making references to geometric concepts in the study of linear algebra would help the university freshman, but it could be useful to the same student a few years later, when, as a young teacher, they have to teach three-dimensional geometry in high school: In this context, they will have to find a balance between intuition and rigor, between the algebraic reasoning typical of an analytical approach and the synthetic reasoning. In subsequent university courses it happens that a teacher speaks of affine spaces and Euclidean spaces without explaining the links with affine and Euclidean geometry studied in high school. The good student manages to find the connections and to motivate the terminology; for the others, knowledge that would be useful to the future teacher is not acquired. Let's also mention the geometrical *perspective*, which is discussed in high schools when studying history of art, but which is rarely treated in university courses of Mathematics.

The first courses of *Mathematical Analysis* (*Calculus*) are somehow more consistent with high school curriculum. A freshman that has attended a good scientific high school rediscovers concepts and techniques that they have already seen in the last years of high school. Naturally, as it should be, at the university more theorems are proved, more complex examples and exercises are faced, there is greater attention to rigor. A curiosity: In the Volume II of his lectures on *Elementary Mathematics* Klein ironically comments on the fact that, in that period (beginning of the 1900s) in Italy, in "lecture courses for engineers one proved Taylor's theorem first for any number of variables, and only later specialized for one variable" (Klein 2016b, p. 263). Let's hope that habits of this kind are no longer current. What still persists today is an excessive tendency to generalize: It is bizarre to present a more general concept or statement for which the only significant case is the original particular one. We will come back to analysis later on, considering integrals and measure.

In *Algebra* courses, we often talk about arithmetic in secret, that is, neglecting those properties of numbers that are also closely linked to the concepts being studied. We don't know if all the students realize the link between Euclidean or unique factorization domains, and the fundamental theorem of arithmetic on the existence and uniqueness of the decomposition of integers into prime factors. It's worth spending some time on elementary number theory. Speaking of Diophantine equations, we can cite Hilbert's tenth problem, advising that the answer is simple if we limit ourselves to equations of the first degree, or equations of any degree but with only one unknown. Prime integers could be examined from many points of view: Infinity of primes of the form 4n − 1, infinity of primes of the form 4n + 1, open problems, applications, etc.

Probability, mathematical logic and computer science deserve a separate discussion. These are topics that are present today in secondary school programs, but the majority of those who graduated in mathematics 30–40 years ago had not followed university courses in these disciplines and, consequently, found themselves unprepared to teach them. Today almost all recent math graduates have taken university courses in probability and computer science, and many of them have taken a course in mathematical logic. However, it happens that even young teachers do not feel adequately prepared and are not confident with the "new" topics. In fact, the university course often has an abstract approach and is of little help to a teacher. For example: Probability in secondary school refers to events that have only finitely many outcomes, while in university the sample space is mostly infinite. As far as mathematical logic is concerned, we believe that it is important to talk about the "architecture" of a mathematical exposition in high schools: How an axiomatic theory is constructed, with examples also examined from a historical point of view; what is the role of definitions; a continuous reflection on axioms, proofs, theorems, and on the concept of counterexample is particularly important. It is worth comparing different proofs of the same theorem, asking questions about the theoretical importance and the educational benefit of presenting two or more different procedures. Wrong arguments can be discussed, that is, reasoning that seems correct but leads to an unacceptable result. We find it useful to present these "proofs" in a high school, to underline the importance of rigor in reasoning and language. By the way, the teacher must bear in mind a concrete risk: Some students will not grasp the paradoxical aspect in the reasoning in question, because they are used to accepting everything the professor says.

The Attitude of the University Teacher

There is a problem that every teacher must face when starting a course with new students: Connecting what they intend to teach with what the students already know. Each student has a personal "body of knowledge", acquired occasionally in the environment in which they live or more systematically in previous studies. The new concepts exposed by the teacher will inevitably interact and perhaps will come into conflict, with this body of pre-existing knowledge.

In reality, not all teachers clearly address the problem. Some years ago, it could even happen that a teacher warned the new students "forget everything, we start from scratch" (note that this arrogant warning was then immediately contradicted by the same teacher who took for granted theoretical concepts, linguistic knowledge, practical skills). Fortunately, such an attitude is today rare.

However, still today most of the university teachers who teach in a first-year course do not know the syllabi of high schools, neither the official nor the real ones, and are unaware of the school of origin of their pupils (in the sense that the teacher has a very rough idea of the percentages of his students who attended scientific high school, classical high school, technical institute, …). In these conditions, our teacher can only refer to their own personal experience, or that of their children, which do not constitute a reliable neither an impartial sample.

In general, the university teacher supports the importance of mathematics education at all levels, but has little information about it. They distrust "experts" and have their own solution for all school problems, a solution that often boils down to a naive recipe, in a very personal vision of problems. As far as their teaching is concerned, they feel like an excellent teacher, and in fact they devote times and effort to the lessons. On the other hand, they have many shortcomings: For example, they neglect the students' reactions during lessons, believing that in a lesson only what the teacher says is important. Our teacher rarely encourages the reading of books, other than course notes or recent research. The university students of Mathematics thus acquire a bad habit that will accompany them in the years following their graduation.

Intensive Teaching in Advanced Courses

In general, many of the university courses in Mathematics seem to have the training of the *future researcher* as their objective, even if this objective can only be achieved in the Ph.D. In advanced courses, the teacher thinks that it is essential to arrive at the most significant results in the recent development of the treated mathematical area; as a result, he wastes no time framing concepts, explaining why certain ideas were born, discussing examples, links, and applications. In an intensive teaching, advanced techniques are preferred, which give an idea of the development of mathematics, but limited to a very specific sector. Advanced concepts and techniques will soon be forgotten by students. A course taught in this way is completely unrelated to the mathematics taught in secondary schools and, in fact, is almost useless for future teachers. Let us add an apparent paradox: The more recent is the research presented in a course, the faster that course ages; after 20 years that research will be superseded by other even more recent results.

Extensive Teaching in the Curriculum "Mathematics Education"

Let's now talk about the second discontinuity. If we refer again to the Italian situation, the university-to-school discontinuity is attenuated for those who follow a curriculum dedicated to mathematics education during the master's degree. As anticipated at the beginning of this section, most of the Italian universities have, in the master, this specific curriculum. Sometimes, it is called *Education and history*, *Education and communication*, etc. Even in absence of a specific curriculum, in almost all universities there are courses related to the teaching of mathematics. In about half of the universities, there is a course that is explicitly named *Elementary mathematics from a higher standpoint* (with some variations such as algebra from a higher standpoint, geometry from a higher standpoint, complementary mathematics, foundations of complementary mathematics). These courses are generally accompanied by other courses about the history of mathematics and mathematics education. On the other hand, there are no pedagogical courses, which can possibly be followed at a later time in some teacher training paths. These curricula always include also advanced mathematics courses; moreover, the final qualification is a master's degree in mathematics, regardless of the curriculum and exams. But the main problem is that it is not mandatory to follow an educational curriculum to access teaching.

Let us now take a look at the courses of the type "Elementary mathematics from a higher standpoint". First of all, keep in mind that all the courses of the "Mathematics Education" curriculum are aimed at students who have already acquired basic mathematical knowledge. The contents of this kind of courses are often similar to the ones of Klein's volumes: Their aim is to encourage a critical rethinking of what has already been learned, comparing different approaches to a concept, framing it historically, linking different areas of mathematics, making it clear that there are problems with multiple possible solutions. In this type of courses, *extensive teaching* takes place, rather than intensive teaching. In fact, the course is not aimed at the study of specific advanced theorems; instead, it is a question of enabling the student to evaluate strengths and weaknesses of various approaches, so that the future teacher can consciously make didactic choices whenever necessary. Naturally, when critically rethinking a subject, there is often an opportunity to fill gaps in student preparation. Time can also be found for issues directly related to teaching in high school: Discussing didactic transpositions of recent theories and new presentations of classic topics, or choosing traditional teaching topics that can be dropped in today's teaching in favor of new ones. In general, the *Mathematics education* curriculum must aim to complete and develop in students a broad vision of mathematics, promote their interest, motivate them for the activity they will have to carry out.

Mathematical Knowledge to Stand Above

In all mathematical areas, there are non-trivial contents that the future teacher should master. Often these questions are elementary just in the sense that they refer to the elements of the discipline, i.e., to the basic concepts. Let's see some examples in the framework of the *mathematical knowledge for teaching*, that is, that knowledge that enables the mathematics teacher to "stand above".

Area of figures and integral calculus. In classical school textbooks, to introduce the area of polygons we start from the equi-decomposability: Two polygons are said to be equi-decomposable if they can be decomposed into finitely many pairwise congruent polygons. Then, it is proved (or in any case accepted) that for every polygon P there exists a square Q that is equi-decomposable with P. Therefore, the square of the length of the side of Q is assumed as the area of P, avoiding any recourse to the concept of limit. If, on the other hand, the measurement is based on integral calculus of an area, a conceptually different path is followed; at an elementary level analogies with this second method can be found in the introduction of the area of the circle and in the procedures for approximating areas of figures by means of more or less dense grids. It should be emphasized that integral calculus provides both a definition for the area of a plane figure and a method of computation.

Moving on to *space*, arguments become more complicated, both due to the difficulty of visualizing ("seeing" that a cube breaks down into three congruent pyramids is not entirely immediate), and for strictly mathematical reasons. Hilbert's third problem (can two polyhedra of the same volume always be decomposed into finitely many congruent polyhedra?) has a negative answer: Except in particular cases, two tetrahedra with the same base and the same height cannot be decomposed into finitely many pairwise congruent polyhedra.

A general definition of measure is not presented in high school, but measures are found in physics and also in probability. The university teacher will have to enrich and complete the knowledge of students, to avoid that the future teacher, having to talk about perimeters and areas, has only their teacher and high school textbook as a model.

Complex numbers. For the introduction of complex numbers, two paths can be followed, conceptually very different. At school level, we set $\sqrt{-1} = i$ and continue by applying the algebraic properties valid in $\boldsymbol{R}$, without investigating the effective correctness of a procedure of this kind; in a more rigorous context, we present complex numbers as ordered pairs of real numbers, with a strange definition for the product of two pairs. We think it is appropriate to explicitly compare strengths and weaknesses of the two approaches, also because a comparison will in any case be spontaneous for a university student, but the student may not be able to give adequate answers. It is also good for students to know that the historical road to accepting complex numbers has been long; acceptance has been linked to the possibility of solving third degree equations rather than to logical justifications.

Euclidean geometry. Some theorems are too difficult to be treated in high school, but nevertheless they are particularly elegant. Let us think, for example, of Morley's theorem, Euler's straight line, the circle of 9 points, Ceva's theorem, the arbelos and its properties. Some of these theorems can be seen in the context of geometric transformations, including circular inversion. Moving on to the geometry of space, let's consider the regular polyhedrons, which have an undoubted aesthetic value and have links with philosophy. There are many properties worth dwelling on. Here we mention only one: The group of direct isometries of the cube is isomorphic to that of the octahedron (which is almost obvious, because cube and octahedron are dual polyhedra), and also to the group of direct and inverse isometries of the regular tetrahedron. It is the symmetric group S_4, i.e., the group of 24 permutations of four elements. In the case of the regular tetrahedron, the argument is simple: The tetrahedron has four vertices and each function from the set of vertices in itself defines a direct or inverse isometry. Also, the direct isometries (that is, the rotations) of a cube form a group that is isomorphic to S_4: This circumstance can be interpreted geometrically by thinking that every direct isometry determines a permutation of the four diagonals of the cube; conversely, each permutation of the diagonals defines one and only one direct isometry of the cube (Table 1).

Infinity. We cannot fail to mention the concept of infinity in mathematics and its history, starting with the ancient Greeks. The discussion can easily be extended: Galileo's paradoxes, infinity in literature and art, infinity in physics. In general, we are concerned that many mathematics graduates are unclear on issues relating to the foundations and epistemology of mathematics. The point is that this deficiency is perhaps more serious for a teacher than a specialist mathematician.

Conclusions

We would like to highlight one more aspect. The epistemological conceptions of mathematics, such as Platonism, formalism, constructivism, are reflected in the didactic field more than in research, even if often the teacher is not aware of it. For example, when the teacher chooses exercises in "clean" mathematics (problems in which at some point computations are greatly simplified, properties of "perfect" figures) or "dirty" mathematics (with estimates and approximations), they make a choice of epistemological character. Let's see an even clearer example. Geometry can be taught because it is considered an excellent example of an axiomatic-deductive theory, or with a view to practical applications, or to offer a model of algebraic structures, or even thinking about the organization of students' spatial experiences. They are all legitimate choices, but the educational potentials are different: For example, a century ago, in the first perspective, in Italy it was recommended not to teach geometry in primary and even middle school, while today, speaking of orientation skills and spatial organization, the study of geometry is recommended from the very first classes of primary school (and even in Kindergarten).

A final observation concerns the methods of carrying out the courses in the "Mathematics Education" curriculum. To some extent, university teachers become "models" for their students. If the university teacher manages to involve students in their lessons and, at the same time, highlights the cultural aspects of mathematics, we can hope that the students, when becoming in turn professors, behave in a similar way. From a not too different point of view, there are useful activities that can be carried out alongside a course in elementary mathematics, e.g., reports presented by students, guided discussions, problem-solving sessions, reflections on possible internship experiences. These activities are fully part of the "free excursions" mentioned in Klein's preface. Indeed, the "free excursions" suggest a very suitable teaching methodology for teacher training.

Table 1 Isomorphism between the group of rotations of a cube and the group of direct and inverse isometries of a regular tetrahedron

To get a one-to-one correspondence between the group of rotations of the cube and the group of direct and inverse isometries of the regular tetrahedron, we inscribe a tetrahedron in the cube, so that the four vertices of the tetrahedron are also vertices of the cube. As can be seen in Fig. *a*, the edges of the tetrahedron are diagonals of the faces of the cube and are therefore all equal to each other

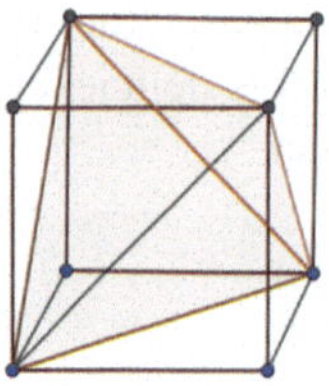

fig. *a*

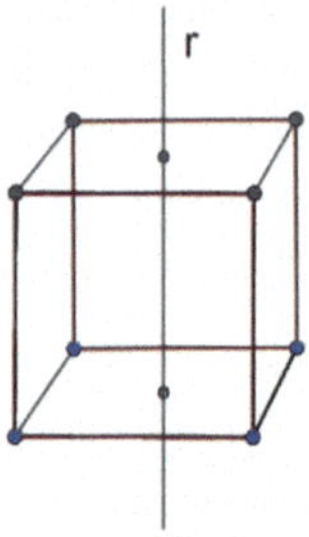

fig. *b*

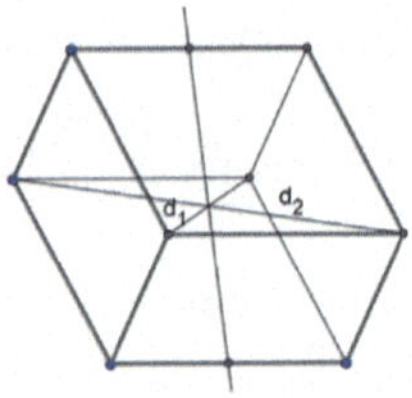

fig. *c*

Let us consider a rotation of the cube; to this we associate an isometry of the tetrahedron as follows

- If the rotation of the cube maps the vertices of the tetrahedron into the vertices of the tetrahedron (like the rotation of π around the straight line *r*, Fig. *b*), it is also a rotation of the tetrahedron
- If the rotation of the cube maps the vertices of the tetrahedron into the other four vertices of the cube (like the rotation of $\pi/2$ around the straight line *r*, Fig. *b*), we compose that rotation with the reflection through the center of the cube, thus obtaining an inverse isometry of the tetrahedron.

In this way, all the isometries of the tetrahedron are obtained. The described one-to-one correspondence is in fact a group isomorphism

We get an algebraic proof of the fact that the group of rotations of the cube is isomorphic to S_4 by considering the permutations of the four diagonals of the cube

Let us define a one-to-one correspondence between the group of rotations of the cube and the group of permutations of its diagonals

- If the permutation of the diagonals is the identity, the rotation is also the identity
- If the permutation leaves only one diagonal fixed, then the corresponding rotation is a rotation about that diagonal by $\pm\,\pi/3$
- If two diagonals d_1 and d_2 are fixed (while the other two are exchanged), then this permutation corresponds to the rotation of π about the line perpendicular to the plane of the two diagonals d_1 and d_2, passing through the center of the cube (see Fig. *c*)
- If no diagonal is fixed in the permutation, then we are dealing with a rotation around the straight line joining the centers of two opposite faces (see Fig. *b*). More precisely, if there is just one 4-cycle, then the rotation is by a right angle; if there are two 2-cycles (that is, the diagonals exchange pairwise), then the rotation is by a straight angle

Images are drawn by the authors

References

Bass, Hymann. 2005. Mathematics, mathematicians, and mathematics education. *Bulletin of the American Mathematical Society* 42(4), 417–430.

Bernardi, Claudio, and Menghini, Marta. 2003. Mettiamo ordine fra le geometrie: il programma di Erlangen e altri approcci. *L'insegnamento della Matematica e delle Scienze Integrate*, 26 A-B, 271–292.

Castelnuovo, Guido. 1913/14. *Matematica di precisione e matematica di approssimazione.* Quaderno delle lezioni dell'A.A 1913/1914. http://operedigitali.lincei.it/Castelnuovo/Lettere_E_Quaderni/quaderni/nC913_14A/mostracom.htm.

De Marchis, Martina; Menghini, Marta and Rogora, Enrico. 2020. The importance of "extensive teaching" in the education of prospective teachers of mathematics. In D. Szarková, D. Richtáriková, M. Prášilová (Eds.) *Proceedings of the 19th Conference on Applied Mathematics (APLIMAT 2020)*, 344–353. New York: Curran Associates.

Kilpatrick, Jeremy. 2008. A higher standpoint. In ICMI (Ed.), *Proceedings of the 11th International on Mathematical Education* (pp. 26–43). Monterrey, Mexico, https://www.mathunion.org/icmi/publications/icme-proceedings-and-publications.

Kilpatrick, Jeremy. 2019. A double discontinuity and a triple approach: Felix Klein's perspective on mathematics teacher education. In H. G. Weigand, W. McCallum, M. Menghini, M. Neubrand, G. Schubring (Eds.) *The Legacy of Felix Klein*, ICME-13 Monographs, 215–225. Berlin: Springer.

Klein, Felix. 1928. *Elementarmathematik vom höheren Standpunkte aus, 3: Präzisions- und Approximationsmathematik.* In C. H. Müller (Ed.), Series: Die Grundlehren der mathematischen Wissenschaften, 16 (3rd ed.), completed for the print and with additions by Fr. Seyfarth. Berlin: Springer.

Klein, Felix. 2016a. *Elementary Mathematics from a Higher Standpoint. Vol. I. Arithmetic, Algebra, Analysis* (Translated by Gert Schubring). Berlin: Springer.

Klein, Felix. 2016b. *Elementary Mathematics from a Higher Standpoint. Vol. II. Geometry* (Translated by Gert Schubring). Berlin: Springer.

Klein, Felix. 2016c. *Elementary Mathematics from a Higher Standpoint. Vol. III. Precision Mathematics and Approximation Mathematics* (Translated by Marta Menghini). Berlin: Springer.

Menghini, Marta. 2019. Precision mathematics and approximation mathematics. The conceptual and educational role of their comparison. In H. G. Weigand, W. McCallum, M. Menghini, M. Neubrand, G. Schubring (Eds.) *The Legacy of Felix Klein.* ICME-13 Monographs, 181–201. Berlin: Springer.

Schubring, Gert. 1989. Pure and Applied Mathematics in Divergent Institutional Settings in Germany: the Role and Impact of Felix Klein. In D. Rowe, J. McCleary (Eds.), *The History of Modern Mathematics. Volume II: Institutions and Applications*, 171–220. Boston: Academic Press.

Schubring, Gert. 2019. Klein's conception of 'Elementary Mathematics from a Higher Standpoint'. In H. G. Weigand, W. McCallum, M. Menghini, M. Neubrand, G. Schubring (Eds.) *The Legacy of Felix Klein.* ICME-13 Monographs, 169–180. Berlin: Springer.

Paulian's *Commentaire* on L'Hospital's *Analyse des infiniment petits*: On How to Teach Differential Calculus in the 1760s

Mónica Blanco

Introduction

In 1696, L'Hospital published the first treatise on differential calculus, *Analyse des infiniment petits pour l'intelligence des lignes courbes.* The publication of the *Analyse* played a fundamental role in the historical development of Leibnizian calculus and, in particular, in its dissemination and teaching.[1] The first section is dedicated to stating the definitions, basic concepts, and rules of differentiation. The other sections gather all the known applications of the differential calculus to solve problems related to curves, whether they are typical ones like finding tangents, maxima and minima, or inflection points and cusps (sections II–IV), or new problems such as the search for evolutes, caustics, envelopes, and other problems (sections IV–IX). In the second case, the new calculus is almost the only one capable of solving them in a general manner. In the last section, L'Hospital shows the generality of Leibnizian calculus over Cartesian method.

L'Hospital's work received a favorable reception by the periodical *Journal des Sçavans.* Among other aspects, the author of the review praised the way in which the differential calculus was extensively displayed and clearly demonstrated, along with its generality and conciseness.[2] Despite acknowledging the innovations of the *Analyse*, the author of the review considered L'Hospital's work difficult and only apt for "connoisseurs" (Bella 2022, p. 301). As a supplement for the understanding of L'Hospital's work, three commentaries appeared in France in the eighteenth century, namely, the *Commentaire sur l'Analyse des infiniment petits* (Paris, 1721)

M. Blanco (✉)
Universitat Politècnica de Catalunya, Barcelona, Spain
e-mail: monica.blanco@upc.edu

[1] See, for instance: Bella (2022), pp. 251–303; Blanco (2008a); Bradley et al. (2015), pp. v–xlii; Mancosu (1989).

[2] *Journal des Sçavans* (September, 1696), p. 414–418.

A. Verdugo Rohrer and J. Zender (eds.), *History of Mathematics and Its Contexts*,
Trends in the History of Science, https://doi.org/10.1007/978-3-031-60828-5_5

by Jean-Pierre de Crousaz (1663–1750), the *Éclaircissemens sur l'Analyse des infinimentpetits* (Paris, 1725) based on the notes written by Pierre Varignon (1654–1722), and the *Analyse des infiniment petits, suivie d'un nouveau commentaire pour l'intelligence des endroits les plus difficiles de cet ouvrage* (Avignon, 1768) by Aimé-Henri Paulian (1722–1802).[3] They commented on L'Hospital's work point after point. Though they thought highly of L'Hospital's work in general, Crousaz and Paulian objected that the *Analyse* was intended for *savants*, and their commentaries aimed at filling in the gap in the teaching of the subject at the time. In fact, Paulian also considered that Varignon's text was not suitable for a beginner either.

The analysis of textbooks is a key idea in Schubring's approach to the history of mathematics (Schubring 2005, pp. 2–3; Schubring 1996, pp. 363–365). Teaching is evidently an act of communication. According to Secord (2004, pp. 663–664), science can be regarded "as a form of communicative action—to recognize that questions of *what* is being said can be answered only through a simultaneous understanding of *how*, *where*, *when*, and *for whom*.

The writing of educational books[4] represents a practice of communication that contributed to the formulation of the ultimate discourse of the differential calculus. Taking Schubring's approach as starting point, the analysis of such communicative practices should be based on much more material than a history of great men and the diffusion of their ideas (Schubring 1996, p. 363). In this sense, the study of Paulian's *commentaire*, a rather lesser work, allows reconstruction of the discussion on how to teach the differential calculus. What prompted Paulian to publish his *commentaire* more than seventy years after the release of the *Analyse*? Paulian's commentary suggests that not only was L'Hospital's approach still in use in the 1760s, but also that concerns existed about how to teach differential calculus to beginners. The aim of this paper is to analyse Paulian's *commentaire* in order to gain further insights into *how* to teach differential calculus (*what*) in the 1760s in France (*when* and *where*), specifically targetting beginners (*for whom*). The present paper opens by introducing Paulian and his *commentaire*. The subsequent section focuses on the study of some notes on fundamental concepts. Lastly, the final section delves into Paulian's use of the question & answer format in certain notes, as a pedagogical tool, along with the description of the prerequisite knowledge required to grasp L'Hospital's *Analyse*, as explained by Paulian.

---

[3] In 1781, a revised and expanded edition of the *Analyse* was published, edited by Louis Lefèvre-Gineau (1751–1829). However, in his notes, Lefèvre departs from L'Hospital's approach and adopts Euler's approach instead. As a result, the study of Lefèvre's notes goes beyond the scope of this paper.

[4] Since the word "textbook" was not defined in the eighteenth century yet, my preference here is to consider them as educational books or "books employed for educational purposes", as they are referred to in Bertomeu Sánchez et al. (2006, p. 658).

Paulian and His *Commentaire* on L'Hospital's *Analyse*

A Newtonian Jesuit, Aimé-Henri Paulian taught physics at the Jesuit Collège in Avignon.[5] Born in Nimes in 1722, Paulian studied with the Jesuits and soon entered the Society. The study of physical sciences became his main occupation, which he pursued until the suppression of the order. He then went on to write and publish several works on physics, among which the *Dictionnaire de physique* (1761) stands out, having been published at least nine times. Another notable work is the *Traité de paix entre Descartes et Newton* (1764), which arose from Paulian's interpretation of Cartesian and Newtonian physics.[6] Other works include *Nouvelles conjectures sur les causes des phénomènes électriques* (1762), *Système général de philosophie* (1769), *Véritable système de la nature* (1771), *Dictionnaire philosophico-théologique* (1774), and *Dictionnaire des nouvelles découvertes faites en physique* (1787).[7] Regarding mathematics, particularly in the context of teaching, in 1765 Paulian published *Le Guide des jeunes mathématiciens dans l'étude des Leçons de Mathématiques de M. l'Abbé de la Caille*, which was a commentary on the 1764 edition of the *Leçons élémentaires de mathématiques* by the Abbé Nicolas Louis de La Caille's (1713–1762), professor of mathematics at the Collège Mazarin. The first edition of La Caille's work (1741) only covered geometry and algebra, but starting from the fourth edition (1756), the foundations of differential and integral calculus were added (Schubring 2005, pp. 213–214; Lacoarret and Ter-Menassian 1986, pp. 144–145, 156–157).

As mentioned above, in 1768, a new commentary on the *Analyse des infiniment petits* was published. This commentary focused on the third edition of L'Hospital's work and was written by the author of the *Guide des jeunes mathématiciens dans l'étude des Leçons de Mathématiques de M. l'Abbé de la Caille*. Remarkably, Paulian's name is not explictily mentioned on the front page, as shown in Fig. 1. When examining the front page of *Le Guide des jeunes mathématiciens* (1765), Paulian's name does not appear either. Instead, it states: "Par un ami de l'Auteur de ces Leçons". However, on the front page of the *Guide des jeunes mathématiciens, ou*

[5] On the teaching of sciences at the Jesuit *collèges*, see Dainville (1986). In eighteenth-century pre-Revolutionary France university education was mainly restricted to the *collèges*, run occasionally by religious orders, where mathematics was taught at a rather elementary level. This system coexisted with some others, where the study of mathematics was more essential. By the 1750s a well-developed network of military schools had been established, with navy, artillery, and engineering as major fields of education (Schubring 1996, p. 365).

[6] For a detailed analysis of Paulian's work on physics, see Mulard, Pascal (2015). *Aimé-Henri Paulian (1722–1802) et sa physique «Newto-cartésienne». L'optique Newtonienne revisitée.* Thèse doctorat, Université Lille. In particular, Chaps. "The Dissemination of Cartesian Ideas. An Analysis of Practice with Diagrams in Van Schooten's Reading of Descartes", "Women's Access to Mathematical Research in Spain (1868–1936): The State of the Art" and "The Training of Mathematics Teachers: A Higher Standpoint" focus on Paulian's *Traité de paix entre Descartes et Newton.*

[7] See the article on Paulian in the *Biographie Universelle Ancienne et Moderne* (1823), tome 37, pp. 196–197.

Commentaire des Leçons de Méchanique de M. l'Abbé de la Caille, published in 1771, it reads: "Par M. Paulian, Prêtre, Associé à l'Académie Royale de Nîmes". In the preface of this work, Paulian acknowledges having published the *commentaire* on La Caille's lessons on algebra and geometry in 1765.

In his preface (*Où l'on trouvera ce qu'on doit penser de l'Analyse des infiniment petits, & des divers Commentaires qui en ont été faits*), although Paulian refers to the *Analyse* as a masterpiece, he mentions the criticism directed at Marquis de L'Hospital: "on a reproché à M. Le Marquis de l'Hospital de n'avoir écrit que pour les Sçavans, tellement rompus dans le calcul, qu'ils entendent tout à demi mot" (Paulian 1768, p. ij). It was to render the *Analyse* accessible to ordinary beginners that Crousaz published his commentary in 1721 (Paulian 1768, preface,

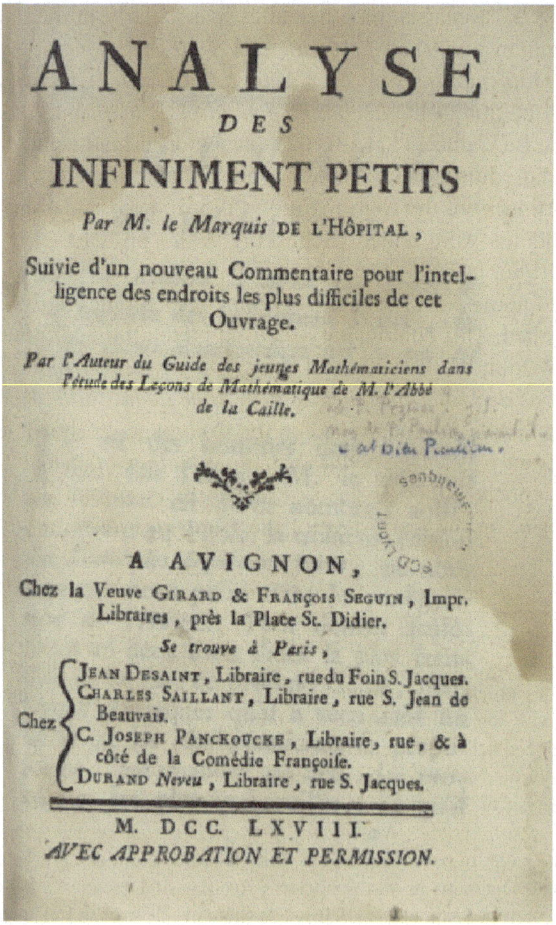
ANALYSE
DES
INFINIMENT PETITS
Par M. le Marquis DE L'HÔPITAL,
Suivie d'un nouveau Commentaire pour l'intelligence des endroits les plus difficiles de cet Ouvrage.
Par l'Auteur du Guide des jeunes Mathématiciens dans l'étude des Leçons de Mathématique de M. l'Abbé de la Caille.

A AVIGNON,
Chez la Veuve GIRARD & FRANÇOIS SEGUIN, Impr. Libraires, près la Place St. Didier.
Se trouve à Paris,
Chez JEAN DESAINT, Libraire, rue du Foin S. Jacques.
CHARLES SAILLANT, Libraire, rue S. Jean de Beauvais.
C. JOSEPH PANCKOUCKE, Libraire, rue, & à côté de la Comédie Françoise.
DURAND Neveu, Libraire, rue S. Jacques.
M. DCC. LXVIII.
AVEC APPROBATION ET PERMISSION.

Fig. 1 Front page of Paulian's *commentaire*

p. iij). However, this commentary had been strongly criticized, according to Paulian (1768, preface, pp. iij–iv). On the other hand, while Paulian recognized the value of Varignon's clarifications, he believed they were not suitable for beginners: "Cet ouvrage, tout excellent qu'il est, ne peut guere être mis entre les mains d'un Commençant; M. Varignon n'y éclaircit pour l'ordinaire que les points qui ont été capables de l'arrêter lui-même" (Paulian 1768, p. v). Varignon's work not only contained explanations of what he considered to be obscure points and a revision of L'Hospital's text, but also contributed new propositions, additional problems, construction rules, and different methods that he had worked out. Yet, in Paulian's opinion, Varignon had only attempted to clarify those aspects that he himself had not fully grasped, and his *éclaircissemens* failed to introduce beginners to the subject. Paulian's goal was to help "les Commençans à se passer de guide dans la route épineuse du calcul différentiel" (Paulian 1768, pp. vj–vij). Nevertheless, he acknowledged the positive contributions of both Varignon and Crousaz: "la lecture des *éclaircissemens* de M. Varignon nous a fait naître la plupart des idées que nous avons mis en oeuvre; & nous ajouterons que nous avons profité de qualques bons endroits qui se trouvent dans le *commentaire* de M. Crouzas" (Paulian 1768, p. xiij).

Paulian's work received positive reviews published by several journals, such as the *Journal Encylopédique* (February 1768, tome II, p. 128) and the *Journal de Sçavans* (May 1768, vol I, tome XXX). Nonetheless, there were also rather negative opinions. Hence, according to Montucla, Paulian's commentary: "n'est guère propre qu'à guider des foibles commençans" (Montucla 1799, tome II, p. 398). It is worth adding that the format of the book was also chosen with the potential readers' benefit in mind, as emphasized by the author of the review in the *Journal Encyclopédique*. Unlike the first two editions, which were printed in 4° format, the third edition was printed in 8° format, specifically for the sake of the readers, as highlighted by Paulian: "c'est rendre un veritable service au public, que de lui présenter à un prix très modique, en un volume d'environ 500 pages" (Paulian 1768, p. xvj). While L'Hospital's text was presented at the beginning for the sake of the *savants*, Paulian's notes were annexed separately at the end for the sake of the *commençans* (Paulian 1768, p. vj). Paulian's *commentaire* consists of 55 notes (distributed as shown in Table 1), and nine extra figures included after the *Analyse*'s figures.

Although Paulian's *commentaire* can be considered as a relatively lesser work in the context of the differential calculus, it is still worth studying it to gain insights into the pedagogy of differential calculus from the perspective of a practitioner, of a contemporary actor. Some of his notes elucidate fundamental concepts, while others develop specific computations with meticulous detail. Furthermore, Paulian's notes also include numerous references, aiming to provide the necessary prerequisite knowledge for understanding differential calculus. These aspects are examined in the following sections.

Table 1 Contents of Paulian's *commentaire*

Sections in L'Hospital's *Analyse*	Paulian's notes
I. Où l'on donne les règles du calcul différentiel	I–IV
II. Où l'on fait usage du calcul des différences pour trouver les tangentes de toutes sortes de lignes courbes	V–XXV
III. Où l'on fait usage du calcul des différences pour trouver les plus grandes ou les moindres appliquées	XXVI–XXXVII
IV. Où l'on fait usage du calcul des différences pour trouver les points d'infléxion & de rébroussement	XXXVIII–ILIX
V. Où l'on fait usage du calcul des différences pour trouver les développées	L
VI. Où l'on fait usage du calcul des différences pour trouver les caustiques par réflexion	LI
VII. Où l'on fait usage du calcul des différences pour trouver les caustiques par réfraction	LII
VIII. Où l'on fait usage du calcul des différences pour trouver les points des lignes courbes touchent une infinité de lignes données de position, droites ou courbes	LIII
IX. Où l'on trouve la solution de quelques problèmes qui dépendent des méthodes précedentes	LIV
X. Où l'on trouve une nouvelle manière de se servir du calcul de différences dans les courbes géométriques, d'où l'on déduit la méthode de Mrs Descartes & Hudde	LV

Some Notes on Fundamental Concepts

As mentioned above, Paulian's criticism of L'Hospital's work turned essentially to the too concise rules and rather complicated developments. That he was very concerned about conciseness, it is evident from his preface to the *Guide des jeunes mathématiciens*, where with regard to La Caille's *Leçons*, Paulian wonders "comment il faut s'y prendre pour profiter, sans le secours d'aucun maitre, des leçons laconiques de ce grand homme" (Paulian 1765, p. iij). Despite being an excellent work, it was not made "pour être mis entre les mains des commençans abandonnés à eux-mêmes" (Paulian 1765, p. iv). As we will see, this is a recurrent argument in Paulian's *commentaire* on the *Analyse*.

The aim of Notes I-IXL was to "mettre à la portée des Commençans ordinaires les quatre premieres sections dut Traité de l'*Analyse des Infiniment Petits*" (Paulian 1768, p. xij). The first section of the *Analyse* is dedicated to stating the definitions, requirements, and rules of differentiation. L'Hospital opens this section with the following definitions:

> DÉFINITION I: On appelle quantités *variables* celles qui augmentent ou diminuent continuellement; & au contraire quantités *constantes* celles qui demeurent les mêmes pendant que les autres changent. (L'Hospital 1696, p. 2)

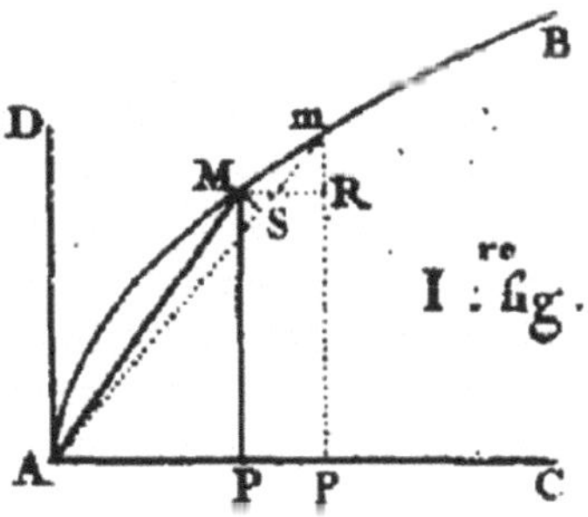

Fig. 2 Diagram used by L'Hospital (1696, Fig. 1) to illustrate the first postulate

> DÉFINITION II: La portion infiniment petite dont une quantité variable augmente ou diminue continuellement, en est appellée la *Différence*. (L'Hospital 1696, p. 3)

Next, L'Hospital formulates the first postulate, without further explanation (Fig. 2):

> I. DEMANDE OU SUPPOSITION. On demande qu'on puisse prendre indifféremment l'une pour l'autre deux quantités qui ne diffèrent entr'elles que d'une quantité infiniment petite: ou (ce qui est la même chose) qu'une quantité qui n'est augmentée ou diminuée que d'une autre quantité infiniment moindre qu'elle, puisse être considérée comme demeurant la même. On demande par éxemple qu'on puisse prendre *Ap* pour *AP*, *pm* pour *PM*, l'espace *Apm* pour l'espace *APM*, le petit espace *MPpm* pour le petit rectangle *MPpR*, le petit secteur *AMm* pour le petit triangle *AMS*, l'angle *pAm* pour l'angle *PAM*, &c. (L'Hospital 1696, art. 2)

To make this first postulate more comprehensible, in the first note Paulian employs several comparisons. To emphasize that the postulate is not fundamentally unreasonable, he draws a parallel with the omissions made by a geometer who, while measuring the height of a mountain, does not concern himself with a grain of sand being carried away from its summit by the wind. Similarly, he points out how astronomers, when discussing fixed stars, may overlook the Earth's diameter of approximately three thousand leagues. Furthermore, he questions whether they consider the Earth as a spherical object when calculating lunar eclipses, taking into account the buildings, towers, and mountains present on its surface:

> Lorsqu'un Géométre, par exemple, prend la hauteur d'une montagne, fait-il attention à un grain de sable que le vent enlève de dessus son sommet? Lorsque les Astronomes nous parlent des étoiles fixes, ne négligent-ils pas le diamètre de la Terre dont la valeur est d'environ trois mille lieues? Lorsqu'ils calculent les éclipses de Lune, ne regardent-ils pas la Terre comme sphérique & par conséquent ont-ils égard aux maisons, aux tours, aux montagnes qui se trouvent sur sa surface? Or tout cela est beaucoup moins à négliger que *dx*, puisqu'il faut un nombre infini de *dx* pour faire *x*. (Paulian 1768, pp. 257–258)

Paulian asserts that all these comparisons were inspired by the first volume of Wolff's mathematics course.[8] It is worth pointing out that one can find the same kind of analogies in Crousaz's commentary (Crousaz 1721, p. 4).

L'Hospital relies on his first postulate to formulate the rule of differentiation of the product *xy* (L'Hospital 1696, art. 5). As the quantity x increases and becomes $x + dx$, the quantity y becomes $y + dy$, and the difference of the product is $ydx + xdy + dxdy$. If we divide *ydx* and *dxdy* by *dx*, we get y and *dy*, respectively. Since *dy* is infinitely smaller than y, the term *dxdy* can be neglected because it is infinitely smaller than *ydx* and *xdy*. Therefore, the difference of the product is $ydx + xdy$. According to Paulian, L'Hospital presents the rules of differential calculus "d'une maniere si concise, qu'il est presque impossible qu'un homme qui les lit pour la premiere fois, apprenne, sans le secours d'un habile maître, à differentier des produits compliequés, des quantités fractionnaires, des nombres affectés d'un ou plusieurs signes radicaux, &c" (Paulian 1768, p. viij). In his second note, Paulian offers what he considers a rigorous proof to explain the elimination of the expression *dxdy*: "On convient que la différence de *xy* est $ydx + xdy + dxdy$; mais on ajoute qu'on peut sans erreur sensible omettre dans la pratique *dxdy*. L'on a raison; en voici la démonstration la plus rigoureuse" (Paulian 1768, p. 258).

The differentiation of the product was the cornerstone for the question of how to avoid the vanishing infinitesimal quantities. It is necessary to justify why certain quantities can be neglected in relation to others, which is not evident to contemporaries (Bella and Blanco 2017). Paulian proposes eleven steps to demonstrate the rule (Paulian 1768, pp. 258–261). First, he presents the notation and the orders of infinite quantities (steps 1–4). Then he defines an infinitely small quantity as a fraction where the numerator is a finite quantity and the denominator is an infinite quantity, for example: 1 (step 5). Based on the orders of infinite quantities, we can infer the orders of infinitely small quantities (steps 6–7). Paulian compares infinitely small quantities of different orders: "un infiniment petit du second ordre représente une grandeur infiniment plus petite qu'un infiniment petit du premier ordre" (Paulian 1768, p. 259). Finally, steps 8–11 deal with the differentiation of the product, first of *xy*, then of *xyz*. In the tenth step, Paulian declares that the infinitely small rectangles *CDop* (*ydx*) and *BmDn* (*xdy*) are infinitely greater than the rectangle *Dnpr* or *dxdy* (Fig. 3a). Consequently, the quantity *dxdy* can be neglected "sans erreur sensible" (Paulian 1768, p. 260). Here again we find a similar line of reasoning in Crousaz (1721, p. 21), as illustrated in Fig. 3b. In fact,

[8] Paulian is here referring to Christian Wolff (1679–1754) and his *Elementa Analyseos Mathematicae. Pars Prima, Elementa Analyseos finitorum tradit*, which are part of the *Elementa Matheseos Universae*. The new calculus was presented in Germany for the first time in this work, printed in the Latin version within the first volume of the first edition of 1713 as part of algebra (the German version, first edition, had been printed in 1710). A highly influential work, widely used in Germany for fifty years, whereas the Latin translation was of great influence in Europe, Wolff's presentation of the new calculus lacks profound reflection on its foundations, as noted by Schubring (2005, pp. 239–240). There are numerous editions of the Latin version of Wolff's work from 1713 to 1754. For this contribution, I have used the edition of 1732.

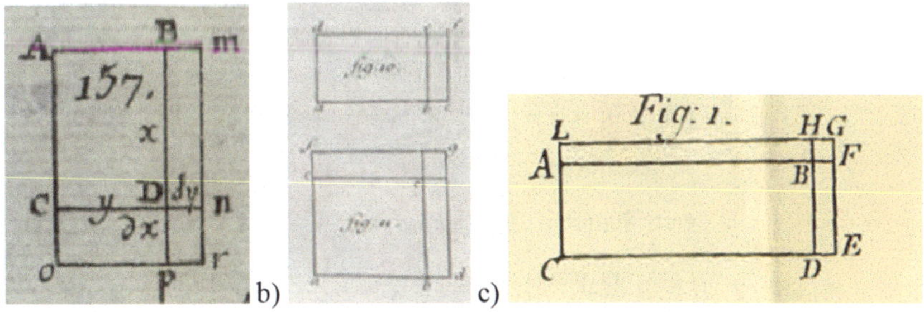

Fig. 3 Diagrams used to illustrate the differentiation of the product: **a** Paulian (1768), Fig. 157; **b** Crousaz (1721), Fig. 11; **c** Wolff (1732), Tab. I, Fig. 1

Wolff (1732, vol. I, p. 419) had turned to the figure of the rectangle to explain the rule (Fig. 3c). We can conclude that the influence of Wolff's work, in his Latin version, can be sensed in the geometric, or rather diagrammatic, justification of the rule of the product.[9]

Even though this type of argumentation had already been discussed in the first half of the eighteenth century, even if it involved the use of resources without reflection, Paulian leaned toward these resources, which were more visual and intuitive, in the teaching of calculus. The study of the prefaces of the commentaries by Crousaz and Paulian helps us understand how, according to these authors, calculus should be taught to beginners, as opposed to Varignon's *Éclaircissemens*, which were more aimed at *savants*. This beginner-*savant* dichotomy is evident in the way Crousaz and Paulian use geometric figures and examples from the physical world as resources to explain the elements and definitions of L'Hospital's *Analyse*. On the other hand, in 1701, Varignon had explained his disagreement with this kind of analogies, like sand-mountain or Earth-stars, to express the nature of infinitesimals, as this could awaken criticisms directed at Leibnizian calculus by its opponents (Bella 2022, pp. 411–412). This is in keeping with the loss of reflection and bowdlerization, regarding the foundations of differential and integral calculus in the context of French universities, as highlighted by Schubring (2005, pp. 213–214). This choice illustrates the tension between the foundations of calculus and their teaching, the more so considering the need to explain the disappearance of infinitesimal quantities.

In the two subsequent notes, Paulian explains the rules for quotients and powers. Remarkably, in Note III Paulian makes a gross mistake. When demonstrating the differentiation of a quotient, it is clear that Paulian, steps 5 and 6 in Fig. 4, simplifies y in the last term of the expression, that is to say:

$$\frac{zdy}{y} = zd$$

[9] For a detailed discussion of Wolff's approach to the differentiation of the product, see Blanco (2008b, pp. 143–144); Schubring (2005, p. 240).

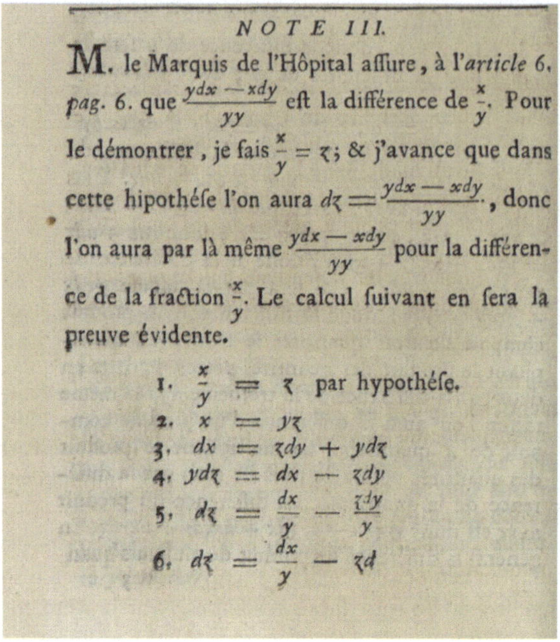

NOTE III.

M. le Marquis de l'Hôpital aſſure, à l'*article* 6, *pag.* 6. que $\frac{ydx - xdy}{yy}$ eſt la différence de $\frac{x}{y}$. Pour le démontrer, je fais $\frac{x}{y} = z$; & j'avance que dans cette hipothéſe l'on aura $dz = \frac{ydx - xdy}{yy}$, donc l'on aura par là même $\frac{ydx - xdy}{yy}$ pour la différence de la fraction $\frac{x}{y}$. Le calcul ſuivant en ſera la preuve évidente.

1. $\frac{x}{y} = z$ par hypothéſe.
2. $x = yz$
3. $dx = zdy + ydz$
4. $ydz = dx - zdy$
5. $dz = \frac{dx}{y} - \frac{zdy}{y}$
6. $dz = \frac{dx}{y} - zd$

Fig. 4 Paulian (1768), Note III

though in the end he gets to the right result.[10]

This is rather striking, the more so since in Note XXXIX, Paulian warns the reader that *dy* is a simple quantity, and not *d* multiplied by *y*:

> L'avertissement qui suit la définition I de la Section IV, fait toujours quelque peine aux commençans. Il s'imaginent que $dy \times dy$ doit donner *ddyy* ou d^2y^2, & que par conséquent le quarré de *dy* doit être d^2y^2, & non pas dy^2... C'est là une erreur dont il est facile de se guerir, lorsqu'on fait attention que *dy* est une quantité très simple, & non pas une quantité composée de *d* multipliant *y*. (Paulian 1768, p. 333)

Paulian's mistakes are not limited to this instance alone. In Note XLVII (Fig. 5), when explaining how to find the difference of $b\sqrt{\frac{a-x}{x}}$, Paulian reminds the reader that:

$$b\sqrt{\frac{a-x}{x}} = \sqrt{\frac{abb - bbx}{\sqrt{x}}}$$

because the denominator x is also affected by the radical sign, like the numerator $a - x$. Was this a genuine mistake? Or was it made on purpose, a visual way

[10] We can find the same mistake in Paulian (1765), Problème 6, lettre 1, livre V.

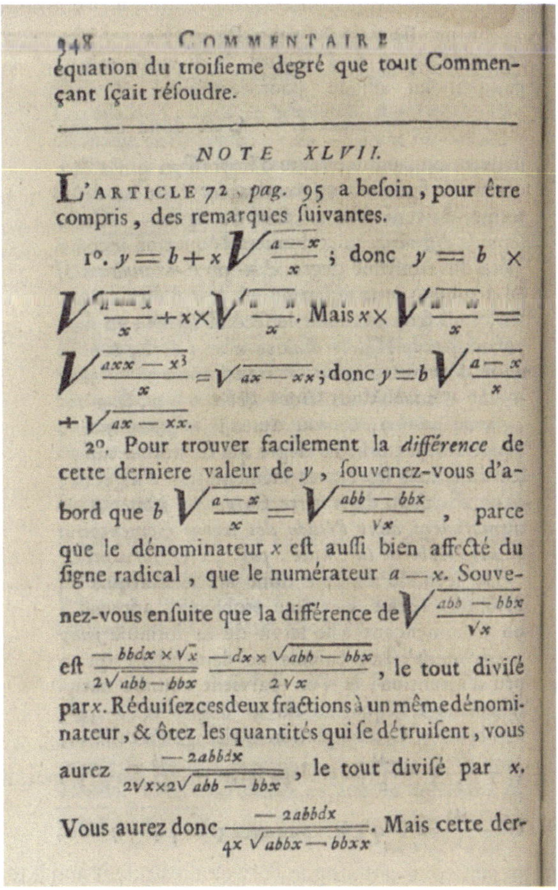
348 COMMENTAIRE

équation du troisieme degré que tout Commençant sçait résoudre.

NOTE XLVII.

L'ARTICLE 72, *pag.* 95 a besoin, pour être compris, des remarques suivantes.

1°. $y = b + x\sqrt{\frac{a-x}{x}}$; donc $y = b \times \sqrt{\frac{a-x}{x}} + x\times\sqrt{\frac{a-x}{x}}$. Mais $x\times\sqrt{\frac{a-x}{x}} = \sqrt{\frac{axx-x^3}{x}} = \sqrt{ax-xx}$; donc $y = b\sqrt{\frac{a-x}{x}} + \sqrt{ax-xx}$.

2°. Pour trouver facilement la *différence* de cette derniere valeur de y, souvenez-vous d'abord que $b\sqrt{\frac{a-x}{x}} = \sqrt{\frac{abb-bbx}{\sqrt{x}}}$, parce que le dénominateur x est aussi bien affecté du signe radical, que le numérateur $a-x$. Souvenez-vous ensuite que la différence de $\sqrt{\frac{abb-bbx}{\sqrt{x}}}$ est $\frac{-bbdx\times\sqrt{x}}{2\sqrt{abb-bbx}} - \frac{dx\times\sqrt{abb-bbx}}{2\sqrt{x}}$, le tout divisé par x. Réduisez ces deux fractions à un même dénominateur, & ôtez les quantités qui se détruisent, vous aurez $\frac{-2abbdx}{2\sqrt{x}\times2\sqrt{abb-bbx}}$, le tout divisé par x. Vous aurez donc $\frac{-2abbdx}{4x\sqrt{abbx-bbxx}}$. Mais cette der-

Fig. 5 Paulian (1768), Note XLVII

to indicate the fact that the radical sign affects both numerator and denominator? In a way, this could convey the idea that, as far as the result is not affected, as far as the operational manipulation does not lead to a wrong result, then it is acceptable. For instance, at the end of the same note, Paulian identifies a sign mistake in L'Hospital's text, a mistake extant in the two previous editions. Since it does not affect the final result, Paulian does not correct it, because "nous sommes fait une loi inviolable de ne rien changer au texte de M. le Marquis de l'Hôpital" (Paulian 1768, p. 350).

Before concluding this section, it is worth pointing out that Paulian expresses a negative opinion about certain passages in L'Hospital's book. For instance, in note XXV, he finds L'Hospital's proposition "assez déplacée", since the general rule explained is not applied. In note XXXVI, Paulian considers article 59 to be "mal énoncé". Paulian's view aligns once more with that of Crousaz, whose opinion regarding the same article was: "l'état de la question (...) me paroit un peu obscurement énoncé" (Crousaz 1721, p. 139). Lastly, in the final note, LV, regarding L'Hospital's tenth section, Paulian deems it to be the least significant among them all:

> La dixieme section est sans contredit la moins importante de toutes. Elle n'apprend que ce que sçavent tous les Mathématiciens, c'est-à-dire, que par le calcul des *différences* on résout beaucoup plus facilement que par toute autre méthode, les problêmes proposés dans les neuf sections précédentes. Pour se convaincre de cette vérité, il ne sera pas nécessaire de lire les 22 *articles* qui composent la dixieme section; on pourra se contenter de la lecture de l'*article* 208; on verra combien compliquée est l'équation que donnent les méthodes qui ne sont pas fondées sur les calcul différentiel. (Paulian 1768, p. 375)

However, this section turned out to be crucial in defending Leibnizian calculus, and its generality, against its opponents in the context of the debate that originated in the Académie Royale des Sciences. Since this section concerned reflections on foundations, could it be that Paulian did not consider it as important for the education of a beginner?

Notes I-ILIX address the fundamental concepts shown in Sections I-IV of L'Hospital's *Analyse*, as well as step-by-step procedures, mainly focusing on the computing mode. Several of these notes are written in what could be called "question & answer format", as we will see in the next section.

On the Question and Answer Format and Prerequisite Knowledge

Several notes are presented as a dialogue between a student and a teacher. In them, Paulian first asks a question, playing the role of a student, and then he gives the answer, as a teacher. Let us discuss some notes to illustrate this specific format. For instance, Note IV (Fig. 6) explains how to operate with negative and rational powers by means of ten questions.

Paulian also uses this format when dealing with fundamental concepts of the differential calculus, such as the determination of the tangent to a curve in Note VI, regarding L'Hospital's article §9. This note opens with the following question: "*Premiere Question.* Comment peut-on démontrer que les triangles *mRM*, *MPT* sont semblables?" (Paulian 1768, p. 283). While L'Hospital simply stated that the triangles were similar, Paulian's *réponse* offers a step-by-step explanation. Yet, he explains that the infinitely small side *Mm* can be taken for the extended line *MT* (Fig. 7), without explicitly mentioning L'Hospital's second postulate: "On demande qu'une ligne courbe puisse être considérée comme l'assemblage d'une infinité de lignes droites chacune infiniment petite" (L'Hospital 1696, §3). Then Paulian moves on to the second question, to explain how to obtain the tangent

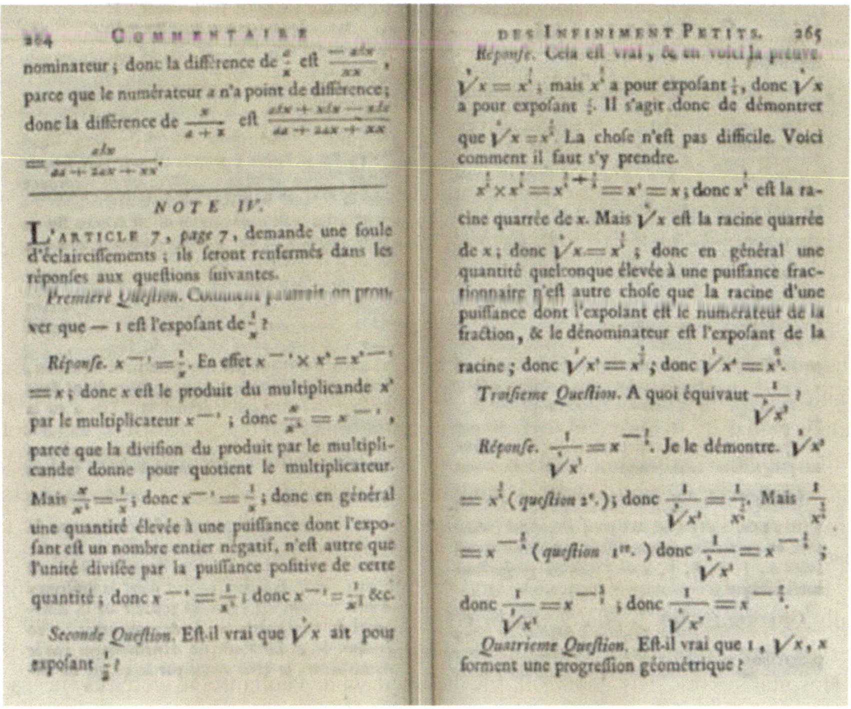

264 COMMENTAIRE

nominateur; donc la différence de $\frac{a}{x}$ eſt $\frac{-adx}{xx}$, parce que le numérateur a n'a point de différence; donc la différence de $\frac{x}{a+x}$ eſt $\frac{adx+xdx-xdx}{aa+2ax+xx}$ $=\frac{adx}{aa+2ax+xx}$.

NOTE IV.

L'ARTICLE 7, *page* 7, demande une foule d'éclairciſſements; ils ſeront renfermés dans les réponſes aux queſtions ſuivantes.

Première Queſtion. Comment pourroit-on prouver que -1 eſt l'expoſant de $\frac{1}{x}$?

Réponſe. $x^{-1}=\frac{1}{x}$. En effet $x^{-1}\times x^2=x^{2-1}$ $=x$; donc x eſt le produit du multiplicande x^2 par le multiplicateur x^{-1}; donc $\frac{x}{x^2}=x^{-1}$, parce que la diviſion du produit par le multiplicande donne pour quotient le multiplicateur. Mais $\frac{x}{x^2}=\frac{1}{x}$; donc $x^{-1}=\frac{1}{x}$; donc en général une quantité élevée à une puiſſance dont l'expoſant eſt un nombre entier négatif, n'eſt autre que l'unité diviſée par la puiſſance poſitive de cette quantité; donc $x^{-2}=\frac{1}{x^2}$; donc $x^{-3}=\frac{1}{x^3}$ &c.

Seconde Queſtion. Eſt-il vrai que $\sqrt[2]{x}$ ait pour expoſant $\frac{1}{2}$?

DES INFINIMENT PETITS. 265

Réponſe. Cela eſt vrai, & en voici la preuve. $\sqrt[2]{x}=x^{\frac{1}{2}}$; mais $x^{\frac{1}{2}}$ a pour expoſant $\frac{1}{2}$, donc $\sqrt[2]{x}$ a pour expoſant $\frac{1}{2}$. Il s'agit donc de démontrer que $\sqrt[2]{x}=x^{\frac{1}{2}}$. La choſe n'eſt pas difficile. Voici comment il faut s'y prendre.

$x^{\frac{1}{2}}\times x^{\frac{1}{2}}=x^{\frac{1}{2}+\frac{1}{2}}=x^1=x$; donc $x^{\frac{1}{2}}$ eſt la racine quarrée de x. Mais $\sqrt[2]{x}$ eſt la racine quarrée de x; donc $\sqrt[2]{x}=x^{\frac{1}{2}}$; donc en général une quantité quelconque élevée à une puiſſance fractionnaire n'eſt autre choſe que la racine d'une puiſſance dont l'expoſant eſt le numérateur de la fraction, & le dénominateur eſt l'expoſant de la racine; donc $\sqrt[3]{x^2}=x^{\frac{2}{3}}$; donc $\sqrt[3]{x^4}=x^{\frac{4}{3}}$.

Troiſieme Queſtion. A quoi équivaut $\frac{1}{\sqrt[2]{x^3}}$?

Réponſe. $\frac{1}{\sqrt[2]{x^3}}=x^{-\frac{3}{2}}$. Je le démontre. $\sqrt[2]{x^3}$ $=x^{\frac{3}{2}}$ (*queſtion* 2e.); donc $\frac{1}{\sqrt[2]{x^3}}=\frac{1}{x^{\frac{3}{2}}}$. Mais $\frac{1}{x^{\frac{3}{2}}}$ $=x^{-\frac{3}{2}}$ (*queſtion* 1re.) donc $\frac{1}{\sqrt[2]{x^3}}=x^{-\frac{3}{2}}$; donc $\frac{1}{\sqrt[3]{x^5}}=x^{-\frac{5}{3}}$; donc $\frac{1}{\sqrt[4]{x^7}}=x^{-\frac{7}{4}}$.

Quatrieme Queſtion. Eſt-il vrai que 1, $\sqrt[2]{x}$, x forment une progreſſion géométrique?

Fig. 6 Paulian (1768), Note IV

from the subtangent: "*Seconde Question.* Comment la connoissance de la soutangente *PT*, peut-elle conduire à la connoissance de la tangente *MT*" (Paulian 1768, p. 283).

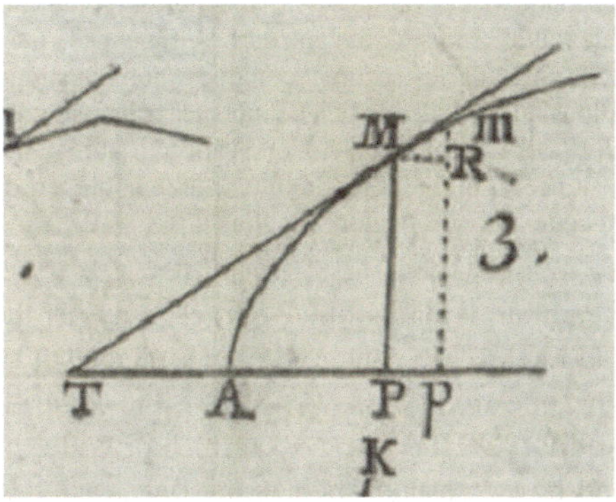

Fig. 7 Diagram used by L'Hospital (1696, Fig. 3) to illustrate the second postulate

The *Guide des jeunes mathématiciens* had been written in this style. It is evident from the preface that Paulian has a penchant for the question & answer format, as a means to make the text more understandable:

> Pour en rendre la lecture supportable, nous nous sommes déterminés à le présenter sous la forme d'un commerce épistolaire. Les difficultés que nous ont proposé, pour l'ordinaire par écrit, ceux à qui nous avons conseillé la lecture des ouvrages de M. l'Abbé de la Caille, seront la matière de toutes les *lettres* quie composent ce recueil; c'est dans les *réponses* que l'on trouvera les éclaircissemens des points que nous croyons être au-dessus de la portée des commençans, ou pour mieux dire, c'est dans les *réponses* qu'est contenu ce qui forme proprement le commentaire que nous donnons au Public. (Paulian 1765, p. 2)

The questions had been allegedly asked by real students, in their attempt to read La Caille's work. Being aware of the difficulties posed by La Caille's lectures, Paulian believed that his *Guide* could be used to complement them, since they were not well adapted for studying "sans le secours d'aucun maitre" (Paulian 1765, p. iij). Therefore, it is not unlikely that Paulian's *commentaire* addresses those learners who intended to read the *Analyse des infiniment petits* without the assistance of a teacher.

Paulian also covers Notes V and VII-X in this fashion, where he specifically elucidates the fundamental principles concerning conic sections and other curves. This serves as prerequisite knowledge, necessary for the understanding of L'Hospital's work. His *commentaire* included what Paulian considered as necessary prerequisite knowledge for comprehending the propositions of the *Analyse des infiniments petits*: arithmetics and algebra, analysis or the sciences of equations, geometry (speculative and practical), and trigonometry (Paulian 1768, p. xiv). Paulian emphasizes the importance of such explanations throughout his work. In particular, before tackling the study of tangents, Note V focuses on conic sections, following the question & answer format discussed above. According to Paulian, this is absolutely crucial for the understanding of L'Hospital's second section:

> Dans toute la section seconde M. Le Marquis de l'Hospital se sert du calcul différentiel pour trouver les tangentes de toutes sortes de lignes courbes. Il suppose que le Lecteur a étudié avec attention tout ce qui regarde les sections coniques; nous le supposons aussi. Malgré cela cependant nous allons lui rappeller en peu de mots les principales propriétés deu cercle, de la parabole, de l'ellipse & de l'hyperbole. Cette espéce d'abrégé du Traité des sections coniques est absolument nécessaire pour rendre intelligible la plupart des problêmes & des exemples que contient cette seconde section. (Paulian 1768, p. 273)

In subsequent notes, Paulian explains other curves, such as the cycloid (XI–XII), the Archimedean spiral (XIV), the conchoid (XV), the cissoid (XVII), the quadratrix (XVIII, also in the question & answer style), the logarithmic curve (XXIII) and the logarithmic spiral (XXIV). These examples provide the description of the curves under study and how to assign their equation.

In order to provide additional support, Paulian recommends the reading of his own *Dictionnaire de Physique*, of the treatise on conic sections by the Abbé de la Caille (notes V, XI, XII) and of his commentary on the latter, namely, the *Guide des jeunes mathématiciens*. He also refers to his *Traité de paix entre Descartes & Newton* on account of the Archimedean spiral (note XIV), L'Hospital's *Traité de sections coniques* and the first volume of Wolff's course concerning the hyperbola (note XXXV).

Another aspect that needs further explanation, according to Paulian, is the resolution of equations (on 3rd degree: note XLVIII; on 4th degree: note XXXVI; on 5th degree: note XLVIII). Once more, Paulian refers to other works of reference, not only his own *Guide*, but also Wolff's course, Clairaut's elements of algebra (first published in 1746), Rabuel's edition of Descartes' *Géométrie* and La Caille's elements of mathematics. In general, Paulian's *commentaire* is peppered with references to the latter work (notes V, XI, XII, XXXII, XLVIII, LIII), together with his own *Guide des jeunes mathématiciens* (notes XLVI, XLVI), but also the works of Wolff and Rabuel are mentioned upon several occasions. In addition, we also find references to the commentaries of Varignon and Crousaz, in line with what was announced in the preface (see for instance notes XIII, XLI, XLV, LIII).

Before concluding, it is worth highlighting the shift in the central concept of differential calculus that occurred in the eighteenth century. In all the works mentioned in Paulian's *commentaire* (eg: L'Hospital, La Caille, Wolff), the difference is considered to be the fundamental concept of calculus. In the theory developed by Leonhard Euler (1707–1783), differential calculus was part of a much broader theory, that of functions. For the first time, functions played a fundamental role in the development of calculus, a fact which enhanced the systematic analysis and classification of elementary functions. In his *Institutiones calculi diferentialis* (1755), Euler regarded differential calculus as a particular case of the method of finite differences, and the infinitely small or evanescent quantities as absolute zeros. The ratio of two zeros could be determined, as approaching a certain limit, only attained when the increments were zero. This ratio, dy:dx, becomes the main object of differential calculus, hence moving the difference away from its central position. However, even after more than a decade, the difference continues to hold a fundamental role in Paulian's approach to the teaching of differential calculus. This is especially noteworthy given that Paulian does not allude to functions; instead, he sticks to the concept of quantity. This parallels the absence of the concept of function within the university context in France, as noted by Schubring (2005, p. 217).

Final Remarks

The preface of Paulian's *commentaire* on the *Analyse des infiniment petits* has allowed us to explore how to teach differential calculus to beginners in 1760s France. Generally speaking, an educational book on differential calculus should avoid complex developments and too concise explanations. Furthermore, it was highly desirable for the book to include references to other works, as well as some hints on the prerequisite knowledge required to fully understand it. It is also remarkable the question and answer format used by Paulian in some of his notes. Was it intended to make the teaching–learning process more vivid?

Beyond these formal aspects, what questions did teachers and *savants* raise when they write educational books on differential calculus, and what choices did they make in the face of discussions about the nature of calculus at that time? The publication of L'Hospital's *Analyse* and the resulting commentaries allow us to reconstruct the discussion on teaching the differentiation of the product, especially when explaining the removal of infinitesimal quantities to beginners. The debates surrounding the nature of infinitesimal quantities and the justification of the differentiation of the product had a significant impact on the teaching of differential calculus. This impact is partly due to the use of without-reflection resources and previously debated arguments, which nonetheless prove useful from a pedagogical perspective. These resources are more visual and intuitive in the teaching of calculus, such as those used by Crousaz and Paulian in their commentaries on L'Hospital's *Analyse*. The choice of these pedagogical resources illustrates the tension between presenting the foundations of calculus and teaching its principles. In a way, the pedagogical presentation takes the approach of pacing, or somehow avoiding, the reflections on foundations to ensure understanding.

The stakes of this tension become more evident when we try to understand for whom these educational books were intended for, and who the authors were. Here, we can distinguish between two categories: texts written by and for *savants*, and manuals aimed at beginners, mostly authored by teachers. This raises the question of how the differential calculus should be communicated to beginners, as opposed to *savants*. Crousaz and Paulian considered that the *Analyse* was intended only for *savants*, despite deeming it as a good work in general. Being mathematics teachers themselves, they attempted to make the content of the *Analyse* accessible to non-*savants* through their commentaries, utilizing geometric diagrams and examples from the physical world as pedagogical resources. In particular, Paulian seems to address self-taught readers, those who would attempt to master calculus without the assistance of a teacher, within the context of French universities.

Acknowledgements This paper was written with the support of the project PID2020-113702RB-I00. I am very grateful to Mª Rosa Massa-Esteve (Universitat Politècnica de Catalunya) for her helpful suggestions and comments on earlier drafts of the paper.

References

Primary Sources

Crousaz, Jean-Pierre de (1721). *Commentaire sur l'Analyse des infiniment petits*. Paris: Chez Montalant.

L'Hospital, Guillaume de (1696). *Analyse des infiniment petits pour l'intelligence des lignes courbes*. Paris: Imprimerie Royale.

Montucla, Jean-Étienne (1799). *Histoire des mathématiques*. Paris: Chez H. Agasse.

Paulian, Aimé-Henri (1765). *Le guide des jeunes mathématiciens dans l'étude des "Elémens des mathématiques" de M. l'abbé de La Caille*. Paris: Chez les livraires associés.

Paulian, Aimé-Henri (1768). *Analyse des infiniment petits, suivie d'un nouveau commentaire pour l'intelligence des endroits les plus difficiles de cet ouvrage*, Avignon: Chez la Veuve Girard et François Seguin.

Varignon, Pierre (1725). *Éclaircissemens sur l'Analyse des infiniment petits*. París: Chez Rollin.

Wolff, Christian (1732). *Elementa Matheseos Universae. Tomus Primus*. Geneva: Bousquets.

Secondary Sources

Bella, Sandra (2022). La (Re)construction française de l'analyse infinitésimale de Leibniz (1690 1706). Paris: Classiques Garnier.

Bertomeu Sánchez, José Ramon; García Belmar, A.; Lundgren, A., and Patiniotis, M., eds. (2006). Textbooks in the Scientific Periphery. Science and Education 15 (7–8).

Blanco, Mónica (2008a). On how Johann Bernoulli's lessons on differential calculus were communicated in eighteenth-century France and Italy. In: Simon, J.; Herran, N., with Lanuza-Navarro, T.; Ruiz Castell, P.; Guillem-Llobat, X. (ed.), Beyond Borders: Fresh Perspectives in History of Science. Cambridge: Cambridge Scholars Press, pp. 113–140.

Blanco, Mónica (2008b). Hermeneutics of Differential Calculus in Eighteenth-Century Northern Germany. Sudhoffs Archiv: Zeitschrift für Wissenschaftsgeschichte, 92, pp. 133–164.

Bella, Sandra and Blanco, Mónica (2017). Comentaris sobre l'Analyse des infiniment petits de L'Hospital (1696–1768): interpretació i ensenyament de conceptes fonamentals del càlcul diferencial. In: M. R. Massa-Esteve and P. Grapí (eds.). Actes de la XV Jornada sobre la Història de la Ciència i l'Ensenyament «Antoni Quintana Marí». Barcelona: Societat Catalana d'Història de la Ciència i de la Tècnica, pp. 49–56.

Blay, Michel (1986). Deux moments de la critique du calcul infinitésimal: Michel Rolle et George Berkeley. Revue d'histoire des sciences, 39, pp. 223-253.

Bradley, Robert E.; Petrilli, Salvatore J.; Sandifer, Charles E. (2015). L'Hôpital's Analyse des infiniments petits: an annotated translation with source material by Johann Bernoulli. Cham: Birkhäuser.

Dainville, François de (1986). L'enseignement sicentifique dans les collèges des jésuites. In R. Taton (ed.). Enseignement et diffusion des sciences en France au XVIIIe siècle. Paris: Hermann, pp. 27-65.

Lacoarret, Marie and Ter-Menassian, Taline (1986). Les Universités. In R. Taton (ed.). Enseignement et diffusion des sciences en France au XVIIIe siècle. Paris: Hermann, pp. 125-168.

Mancosu, Paolo (1989). The metaphysics of the calculus: A Foundational debate in the Academy of Sciences, 1700-1706. Historia Mathematica, 16, pp. 224-248.

Schubring, Gert (1996). Changing Cultural and Epistemological Views on Mathematics and Different Institutional Contexts in Nineteenth-Century Europe. In Goldstein, et al., eds. (1996). Mathematical Europe. Myth, History, Identity. Paris: Éditions de la Maison des sciences de l'homme, pp. 361–388

Schubring, Gert (2005). Conflicts between Generalization, Rigor, and Intuition. New York: Springer.

Secord, James A. (2004). Knowledge in Transit. Isis 95 (4), pp. 654-672.

The Brazilian Mathematics Textbook Assessment Program and Its Influence on Academic Research

Joao Bosco Carvalho

Introduction

Brazil's *Programa Nacional do Livro e do Material Didático* (National Textbooks Program, PNLD, from now on), one of the federal government measures to promote school inclusiveness, was instituted in 1985 with the purpose to distribute free didactical material, mainly textbooks for public schools in the country. At first, the national government distributed only mathematics and Portuguese language textbooks for primary education. Slowly, the scope of the program increased till it included textbooks for all school subjects of public elementary, middle and high schools. It also included dictionaries and books for school libraries. These programs make publishers of school materials almost completely dependent of the national government (Cassiano, 2007). In fact, if the government stopped buying their textbooks and other pedagogical materials, they would very quickly close their doors. The details of the program evolution along the years can be found in Carvalho (2018a, b). Here, it is sufficient to say that prior to 1992, publishers just sent to the Ministry of Education a list of the books they had available for sale to the government; public school teachers had to choose the texts they would use from this list, which included unsold books 'stuck' in the publishers warehouses.

Due to educators' complaints about the poor quality of elementary school textbooks, the Ministry of Education promoted, in 1992–1993, a pilot assessment of mathematics, sciences, social sciences and Portuguese language textbooks used in elementary public schools. It examined the ten most used texts in the mentioned subjects. The results were shocking and widely quoted by the press. For example, none of the science textbooks was approved by the assessment group,

J. B. Carvalho (✉)
Universidade Do Estado Do Rio de Janeiro, Rio de Janeiro, Brazil
e-mail: jbpfcarvalho@gmail.com

A. Verdugo Rohrer and J. Zender (eds.), *History of Mathematics and Its Contexts*, Trends in the History of Science, https://doi.org/10.1007/978-3-031-60828-5_6

which included representatives of universities, educational organizations and state and county school boards. As a result, shortly afterward the Ministry of Education started systematically assessing elementary school textbooks.

As described in Carvalho (2018a, b), from 1992 through 2016, Brazil had a very well-structured program for the assessment of mathematics textbooks. The author of the present chapter was involved with this enterprise from its conception till 2015 and kept careful records of the titles assessed, with detailed reports on each, participants and so on.

Publishers sent their textbooks to be assessed in each of PNLD's assessments. As the end-product of the assessment, a book could be approved for use without reservations, with reservations or be excluded. There was written a long careful review of each textbook, which constituted a catalog, distributed to all public schools in the country, from which teachers had to choose the textbooks they were going to use. The list of non-approved textbooks was not made public.

Along the years, the textbooks were assessed by a total of 250 persons, most of them with degrees equivalent to a Ph.D. or a M.Sc degree, chosen from all the five great Brazilian geographic regions. There were also active mathematics teachers from several states with no graduate degrees and some of the assessors connected to universities taught at their laboratory schools.

The purpose of this chapter is to investigate how this participation influenced the production of academic studies on mathematics textbooks for elementary, middle and high schools. This is no idle academic exercise. It shows a specific case of how "political" choices influence academic production. The World Bank stressed the importance of good textbooks in developing countries, "to compensate for poorly trained teachers, mentioning that in many countries these books impose de facto curricula and are very cheap" (Torres 2000; apud Carvalho 2018a, 775). So, steps to assure textbooks of good quality are indeed an important contribution to a good education, which is a very important asset toward human growth and economic development (King 2011).

The Assessment Program

As already mentioned, the mathematics textbooks bought by the Brazilian Government for free distribution in public elementary, middle and high schools were assessed for PNLD related to the years 1997[1] through 2018 by persons from universities from all great Brazilian geographic regions. Details can be found in Carvalho (2018a, b). The first three assessments (for PNLD-97, 98 and 99, respectively) were run directly by Brazil's Ministry of Education (MEC, from now on). In 2000, it set up the *Comissão Técnica do Livro Didático* (Textbooks Technical

[1] The numbering of PNLD—for example, PNLD 1997, PNLD, 2011, PNLD 2017—refers always to the school year in which the books will be used. Their assessment started three years before their actual use in schools. This should always be kept in mind to avoid confusion and misunderstandings.

Committee), with representatives of the school subjects assessed by PNLD and members of MEC, with the mandate to advise the Ministry on the establishing textbooks public policies. This resulted in a fundamental policy paper which was a guideline of the Ministry actions from then on (BRASIL/MEC 2001).

At the first meeting, MEC's representatives recommended that the Ministry itself should not set up the assessments, that it should hire private educational assessment organizations to do so, supervised by the Committee. The representatives of the curriculum areas (mathematics, science, geography, history, languages) strongly opposed this. They argued that these institutions would develop their own assessment methodologies or appropriate the ones already developed by MEC and would not share them. These representatives proposed that MEC delegates the setting up of the assessments to public universities in the various Brazilian regions, to help to create a textbook assessment culture. In mathematics, the Ministry chose the Universidade Federal de Pernambuco (UFPE), which organized and executed all mathematics textbooks' assessment from PNLD 2000 through PNLD 2018.

The same group at UFPE coordinated continuously these assessments; along the years, it incorporated members from other institutions, both as assessors and as coordination members. The competent work of this coordination, always improved, which developed innovative assessment methodologies, in particular for the case of digital textbooks has not been much studied. As already mentioned, the coordination group invited assessors from all over the country, in order to have representatives of the Brazilian cultural diversity and to establish an assessment culture throughout the country. Along the years, a total of 250 assessors took part in the program.

For each PNLD, MEC distributed a catalog with a list of the approved books and a detailed review of each one. This catalog was distributed to all public elementary, middle and high schools in Brazil, and teachers had to choose the textbook they used from the catalog.

The Influence of Participation in PNLD on the Number of Ph.D. and M.Sc Dissertations on Mathematics Textbooks

This chapter is a step to measure the influence of this assessment program in creating a textbook assessment culture in Brazil. This will be done measuring how many PNLD participants advised Ph.D. and M.Sc dissertations on textbooks.

Presently, Brazil has three kinds of graduate degrees: *Doutor*, equivalent to a Ph.D. degree; *Mestre*, equivalent to a M.Sc. There are two type of *Mestre* degrees, the academic ones and the professional ones, most of which are part of a network organized and supervised by the Brazilian Mathematical Society (Sociedade Brasileira de Matemática, SBM). This network is called PROFMAT (Mestrado Profissional em Matemática) and was instituted in 2011.

The names and institutions of the assessors were easy to find, thanks to the records kept by the author of the present paper. On the other hand, to find out which ones were advisors of dissertations about mathematics textbooks was more difficult. On the one hand, the *Instituto Brasileiro de Informação Científica e Tecnológica* (IBICT) has on its site, a digital library with all Doctors' and Masters' dissertations sponsored by accredited programs in the universities, both public and private.[2,3] Besides, author of the present paper wrote to many persons he knew in many institutions and asked them to send lists of dissertations on textbooks. He also accessed these institutions' internet repositories and the site of the PROFMAT program and was able to find and download 283 dissertations, with a single exception (Meksenas, 1992). Obviously, it cannot be guaranteed that all dissertations were found. These data have not yet been fully studied and will be a valuable tool for future researches.[4]

The Studied Dissertations

First of all, let us show the evolution of the number of Ph.D. and M.Sc dissertations about mathematics textbooks per year.

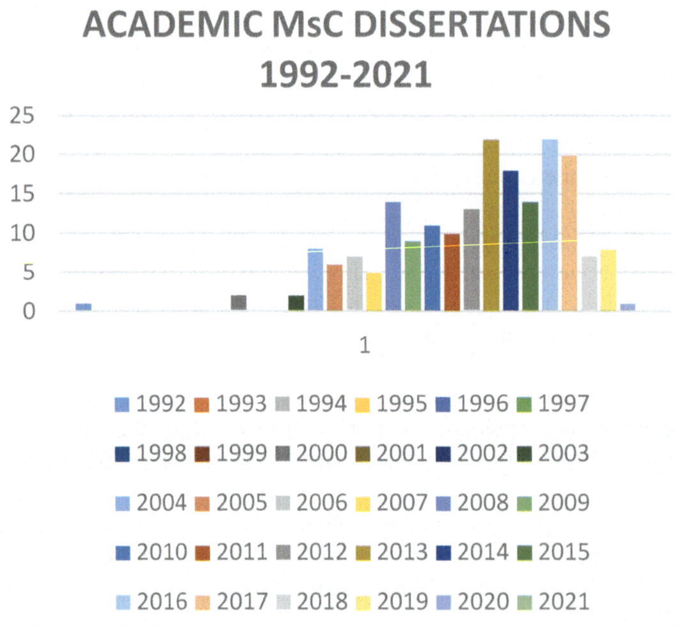

[2] Sometimes, dissertations are included in IBICT's database with a considerable time lag.
[3] The Biblioteca Digital Brasileira de Teses e Dissertações (BDTD), https://bdtd.ibict.br/vufind/Content/whatIs.
[4] The collecting and sorting out of dissertations continue and will be the subject of future papers.

For M.Sc dissertations, we notice a concentration in 2013, 2014, 2016 and 2017. The period 2013–2016 concentrates almost half of the dissertations on mathematics textbooks (96 in a total of 200). The earliest dissertation we could locate was Meksenas (1992).

The following graph shows the distribution of professional M.Sc dissertations (we considered only PROFMAT ones, because of the very small number of degrees received from other professional programs).

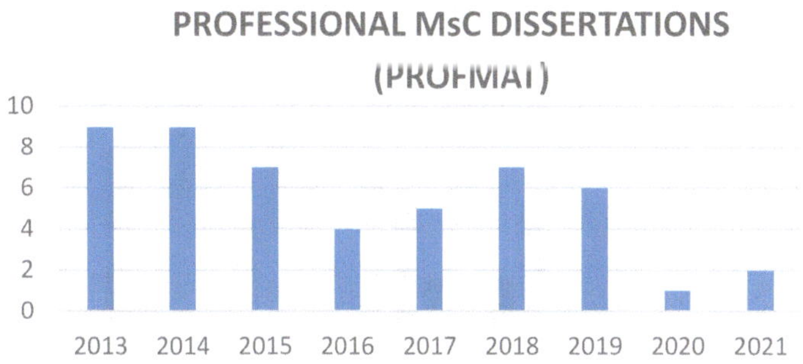

This program started, as we already mentioned, in 2011. There were located 50 professional M.Sc dissertations. The greatest number of dissertations was in 2013 and 2014, probably due to a "repressed demand".

Let us now look at the Ph.D. dissertations.

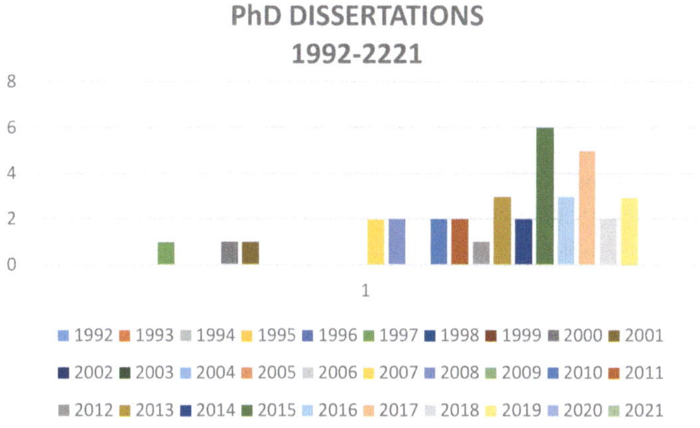

We found Ph.D. dissertations only from 1997 on, with concentrations in 2015 (six dissertations) and 2017 (five dissertations).

Among the 36 Ph.D. dissertations, only six had advisors who were PNLD assessors; this means a modest influence of PNLD on academic production at this level (16.6%). Only five of the 250 PNLD participants were advisors of these six dissertations. For academic M.Sc dissertations, the situation is completely different: 81 of the 200 **(40.5%)** M.Sc *dissertations on mathematics textbooks had advisors who* *participated in* PNLD. These advisors amounted to 29 (14.6%) of PNLD participants (250). *This shows indeed that* PNLD *had a strong influence on the academic production concerning textbook assessment*.

The following graph shows the distribution of these dissertations along the years.

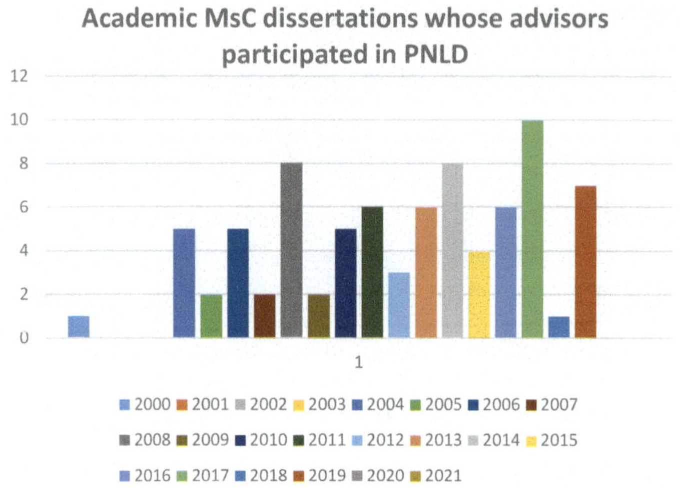

Now, let us look more closely at the dissertations. They were classified as follows:

1. Public policies on textbooks.
2. Influence of PNLD on textbook publishing.
3. Mathematics textbook publishing market.
4. Assessment of mathematics textbooks.
5. Choice and use of mathematics textbooks by teachers.
6. Production and testing of new mathematics textbooks.
7. Other subjects.

If we consider the Ph.D. dissertations, the distribution is the following.

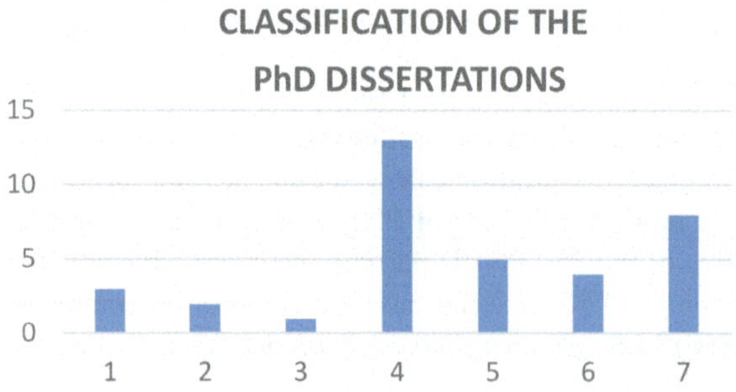

The PROFMAT M.Sc dissertations were distributed as follows.

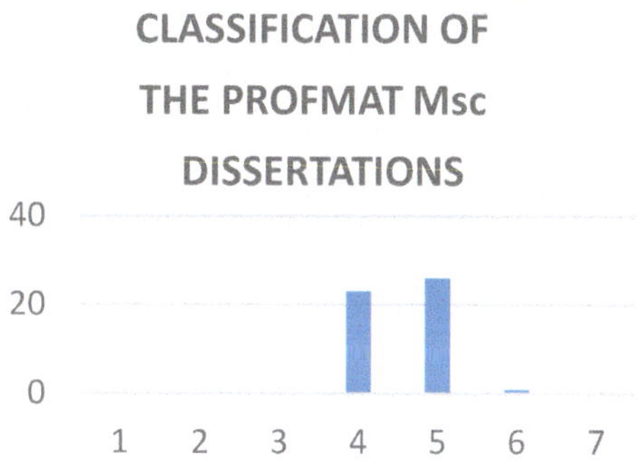

The situation for M.Sc academic dissertations is as shown below.

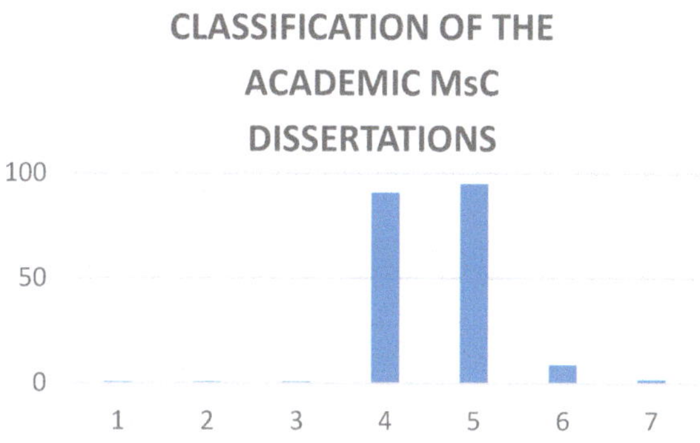

Note: All images were drawn by the author.

References

BRASIL/MEC. 2001. Recomendações para uma Política Pública de Livros Didáticos. Brasília, DF.

Cassiano, Célia Cristina de Figueiredo. 2007. O Mercado do Livro Didático no Brasil: da Criação do Programa Nacional do Livro Did'tico (PNLD) à Entrada do Capital Internacional Espanhol (1985–2007). Tese de Doutorado. Doutorado em Educação. São Paulo, SP: Pontifícia Universidade Católica de São Paulo.

Carvalho, João Bosco Pitombeira Fernandes de. 2018a. The Brazilian Mathematics Textbook Assessments. ZDM, 50, 773–785.

Carvalho, João Bosco Pitombeira Fernandes de. 2018b. Textbooks for Millions: The Brazilian Mathematics Textbook Assessment Program. Proceedings of the Second International Conference on Mathematics Textbook Research and Development, 5–24. Rio de Janeiro: Universidade Federal do Rio de Janeiro.

King, Elizabeth. 2011. Education is Fundamental to Development and Growth. World Bank Blogs. Accessed on April, 21, 2023.

Meksenas, Paulo. 1992. A produção do livro didático: sua relação com o Estado, autor e editor. Dissertação de Mestrado. São Paulo: Universidade de São Paulo (USP).

Torres, Rosa Maria 2000. Melhorar a qualidade da educação básica? As estratégias do Banco Mundial. In Livia de Tommasi, Mirian. J. Warde & Sérgio Haddad (orgs), O Banco Mundial e as Políticas Educacionais (3ed). São Paulo: Cortez/PUC.

Mistaken Identities: Scribal Errors or Conceptual Missteps in Ancient Chinese "Multiplication" Tables?

Joseph W. Dauben

When I last saw Gert Schubring it was in Rio de Janiero, for lunch at a breezy restaurant overlooking the beach at Copacabana, in the spring of 2013. Although Gert Schubring and I have corresponded over many decades, have exchanged letters over matters of mutual interest related to the history of mathematics and mathematics education, have participated in various congresses and meetings at Oberwolfach, among others, the most recent occasion I had to see Gert in person was in Brazil, in the spring of 2013. I was there thanks to an invitation to participate in the inaugural meeting of the CLE Colloquium for Philosophy and History of Formal Sciences at the University of Campinas (Unicamp) in Campinas, Brazil, in March of that year. Gert was also on hand a few days later for the 10th Brazilian National Seminar on the History of Mathematics, which also met at Unicamp, for which he had organized a workshop, "O papel da visualizaçao na história dos números complexos." I had also been invited to speak at the Seminar as well, where I presented an overview of my own recent research: "The Evolution of Mathematics in Ancient China. From the newly discovered 數 *Shu* and 算數書 *Suanshu shu* Bamboo Texts to the *Nine Chapters on the Art of Mathematics*" (Dauben 2020: 25–78).

I'm not sure if anything I said in that lecture sparked Gert's interest in particular problems in the history of Chinese mathematics, but several years later he wrote to me about the *Nine Chapters* and a matter that arose there concerning negative numbers that were introduced for the first time by Chinese mathematicians in the course of their attempts to find solutions of simultaneous linear equations in which negative numbers arose quite naturally in the course of their algorithmic solution

J. W. Dauben (✉)
City University of New York, New York, NY, USA
e-mail: jdauben@gc.cuny.edu

A. Verdugo Rohrer and J. Zender (eds.), *History of Mathematics and Its Contexts*, Trends in the History of Science, https://doi.org/10.1007/978-3-031-60828-5_7

procedures, and to which an entire chapter of the *Nine Chapters* is devoted, namely Chapter 8 on the *fangcheng* method. Gert later mentions this in his recently published *Analysing Historical Mathematics Textbooks* (2022). There he devotes an entire section to textbooks in China and offers a succinct yet informative account of the role textbooks played in education from early times in the Han dynasty down to the Song dynasty (206 BCE–1279 CE), roughly the period he covers until the end of the thirteenth century (Schubring 2022: 27–31). Of particular interest to Gert was the discovery Chinese mathematicians made of negative numbers. As he explained:

> This chapter, named *fangcheng* (square arrangement), is valued as the mathematically most significant one. It shows solving problems by manipulating numbers arranged in table form in parallel columns. It corresponds to the solution of linear systems in n unknowns (≤ 6), which has been interpreted as using matrices. This chapter presents major challenges to understand and translate it (see Table 2.2). For instance, regarding what one can understand as a rule of signs, Chemla and Guo Shuchun translate modernising a term which commentator Liu Hui gives as quantities (gains and losses represented by red and black counting rods), as "numbers" (Nombres de même nom…) (Chemla and Guo Shuchun 2004, p. 627), while Dauben and [Xu] Yibao just refer to quantities ("If the signs are the same …"), (Dauben et al. 2013, vol. 1, p. 248). (Schubring 2022: 28)

As early as 2005 Gert had pointed out the distinction Chinese mathematicians made from the earliest times between positive and negative quantities in his book, *Conflicts Between Generalization, Rigor, and Intuition. Number Concepts Underlying the Development of Analysis in 17th-19th Century France and Germany*. There he cites (Tropfke 1980: 145) and (Sesiano 1985: 107) in explaining that:

> It is well known that red rods were used for calculating with positive quantities and black rods for calculating with negative quantities in Chinese mathematics. Negative values were permissible as intermediate values during calculation, but not as solutions of systems of equations. The literature refers to a typical example, one of the tasks contained in the classical mathematics textbook *Chiu chang suan shu* [*Jiuzhang suanshu*] (mathematics in nine chapters, approx. 250 BCE). In modern terms, it can be written as the system of equations
>
> $$\begin{aligned} 2x + 5y - 13z &= 1000 \\ 3x - 9y + 3z &= 0 \\ -5x + 6y + 8z &= -600 \end{aligned}$$
>
> having the solutions $x = 1200$, $y = 500$, $z = 300$ (Schubring 2005: 36).

But he cautions in a footnote:

> [Lam] Lay-Yong and [Ang] Tian-Se strove to infer from the practice of such examples that the Chinese were the first to have had "the earliest negative numbers" and "the concept of negative numbers" ([Lam] Lay-Yong, [Ang] Tian-Se 1987, p. 222 and 236). They did not see that operating with subtractive quantities as intermediate entries means just one step in a long process of conceptualization (Schubring 2005: 36).

As Gert later explained to me, this was a reflection of the careful thought he had given to the important distinctions, conceptually, that needed to be drawn among the concepts of number, quantity, and magnitude: "In my 2005 book *Conflicts between* ..., I am analysing the issue of quantities vs numbers as basic dimension for developing general number concepts..." (personal communication dated June 10, 2023). In his related critique of accounts of the development of algebra generally, Gert has been quite specific: "Not to include the difference between quantity (or magnitude) and number into the basic dimensions of the research design means to exclude from the analysis a majority of the most important contemporary problems in the development of the algebraic concept field and thus to miss key historical insights, in particular concerning the extension of the multiplication operation" (Schubring 2005: 35).

It is not my intention here to go further into the details of the history of algebra or analysis and the importance of these distinctions for properly appreciating their development, particularly in the modern period, to which Gert has produced his own careful study in *Conflicts Between Generalization, Rigor, and Intuition.* But in appreciating the importance he has placed on distinctions concerning different concepts of number, it is also worth considering the different ways in which operations with numbers can be treated, as in the case of early Chinese mathematics, especially with respect to the close connections between addition and multiplication. In fact, in what follows the focus will be specifically on the case of elementary arithmetic in ancient Chinese mathematics—in particular the relationship between addition, multiplication, and the appearance very early in Chinese history of what have often been described as nine-nines multiplication tables.

Beginning with the Basics

As Jacques Sesiano notes in a paper devoted to "A Reconstruction of Greek Multiplication tables for Integers":

> A fundamental step for beginners in mathematics is to know about multiplication. Thus people had either to know the multiplication tables or to have them readily at hand. Such tables took various forms, depending on the base of the numerical system and the symbols used for the numerals. (Sesiano 2002: 45)

But Sesiano adds an important caveat at the beginning of his article about such tables:

> Greek multiplication tables for integers have come down to us in fragmentary form only. An additional problem is that editors of such fragments generally transcribe them without any attempt to look at the structure of the tables and thus often without drawing parallels with other related fragments. (Sesiano 2002: 45)

The same, it turns out, is true in the case of tables that are known in China since ancient times as *nine-nines* tables, which generally begin (until the Tang dynasty or so, 618–905 CE) laconically with "nine-nines—eighty-one" and generally proceed in descending order from there. Here the grammar, or lack of any verbs or other connectives, will play a significant role in interpreting exactly what these nine-nines tables were meant to convey.

Edouard Chavannes and the First Published Han Dynasty Nine-Nines Table

Émmanuel-Édouard Chavannes (1865–1918) was an accomplished French sinologist whose interest in China was first awakened as a student at the École Normale Supérieure in Paris and reinforced after he served briefly as an attaché with the French delegation to Peking in 1889. Later that year he assumed the Chair of Chinese and Manchu Tartar Languages and Literatures at the Collège de France. In 1903 he was elected a member of the Institute de France, and the following year assumed the editorship of 通報 *T'oung Pao*, the world's oldest international journal of sinology, founded in 1890. For complete biographical details of Chavannes and his several expeditions in China as well as his many publications, see (Cordier 1917).

Of the fifteen or so pages that Cordier devotes to Chavannes's bibliography (Cordier 1917: 131–147), the item of greatest interest to historians of mathematics is Chavannes's publication, in 1913, of the works mostly on wooden slips and boards, as well as some on paper, that Aurel Stein had acquired during his second expedition to China (1906–1908) and had entrusted to Chavannes for study, translation, and commentary. Remarkably, although Stein was a gifted linguist and counted Sanskrit, Kashmiri, Persian, and Punjabi among the oriental languages in which he was fluent, Chinese was not one of them, and so he often relied on Chavannes for translations of material in which Stein was particularly interested (see, for example, Stein's publication of material related to what he termed "relics of China's ancient silk trade," for which he cited Chavannes's translations of passages in Chinese from the sources Stein wished to discuss (Stein 1920–1921)) (Figs. 1 and 2).

Sir Mark Aurel Stein (1862–1943) was a highly respected orientalist and intrepid archaeologist/anthropologist/adventurer, whose accomplishments were uniquely the product of Britain and British India in the Victorian Age, as Frank Ikle has amply demonstrated in his brief biographical sketch of Stein (Ikle 1968). Stein, Jewish by birth but raised as a converted Lutheran, was born in Budapest, educated in Hungary and Germany, pursued Indian and Iranian studies in Vienna and Leipzig, but received his doctorate from Tübingen in 1883, *cum laude*. Following further studies at Oxford and the British Museum, Stein obtained a position in India as Registrar of Punjab University in 1887 and then, the following year, was appointed Principal of the Oriental College in Lahore, a position he retained until 1899. This gave him the opportunity to emulate the travels of two of his personal heroes, Alexander the Great and 玄奘 Xuanzang. The latter, a Tang dynasty Chinese Buddhist monk famous for his journey to India in the early seventh century,

Fig. 1 Émmanuel-Édouard Chavannes (1865–1918). Wikimedia Commons

Fig. 2 Mark Aurel Stein (1862–1943). Photograph by John Thomson, The Grosvenor Studios, 1909, courtesy of the Wellcome Collection

spent fifteen years collecting Indian manuscripts which he translated into Chinese upon his return to China in 643 CE.

In 1899, Stein secured funding from the British government for an expedition into Chinese Turkestan, in the course of which he retraced in part the same path Xuanzang had taken along one of the southern trails of the silk route. Stein used his expertise as a land surveyor (acquired during a period of obligatory military service in Hungary where he was trained in geography and land surveying) to draw up detailed maps of the regions he later traversed, travelling as far as Khotan, but all the while he was also looking for artefacts of archaeological and ethnographic interest. It was on the second of Stein's expeditions (1906–1908) that he made his way through 新疆 Xinjiang (尼雅 Niya and 樓蘭 Loulan) into 甘肅 Gansu Province (敦煌 Dunhuang), where the northern and southern Silk Routes meet.

Stein had heard of the discovery at Dunhuang not long before of a trove of documents in Buddhist caves, the Mogao Grottoes. In particular, numerous manuscripts had been discovered in a small cave, No. 17, the "Library Cave." There, Stein was able to negotiate the purchase of thousands of works on scrolls and paper written in such languages as Chinese, Khotanese, Sanskrit, Sogdian, and Tibetan, among others, along with textiles and other artefacts. What is of particular interest here is that he found a Chinese nine-nines table written on a wooden slip (see Fig. 4) in an abandoned watchtower at 萬歲顯武燧 Wansui xianwusui, along a portion of the Great Wall just to the north of Dunhuang (Luo and Wang 1993: 92), and others, at Dunhuang cave No. 17, written on paper (as one depicted below, Fig. 10) (Lam and Ang 2004: 75).

Some idea of what Stein, Chavannes, and others faced in dealing with the vast collection of materials obtained as a result of their travels in Turkestan may be gathered from photographs taken when Chavannes's protégé, Paul Pelliot (1878–1945), visited Dunhuang and, like Stein, came away with another trove of manuscripts. Educated at the Sorbonne and the École des Langues Orientales Vivantes, where he mastered Mandarin Chinese, Pelliot accepted a research position at the École Française d'Extrême Orient in 1900 in Hanoi, where later he became Professor of Chinese. But it was his expedition to the Dunhuang caves and the many manuscripts he was able to obtain there, in 1908, that established his reputation and brought him into the same circle of scholarly adventurers as Stein and Chavannes. A photograph of Pelliot in the "Library Cave" at Dunhuang gives some idea of what everyone who dealt with such materials had to confront (Fig. 3).

Working on the Stein Collection of Material from Turkestan

Chavannes himself describes what working on the material Stein had entrusted to him was like as follows:

Fig. 3 Paul Pelliot sorting material stored in the "Library Cave" (Cave 17) at Dunhuang, 1908. Photograph in the collection of the Musée Guimet, archives photographiques, AP8186 (Dischner 2023)

> M. Stein a bien voulu me confier l'étude des documents chinois, pour la plupart écrits sur bois, qu'il a retirés des sables de l'Asie centrale. Cet honneur était redoutable; je n'ai pas tardé à m'en apercevoir lorsque je me suis trouvé en présence de deux mille pièces environ qu'il a fallu d'abord examiner à la loupe une à une pour faire le départ entre celles qui étaient inutilisables et celles qui étaient susceptibles d'être déchiffrées. Une moitié des fiches qui constituent la grosse masse de ces textes ayant été ainsi éliminée, j'ai dû lire celles qui restaient, les classer par groupes et les traduire. Pour mener à bien cette entreprise j'ai profité de diverses aides que je suis heureux de mentionner. (Chavannes 1913: i)

As Chavannes notes, the task he faced was a formidable one, more than 2000 items, mostly on wood, that he first examined with a magnifying glass to determine which were most likely to be of interest and sufficiently intact to be transcribable with reasonable accuracy. This reduced the lot by about half, which he then proceeded to classify into groups and began their transcription and subsequent translation. This work was undertaken, he gratefully acknowledges, with the help of several Chinese colleagues. One of those was in fact Stein's "lettré"

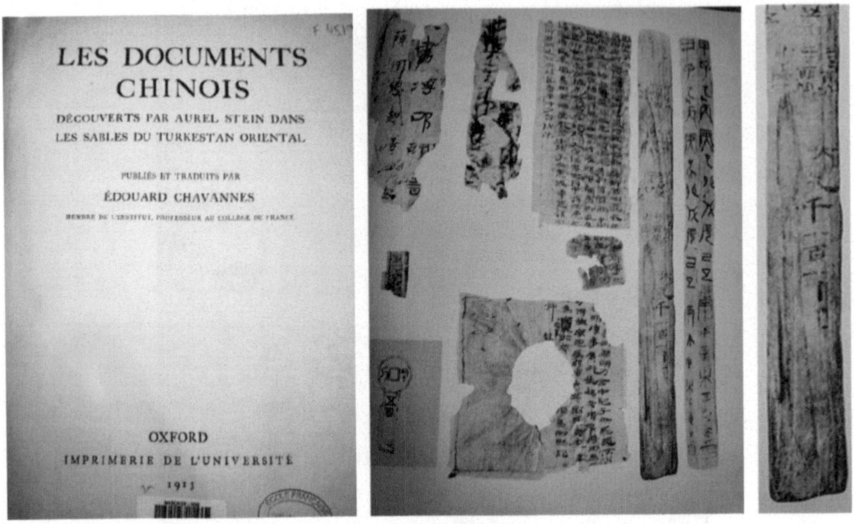

Fig. 4 Left: title page of (Chavannes 1913). In the middle is Plate XX on which the nine-nines wooden slip is second from the right. Right: an enlarged view of the bottom portion of the nine-nines table

(scholar-assistant), 蔣孝琬 Jiang Xiaowan (Tsiang Hiao-yuan), who had been on-site with Stein at Dunhuang and served as his amanuensis in transcribing hastily the documents in which Stein was most interested but had only had time to give "une lecture provisoire." Jiang's notes in fact were only of limited help, "laissait subsister presque toutes les difficultés réelles" (Chavannes 1913: i).

Fortunately, at just that moment Chavannes happened to meet two Chinese students in Paris who proved immensely helpful. The first, 吳勤訓 Wu Qinxun (Wu K 'in-hiun), helped him to collate and arrange the various fragmentary pieces into groups and then aided Chavannes in the tedious task of trying to determine the meaning of specific problematic characters, while the second student, 委懷 Wei Huai (Wei Houai), worked with Chavannes to improve his translations.

Chavannes subsequently had to decide how much of Stein's material to prepare for publication—only the parts of which he was certain, or the much larger portion that he was sure must contain numerous errors? In the end, Chavannes decided that it was better to provide as much access to the diverse materials Stein had collected as possible, and leave to others the matter of corrections "qui seront proposées par les érudits soucieux de collaborer au développement de nos connaissances sinologiques" (Chavannes 1913: i).

Chavannes's Transcription and Interpretation of the Han Dynasty Nine-Nines Table

As noted, Chavannes's 1913 publication was the first of any of the nine-nines tables. Working with the material Stein had either already published or had sent to Chavannes for study, given that Stein was overwhelmed by the amount of material he had retrieved, Chavannes limited his own publication and discussion of the material from Turkestan to works mostly on wood but some on paper, including what Chavannes listed as No. 702, a wooden slip on which is written a portion of a nine-nines table (Fig. 4):

Chavannes transcribed the text on this wooden slip as follows (Fig. 5):

He then translated the nine-nines table in symbolic terms (Fig. 6):

Note that Chavannes here indicates that the total of all the multiples on the board is said to be 1113, but he was aware that the total of all the multiples

XXII

SÉRIE T. xxvi

N° 702. — T. xxvi. 1. 702*

Fiche incomplète dans le sens de la largeur; 260 mm. de haut; 24 mm. de large. Nous sommes ici en présence d'une table de multiplication dont on peut reconstituer la disposition primitive de la manière suivante:

Lisez chaque registre de haut en bas et de droite à gauche:

九九八十一
八九七十二
七九六十三
六*九*五*十*四*
五*九*卌*五*
四*九*卅*六*
三*九*廿*七*
二*九*十*八*

八八六十四
七八五十六
六八卌八
五八卌
四*八*卅*二*
三*八*廿*四*
二*八*十*六*
七*七*卌*九*
六*七*卌*二*

五七卅五
四七廿八
三七廿一
二*七*十*四*
六*六*卅*六*
五*六*卅*
四*六*廿*四*
三*六*十*八*

二六十二
五五廿五
四五廿
三五十五
二*五*十*
四*四*十*六*
三*四*十*二*
二*四*八*
三*三*九*

二三而六
二二而四
一一而一

大凡千一百一十三

Fig. 5 Note that the characters to the left of the fourth column from the right, all denoted with an asterisk, provided what Chavannes surmised was missing from the wooden slip that had survived with only a portion of what should have been a complete nine-nines table (Chavannes 1913: 149)

he has transcribed only comes to 1111, and he was at a loss to account for this discrepancy. Why should this nine-nines table end with a "Total général" of 1113? He summarized the situation as he then understood it as follows:

> C'est le nombre 1111 qui représente la somme totale de tous les produits des multiplications énoncées dans la table ci-dessus. Pour obtenir le nombre 1113 qui est indiqué par la fiche, il faudrait, entre la multiplication 2 × 2 = 4 et la multiplication 1 × 1 = 1 , admettre la multiplication 2 × 1 = 2.. Mais cette multiplication n'est pas conforme au système qui n'admet dans aucun autre cas la multiplication d'un nombre par l'unité; d'ailleurs on voit nettement sur la fiche que la multiplication 1 × 1 suit immédiatement la multiplication 2 × 2. Il n'y a donc pas place pour la multiplication 2 × 1 et je ne m'explique pas comment on obtient le total 1113. (Chavannes 1913: 150)

In order to account for the apparent anomaly Chavannes mentions here, namely that the total given on the nine-nines table comes to 1113—and his lament that there was no room for the multiplication 2 × 1 on the board and he could not explain how the total of 1113 was obtained—we must consider the chronology of archaeological excavations of such documents as the nine-nines tables beginning in the early twentieth century, of which this was the first to be published. In fact, it is remarkable that Chavannes was able to account so clearly for all but a small part of the nine-nines document that he reproduced in his compendium of the Stein material from Dunhuang and nearby sites. It would not be until three more nine-nines tables were discovered in 2002 that further light would be shed on the situation of the nine-nines table for which Chavannes was trying to find an explanation but could not (Fig. 6).

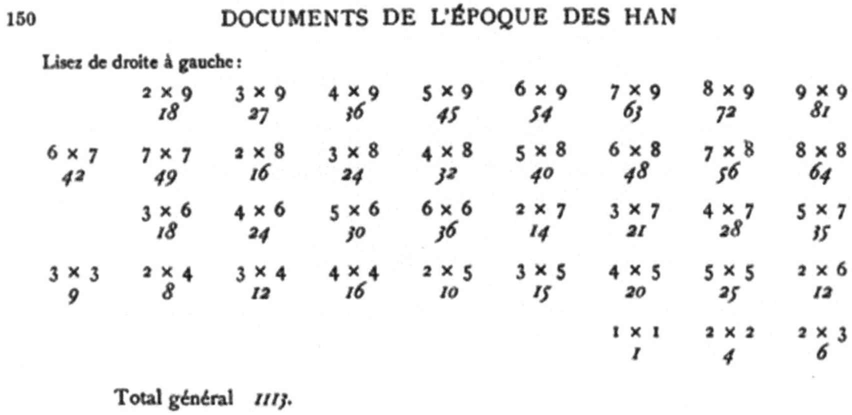
150 DOCUMENTS DE L'ÉPOQUE DES HAN

Lisez de droite à gauche :

	2 × 9 18	3 × 9 27	4 × 9 36	5 × 9 45	6 × 9 54	7 × 9 63	8 × 9 72	9 × 9 81
6 × 7 42	7 × 7 49	2 × 8 16	3 × 8 24	4 × 8 32	5 × 8 40	6 × 8 48	7 × 8 56	8 × 8 64
	3 × 6 18	4 × 6 24	5 × 6 30	6 × 6 36	2 × 7 14	3 × 7 21	4 × 7 28	5 × 7 35
3 × 3 9	2 × 4 8	3 × 4 12	4 × 4 16	2 × 5 10	3 × 5 15	4 × 5 20	5 × 5 25	2 × 6 12
						1 × 1 1	2 × 2 4	2 × 3 6

Total général 1113.

Fig. 6 Chavannes (1913: 150)

Publication of the Nine-Nines Table in Chinese

The first publication in Chinese of the Han dynasty nine-nines tables that Stein had discovered not far from Dunhuang was the result of research by 羅振玉 Luo Zhenyu and 王國維 Wang Guowei published in 1914 (later reprinted in 1993). Remarkably, this work was produced and printed in Japan, where as a Qing loyalist following the collapse of the empire in 1911, Luo had connections (counting among his friends both French and Japanese scholars like Édouard Chavannes, Paul Pelliot, Count Ōtani Kōzui (大谷 光瑞), and Naitō Torajirō (内藤 虎次郎)) as well as the means to publish (see Yang and Whitfield 2013). Following the Xinhai Revolution, Luo took up residence in Kyoto, where he did his most intensive work related to Chinese archaeology, culminating in the book of 1914.

Wang Guowei 王國維 was a student of Luo Zhenyu at the 東文學社 *Dongwen xueshe* (Japanese Language Teaching School), a school Luo had established in Shanghai for teaching Japanese. In 1911, Wang also fled China with Luo following the Xinhai Revolution. Working together with Wang in Kyoto, the two completed an abridged volume devoted to the material collected by Ariel Stein in Turkestan and published by Chavannes (Fig. 7). For details, see (Bonney 1986).

But when Luo and Wang published their Chinese version of Chavannes transcription of the nine-nines table, theirs differed in significant ways from the reading Chavannes had given to the numbers he had found:

Where Chavannes thought he could clearly read 二六十二 (2–6–12) in column *d* (see Fig. 8) and 一一而一 1-1-*er*-1 (1–1–is–1) in column *f*, Luo and Wang were not so sure, and instead indicated these were too indefinite to identify clearly by signifying their positions, but not the characters, with empty square boxes (column *c*). Note that the transcription Luo and Wang provide (bottom of column *b* in Fig. 8) stops at 二二而四 2-2-*er*-4 (2–2–are–4) and gives a "grand total" (大凡千一百一十 *dafan qian yibai yishi*) for the sum of all the entries in the nine-nines table as 1110 (bottom of column *g*) rather than the total of 1113 given by Chavannes. It is this version of the Han dynasty nine-nines table showing a total of 1110 that has later been republished on several occasions, including (Li and Du 1963: 20, 1987: 14; Lam and Ang 2004: 74; Martzloff 2006: 219) (Fig. 9).

According to Li Yan and Du Shiran, in their *Concise History of Chinese Mathematics*, this example of a nine-nines table was just one among many excavated in North-Western China since the end of the nineteenth century. They don't identify its provenance but reproduce this transcription and note that it gives a total of 1110. This is the same wooden strip that Lam and Ang identify as "a multiplication table tablet unearthed at Liu Sha 流沙 and its translation" (Lam and Ang 2004: 74). The designation of Liu Sha here as a place name however is mistaken; 流沙 *liusha* is really a somewhat poetic reference to the "shifting sands" of the desert terrain where Stein had found this multiplication table in an abandoned watchtower along a section of what survives of the Great Wall just north of Dunhuang.

Lam and Ang go on to cite (Li 1954: 16–18) as the source of their information and (Li and Du 1963: 20) for the illustration, noting further that Li Yan reproduced the "multiplication table" from "Studies of the strip by Luo Zhenyu and Wang Guowei," but without referring to Wang and Guo's *Liusha zhui jian* of 1914. They also incorrectly describe the number 1110 as the total number of individual

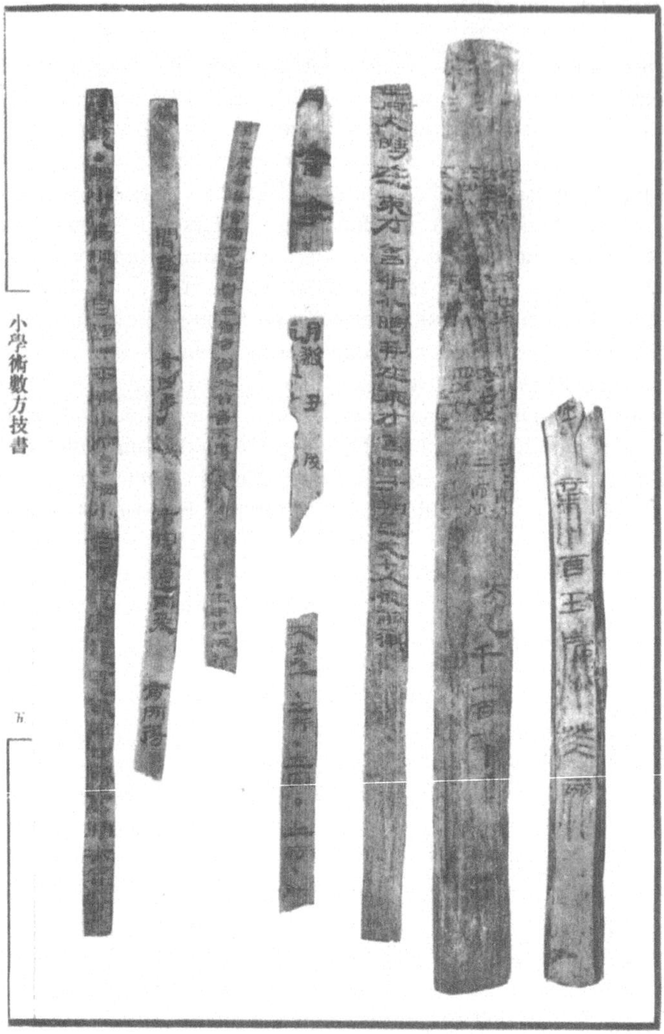

Fig. 7 Plate from the publication in Chinese of selected items from Chavannes's book of 1913, showing the nine-nines table second from the right, in (Luo and Wang 1914: 21)

characters written on the strip (Lam and Se 2004: 75), rather than as the total of the presumed multiples to be found on the slip, as Chavannes had surmised in his publication of the nine-nines table in 1913. They further note (Fig. 10):

> Li Yan (1954, pp. 16–18), who quoted from studies of the strip by Luo Zhenyu 羅振玉 and Wang Guowei 王國維, confirmed that the table ended with two twos. However, since the lower portion of the strip was partially putrefied, the table was tantalizingly incomplete. The legible characters of the multiplication table are reproduced in Fig. 3.1 (here Figure 9), with their translation on the right. (Lam and Ang 2004: 75)

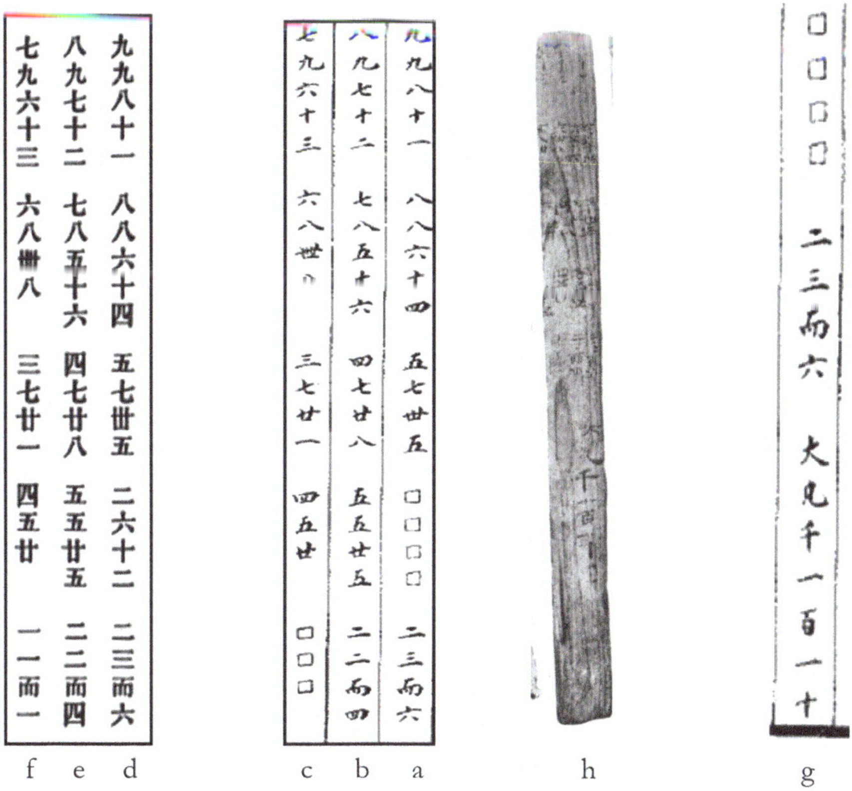

Fig. 8 Leftmost columns d, e, f excerpted from Chavannes (1913). Rightmost three excerpted from Luo and Wang (1914). Bottom seven characters (g): 大凡千一百一十 *dafan qian yibai yishi* (grand total 1110)

They go on to explain:

> The table was also carefully copied on scrolls and preserved in the Dunhuang Caves during the Tang dynasty (Libbrecht 1982, pp. 212 and 218). This type of multiplication table has a special feature: each product displayed is repeated in rod numerals. According to Li Yan and Du (1987, p. 15), this is the earliest extant material displaying rod numerals in a written form (See Fig. 3.2) (here Figure 10). (Lam and Ang 2004: 76)

The Evidence from Liye, Three Han Dynasty Nine-Nines Tables Excavated in 2002

In April of 2002, with the immanent construction of the Wanmipo Hydropower Station at Liye, Longshan County, Hunan Province, the Hunan Institute of Archaeology and Cultural Relics was called in to excavate a site of approximately 20,000 square metres just to the east of the Youshui River (IAHP 2003). In addition to an ancient city wall, the site included wells and tombs from which more than 30,000

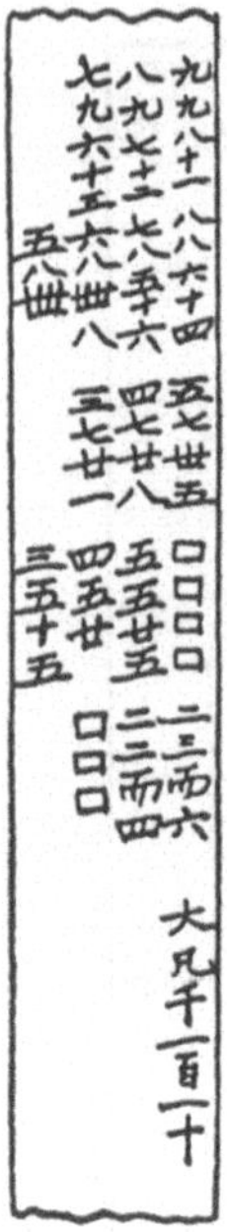

Fig. 9 Example of a nine-nines verse, among many excavated in North-Western China since the end of the nineteenth century. From p. 74 of *Fleeting Footsteps: Tracing the Conception of Arithmetic and Algebra in Ancient China* (Revised Edition), Lam Lay Yong and Ang Tian Se, Copyright @ 2004 by World Scientific Publishing Co. Pte. Ltd., Singapore, by permission of the publisher

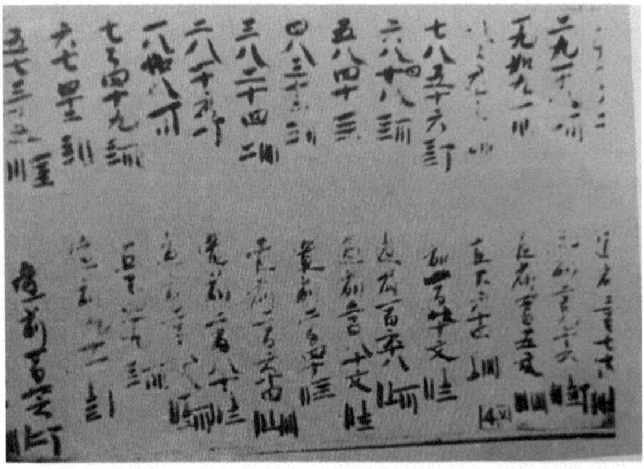

Fig. 10 Section of the multiplication table in the Tang dynasty Dunhuang scroll: 立成 算法 *Li cheng suanfa* (A Short Course on Calculation Techniques). British Library Or. 8210, Stein 930. Reproduced with permission of the British Library, London, UK; from p. 75 of *Fleeting Footsteps: Tracing the Conception of Arithmetic and Algebra in Ancient China* (Revised Edition), Lam Lay Yong and Ang Tian Se, Copyright @ 2004 by World Scientific Publishing Co. Pte. Ltd., Singapore, by permission of the publisher

bamboo strips and wooden tablets have been recovered. Among these, Ancient Well No. 1 was particularly valuable (Fig. 11) and yielded documents recording political, economic, military, cultural, and social events from the Warring States period through the Qin (221–201 BCE) and Han (206 BCE–220 CE) dynasties. In particular, the Qin dynasty annual calendars found at Liye provide fairly complete records, day by day, for the years 221–206 BCE.

Of particular interest for answering the questions Chavannes raised about the nine-nines table he published in 1913 but could not resolve are three Qin dynasty nine-nines tables found in Well No. 1 at Liye. These may be compared with similar nine-nines tables that comprise the opening portion of the *Suanshu shu*. For details, see (Peng 2000, 2001, Cullen 2004, Dauben 2008).

The nine-nines tables from Liye now offer a possible solution to the conundrum faced by Chavannes in accounting for why the nine-nines board he had transcribed from the Han dynasty nine-nines table Aurel Stein had found to the north of Dunhuang should have given the "incorrect" total of 1113, because this is exactly the same result that is found on the nine-nines boards excavated at Liye in 2002. Unlike the works found on hundreds of bamboo strips that comprise the *Shu* and *Suanshu shu*, the nine-nines tables recorded on wooden boards did not require the reconstruction of a jigsaw-like puzzle of disparate pieces, and in fact, on two different wooden boards, three nine-nines tables were found to be almost entirely intact. Thus the examples of nine-nines tables from Liye represent the best preserved and most complete examples of such tables found to date.

Fig. 11 Well No. 1 at Liye, where three versions of nine-nines tables were found on two wooden boards, one with tables on both the front and back. Photo courtesy of Zhou Xulin, Southwest Minzu University, PRC

What the Liye tables reveal, however, is that the assumptions of the early researchers like Édouard Chavannes were wrong in taking for granted that these were multiplication tables. For all three examples of the nine-nines tables from Liye clearly end as follows (HPICRA 2012: Plate 1) (Fig. 12, transcribing the bottom four lines of the right-most nine-nines table, columns *a* through *d*, reading right to left)[1]:

a: 二二而四 2–2–*er*–4 (2–2–are–4)

b: 一一而二 1–1–*er*–2 (1–1–are–2)

c: 二半而一 2–½–*er*–1 (2–½–are–1)

d: 凡千一百一十三字 *fan qian yibai yishi san zi* (total 1113)

Had Chavannes not misread the numbers on the wooden board Stein had sent him to decipher as a multiplication table, he might have realized that it was in fact a table recording the *addition* of numbers. Thus, when adding up the numbers on both the wooden slip, Stein 702, and the three from Liye, they indeed total 1113. As Jacques Sesiano rightly warned in interpreting material from ancient sources, one should not bring contemporary assumptions or prejudices into the historical laboratory (Sesiano 2002: 45). New material should be judged on its own terms, and even if the format and numbers look familiar, that familiarity can be quite misleading, as has been the case for everyone who has interpreted these nine-nines tables from ancient China as multiplication tables, for at least in the case with which Chavannes was confronted, and as the three wooden tables from Liye make clear, this was not always the case (Fig. 13).

And yet we are still left with a further mystery, because the oldest example of such tables also belongs to this arithmetic family—namely the Tsinghua University "computational" table—which is clearly about multiplying and not just from 9 × 9 to 1 × 1, but from 90 × 90 to ½ × ½. For details about this Warring States (475–221 BCE) table and how it could have been used to carry out divisions and possibly even the extraction of roots, see Berlin (2014), Li and Feng (2014), Feng (2017), and Qiu (2014). In fact, how can we account for this much earlier and even more extensive example of a multiplication table that goes well beyond the nine-nines tables we have been examining here? And why do so many of the nine-nines tables, whether of addition or multiplication, often (but not always) end displaying the "grand totals" of 1110 or 1113? These are questions that Zhou Xulin and I intend to answer in a more extensive account than is possible here of the early history of nine-nines tables in ancient China.[2]

---

[1] Note at the top of column *a*, the character for the numeral 9, i.e. 九 *jiu*, is followed by a duplication mark: =, meaning that the character just prior to this mark should be duplicated. In this case, the nine-nines table begins as one would expect with 九九八十一 *jiu jiu bashiyi*, i.e. 9–9–81. At the bottom of column *a*, the duplication mark appears again, 二=而四, meaning 二二而四 as transcribed here.

[2] This larger, more detailed study of nine-nines tables from ancient sites currently in progress, co-authored with Professor 周序林 Zhou Xulin of the School of Chinese Languages and Literatures at Southwest Minzu University in Chengdu, PRC, will encompass not only all of the currently known

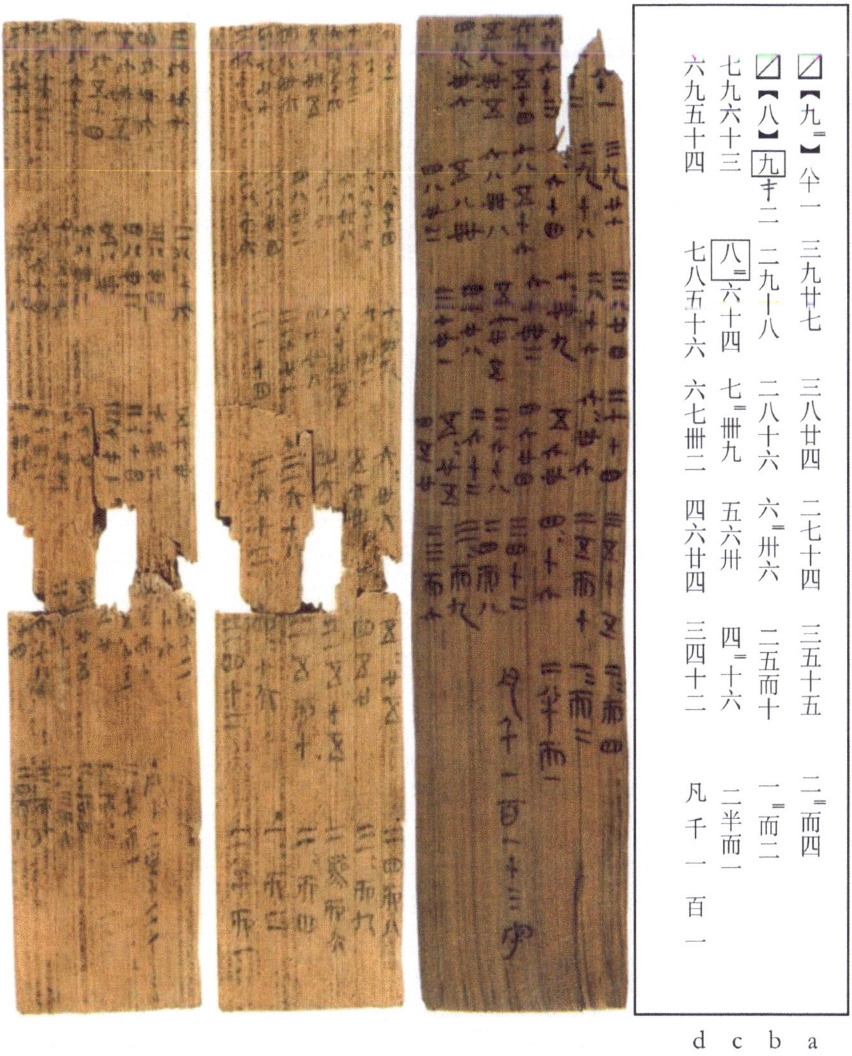

Fig. 12 The two left-most images are the recto and verso of a single wooden board from Liye, published in (Liye Qinjian Museum 2016: 63), reproduced here courtesy of the Liye Qinjian Museum and the Unearthed Documents and Ancient Chinese Civilization Research Collaborative Innovation Center of the Renmin University of China, and with permission of the Zhongxi Book Company, Shanghai, PRC. The third image from the left is a wooden board with a nine-nines verse, also from Liye, published in (HPICRA 2012, vol. 1: 1); image courtesy of the Hunan Institute of Cultural Relics and Archaeology and published here with the permission of Cultural Relics Press, Beijing

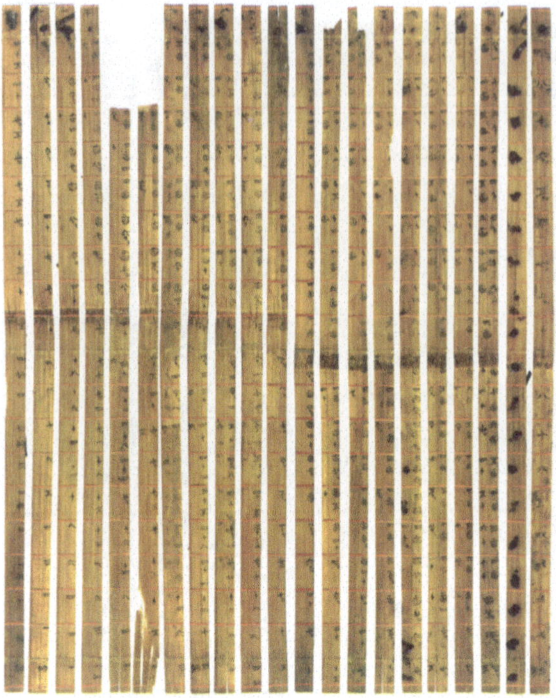

Fig. 13 The Tsinghua University computational table (dated to ca. 300 BCE, Warring States Period), showing multiplications from ½ × ½ to 90 × 90, published in 清華大學藏戰國竹簡 (肆) *Tsinghua daxue cang zhanguo zhujian (si)* (Warring States Bamboo Slips Preserved at Tsinghua University (Beijing), IV), Li Xueqin, ed. Shanghai: Zhongxi Book Company, 2013, vol. 4: Appendix Fig. 3; reproduced here courtesy of the Tsinghua University Center for Research and Protection of Unearthed Documents, and with the permission of the Zhongxi Book Company, Shanghai, PRC

In conclusion, Fig. 14 shows how easily ancient Chinese mathematics has become a part of popular culture. In the case of the nine-nines tables found at Liye, these (along with all of the material recovered from Liye) were almost immediately heralded in local newspapers as "the most important archaeological discovery after the Qin Terracotta Warriors" (*China Daily* 2015). Within just a few years they were featured nationally on government-issued postal stamps, a clear sign of how serious archaeology, elementary mathematics, and something familiar to Chinese

nine-nines tables excavated in China, but also examples of nine-nines tables found in Japan and Korea, as well as examples of Tibetan nine-nines tables from the Dunhuang caves. I am pleased to acknowledge here the collaboration of Zhou Xulin in the development of the basic lines of argument to be found here in this preliminary account of our results to date. For their very helpful comments and suggestions for revisions of earlier versions of the analysis presented here of ancient nine-nines tables, I am grateful to Gregg de Young and Zhou Xulin.

children as part of their earliest education, easily captured the public's attention. Here the Liye nine-nines tables serve as fitting reminders that mathematics and pedagogy have featured significantly in Chinese culture from the most ancient times.

Fig. 14 中國郵政 *Zhongguo youzheng* (China Post). 1.20 yuan stamp issued 13 September 2012, photogravure by Beijing Postage Stamp Printing Works. When issued, this was said "to carry the multiplication table as well as characters on both sides. The earliest calculation table ever found in China, it is of great value to the study of Chinese history of arithmetic." The Chinese at the top of the stamp reads: 里耶秦简·乘法九九口诀 *Liye Qin jian·chengfa jiujiu koujue* (Liye Qin wooden board·nine-nines multiplication rhymes). Reproduced here with the permission of China Post

References

Berlin, Jeremy. "World's Oldest Decimal Times Table Found in China." *National Geographic*, April 5, 2014. https://www.nationalgeographic.com/science/article/140405-chinese-oldest-multiplication-table-decimal (accessed June 25, 2023).

Bonney, Joey. *Wang Kuo-wei: an intellectual biography*. Cambridge, MA: Harvard University Press, 1986.

Chavannes, Édouard, ed. and trans. *Les documents chinois découverts par Aurel Stein dans les sables du Turkistan oriental*. Oxford, UK: Oxford University Press, 1913.

Chemla, Karine, and Guo Shuchun. *Les neuf chapitres: Le classique mathématique de la Chine ancienne et ses commentaires*. Paris: Dunod, 2004.

Chemla, Karine. "L'histoire des sciences dans la sinologie des débuts du XIXe siècle: les Biot père et fils." In *Jean-Pierre Abel-Rémusat et ses successeurs. Deux cents ans de sinologie française en France et en Chine*, edited by Pierre-Etienne Will and Michel Zink, 411–37. Paris: Publications of the Académie des Inscriptions et Belles-Lettres & Collège de France, 2020.

Cordier, Henri. "Nécrologie—Édouard Chavannes." *T'oung Pao* 18 (1917): 114–147.

Cullen, Christopher. *The Suàn shù shū* 筭數書 *'Writings on reckoning': A translation of a Chinese mathematical collection of the second century* BC, *with explanatory commentary*. Needham Research Institute Working Papers, no. 1. Cambridge, UK: The Needham Research Institute, 2004.

Dauben, Joseph W. "算數書 *Suan shu shu* (A Book on Numbers and Computations). English Translation with Commentary." *Archive for History of Exact Sciences* 62, no. 2 (2008): 91–178.

Dauben, Joseph W. "九章算术 *Jiu zhang suan shu* (Nine Chapters on the Art of Mathematics)—An Appraisal of the Text, its Editions, and Translations." *Sudhoffs Archiv* 97, no. 2 (2013): 199–235.

Dauben, Joseph W. "The Evolution of Mathematics in Ancient China. From the newly discovered 數 *Shu* and 算數書 *Suan shu shu* Bamboo Texts to the *Nine Chapters on the Art of Mathematics*." X Seminário Nacional de História da Matemática, *Revista Brasileira de História da Matemática — an international journal on the History of Mathematics* 19, no. 37 (2020): 25–78.

Dauben, Joseph W., Guo Xuchun, and Xu Yibao. *Nine Chapters on the Art of Mathematics*. With the Annotations by Liu Hui [State of Wei] and Notes and Annotations of Li Chunfeng and Associates [Tang Dynasty]. A Critical Edition and English Translation Based upon a New Collation of the Ancient Text and Modern Chinese Translation by Guo Shuchun; English Critical Edition and Translation, with Notes by Joseph W. Dauben and Xu Yibao. Shenyang: Liaoning Education Press, 2013; in 3 volumes.

de la Vallée Poussin, Louis. "Obituary Notice—Édouard Chavannes." *Bulletin of the School of Oriental Studies*, University of London 1, no. 2 (1918): 147–151.

Dischner, Matthew Z. "Marc Aurel Stein." *The Sogdians: Influencers on the Silk Roads*, an on-line exhibition organized by the Freer Gallery of Art and Arthur M. Sackler Gallery, Smithsonian Institution, Washington, DC., 2023. https://sogdians.si.edu/sidebars/marc-aurel-stein/ (accessed June 26, 2023).

Feng, Lisheng 冯立昇. "On the Structure and Functions of the Multiplication Table in the Tsinghua Collection of Bamboo Slips." *Chinese Annals of History of Science and Technology 1*, no. 1 (2017): 1–23.

HPICRA. 湖南省文物考古研究所 *Hunan sheng wenwu kaogu yanjiu suo* (Hunan Province Institute of Cultural Relics and Archaeology). 里耶秦簡 (壹) *Liye Qin jian (I)* (Liye Qin bamboo slips (I)). Beijing: Cultural Relics Publishing House, 2012.

IAHP (Institute of Archaeology of Hunan Province). 湖南龙山里耶战国 — 秦代古城一号井发掘简报 *Hunan Longshan Liye Zhangguo—Qin dai gucheng yi hao jing fajue jianbao* (Hunan Longshan Liye Warring States—Excavation Report of Well No. 1 in an Ancient City of the Qin Dynasty). 文物 *Wenwu* (Cultural Relics) 1 (2003): 4–35.

Ikle, Frank W. "Sir Aurel Stein. A Victorian Geographer in the Tracks of Alexander." *Isis* 59, no. 2 (1968): 144–55.

Lam, Lay Yong, and Ang Tian Se. *Fleeting Footsteps. Tracing the Conception of Arithmetic and Algebra in Ancient China.* Singapore: World Scientific, rev. ed. 2004.

Laufer, Berthold. "Édouard Chavannes." *Journal of the American Oriental Society* 38 (1918): 202–205.

Li, Junming 李均明, and 冯立昇 Feng Lisheng. 清华简《算表》的形制特征与运算方法 *Tsinghua jian "Suanbiao" de xingzhi tezheng yu yunsuan fangfa* (The shape, characteristics, and calculation methods of the Tsinghua bamboo-strip "Calculation Table"). 自然科学史研究 *Ziran kexueshi yanjiu* (Studies in the History of Natural Sciences) 33, no. 1 (2014): 1–17.

Li, Xueqin 李學勤, ed. 清華大學藏戰國竹簡(肆) *Qinghua daxue cang zhanguo zhujian* (si) (Warring States Bamboo Slips from Tsinghua University Collection (IV)). Shanghai: Zhongxi Book Company, 2013.

Li, Yan. 李 儼. 敦煌石室算書 *Dunhuang shishi suanshu* (Arithmetical books from the Dunhuang caves). 中大季刊 *Zhongda jikan* (Chinese University Quarterly) 1, no. 2 (1926): 1–4.

Li, Yan. 敦煌石室立成算經 *Dunhuang shishi Licheng suanjing* (A Short Course on Calculation Techniques from the Dunhuang caves). 圖書季刊 *Tushu jikan* (Books Quarterly) 1, no. 4 (1939): 386–396.

Li, Yan. 中国古代数学史料 *Zhongguo gudai shuxue shiliao* (Historical Materials for the History of Ancient Chinese Mathematics). Shanghai: Shanghai Science and Technology Press, 1954; repr. 1963; in 2 volumes.

Li, Yan 李儼, and 杜石然 Du Shiran. 中国古袋數学简史 *Zhongguo gudai shuxue jianshi* (Chinese mathematics, A concise history). Peking: Zhonghua Shuju, 1963.

Li, Yan and Du Shiran. *Chinese Mathematics, A Concise History.* Translated by John Crossley and Anthony Lun. Oxford, UK: Clarendon Press, 1987.

Libbrecht, Ulrich J. "Mathematical Manuscripts from the Tunhuang Caves." In *Explorations in the History of Science and Technology in China*, edited by Li Guohao et al., 203–229. Shanghai: Shanghai Chinese Classics Publishing House, 1982.

Liu, Ying-hua. "A Study on the Tibetan Arithmetic Works Found at Dunhuang." *Journal of Tibet University* 1 (2015): 74–81.

Liye Qin Bamboo Strips Museum. 里耶秦簡博物館藏秦簡 *Liye Qin jian bowuguan cangQin jian* (Qin Bamboo Slips Preserved at the Liye Qin Bamboo Slips Museum), Shanghai: Zhongxi Book Company, 2016.

Luo, Zhenyu 羅振玉, and 王國維 Wang Guowei. 流沙墜簡 *Liusha zhui jian* (Bamboo strips lost in shifting sands). Kyoto, Japan: Kyoto Publishing House, 1914; repr. Beijing: Zhonghua chuju (Zhonghua Book Co.), 1993.

Martzloff, Jean-Claude. *A History of Chinese Mathematics.* Translated by Stephen S. Wilson. Berlin: Springer, 2006.

Peng, Hao 彭浩. 中国最早的数学著作《算數書》*Zhongguo zuigu de shuxue zhuzuo Suan shu shu* (The Earliest Arithmetic Book in Chinese, Suan shu shu). 文物 *Wenwu* (Cultural Relics) 9 (2000): 85–90.

Peng, Hao. 張家山漢簡《算數書》注釋 *Zhangjiashan hanjian "Suanshu shu" zhushi* (A Collation of Suanshu Shu Excavated from Han Dynasty Tomb 247 at Zhangjiashan). Beijing: 科學出版社 Kexue chubanshe (Science Press), 2001.

Qiu, Jane. "Ancient times table hidden in Chinese bamboo strips." *Nature* (January 7, 2014). https://doi.org/10.1038/nature.2014.14482 (retrieved June 15, 2023).

Schubring, Gert. *Conflicts between Generalization, Rigor and Intuition. Number Concepts Underlying the Development of Analysis in 17th–19th Century France and Germany* (Sources and Studies in the History of Mathematics and Physical Sciences). New York, NY: Springer Science+Business Media, 2005.

Schubring, Gert. *Analysing Historical Mathematics Textbooks* (International Studies in the History of Mathematics and its Teaching). Cham, Switzerland: Springer Nature, 2022.

Sesiano, Jacques. "The Appearance of Negative Solutions in Medieval Mathematics." *Archive for History of Exact Sciences* 32, no. 2 (1985): 105–150.

Sesiano, Jacques. "A Reconstruction of Greek Multiplication Tables for Integers." In *From China to Paris: 2000 Years Transmission of Mathematical Ideas*, edited by Yvonne Dold-Samplonius, Joseph W. Dauben, Menso Folkerts, and Benno van Dalen, 45–56. Stuttgart: Franz Steiner Verlag, 2002.

Stein, Aurel. "Central-Asian Relics of China's Ancient Silk Trade." *T'oung Pao* 20, no. 2 (1920–1921): 130–141.

Tho, Alexis See. "World's Oldest Decimal Multiplication Table Awarded by Guinness World Record." *Tsinghua Newsletter* (Beijing: Tsinghua University), (April 23, 2017). https://www.tsinghua.edu.cn/en/info/1253/2378.htm (accessed June 15, 2023).

Tropfke, Johannes. *Geschichte der Elementarmathematik.* Vol. 1: *Algebra*, 4th ed., newly revised by Kurt Vogel. Berlin: de Gruyter, 1980.

Wang, Helen, ed. *Handbook to the Stein Collections in the UK*. London, UK: British Museum Occasional Papers 129, 1999.

Yang, Chia-Ling, and Whitfield, Roderick, eds. *Lost Generation: Luo Zhenyu, Qing Loyalists and the Formation of Modern Chinese Culture.* Atlanta, GA: Saffron Books, 2013.

Zhang, Xiao-hu 张小虎. 敦煌算经九九表探析 *Dunhuang suanjing jiujiu biao tanxi* (Analysis of Multiplication Tables in Dunhuang Mathematical Classics). 温州大学学报: 自然科学版 *Wenzhou daxue xuebao: zerankexue ban* (Journal of Wenzhou University: Natural Sciences Edition) 32, no. 2 (2011): 1–6.

An Episode from the History of Mathematics at the Beginning of the Twentieth Century (To the Prehistory of the Soviet Mathematical School)

Sergey S. Demidov

The Soviet mathematical school which constitutes one of the leading mathematical schools of the second half of the twentieth century was born in the 1930s. Its formation resulted from the convergence of the ideas of the two leading Russian schools of that time—the Moscow school and the Petersburg one. However, paying tribute to historical reality, we must say Leningrad school—as in these years, the northern capital took the name of V.I. Lenin in order to honor his memory. This fact took place previously the confrontation between mathematicians of the two capitals, which often had a rather violent character (see, for example, (Demidov 2009)). This confrontation ended in 1934 by the transfer of the Presidium of the Academy of Sciences of the USSR and this of the V.A. Steklov Mathematical Institute in Moscow. By the force of the circumstances, the leading mathematicians of the two capitals were obliged to live and work together. The search for a common language laid them to unify their ideas—a fact which determined the birth and the development of the Soviet mathematical school.

P.L. Chebyshev (1821–1894) tried to maintain normal relations between the representatives of the two schools. Graduating from Moscow University, he took there the first steps in science, and until the end of his life, he conserved his love and his respect for his alma mater. However, some of his students had completely different feelings for Moscow, and, first of all, their leader, A.A. Markov (1856—1922). It might be stressed that the confrontation between the mathematicians of the two capitals was formed on ideological contradictions. The Petersburg mathematical circles of the late-nineteenth and early-twentieth centuries dominating by

S. S. Demidov (✉)
Moscow State University, Moscow, Russia
e-mail: serd42@mail.ru

A. Verdugo Rohrer and J. Zender (eds.), *History of Mathematics and Its Contexts*, Trends in the History of Science, https://doi.org/10.1007/978-3-031-60828-5_8

pro-Western aspirations kept anti-monarchist sentiments and militant atheism. The understanding of mathematics and its place in the world was essentially positivist. «We solve concrete problems by rigorous methods[1] and we will not permit any philosophical fog (e.g., in the spirit of G. Cantor)». This orientation was reflected in the nature of the chosen tasks and in the using methods, as well as, ultimately, it was reflected on the topics of their research. The school is characterized from its tendency to applied mathematics,[2] a constant desire for rigor and at the same time an effective tool in order to solve mathematical problems, for the construction of algorithms that allow one to bring it either to an exact numerical result or to a suitable approximate solution. They cultivate also the desire for simplicity and elementary nature of the used means. Such an orientation determined a well-known distrust of the newfangled trends in Western mathematics (in particular, the vanguard Riemann's ideas were considered as mathematical decadence), as well regarding new trends in the foundations of mathematics. In the sharpest form, these ideas were inherent in aforementioned A.A. Markov.

Mathematics in Moscow was developed in a different ideological context than those in St. Petersburg. Although Muscovites, like Petersburgers, had an inherent interest in applied topics (a manifestation of a common desire for all Russian at that time to develop the industrial forces of the Empire), in all the other respects their mathematical predilections were radically diverged: Muscovites had a special interest in geometric studies, had a deep interest in philosophy, as well as in the philosophical comprehension of mathematical activity. From this, latter arose the cause to name the Moscow school of that time philosophical and mathematical (see Nekrasov Pavel 1904)). Moreover, in Moscow mathematical circles followed Leibniz's philosophical ideas, as well as idealistic philosophy and even religious one. The most influential Moscow mathematician of the late-nineteenth–early-twentieth century was N.V. Bugaev (1837–1903), who was a philosopher and more a creator of his own philosophical doctrine—the evolutionary monadology (Nekrasov Pavel 1904; Demidov 1988). His philosophical views were «refracted» in mathematics by a special interest concerning the discontinuous functions, in the development of the theory in which he saw the most important task of modern mathematics.[3] Accordingly, to such worldview directions, their mathematical affinities took shape. The main fields of studies chosen by the Muscovites were differential geometry (it might be stressed that the foundation of the Moscow school of theory of surfaces was laid by F.G. Minding's student, K.M. Peterson (1828–1881), whose ideas were developed in the works of B.K. Mlodzeevsky and D.F. Egorov (1869 - 1931), as well as analytical mechanics and applied mathematics with N.E. Zhukovsky (1847–1921) and S.A. Chaplygin (1869–1942). In the framework of

[1] Rigor was understood in the sense of possibly setting the exact limits of the errors of the used methods.

[2] The only exception was number theory—an area which has become traditional for Petersburgers since L. Euler's epoch.

[3] Solving this problem, Bugaev with his students conceived a special mathematical discipline—the arithmology, which, however, did not seriously developed.

worldview, they radically diverged from the Petersburgers, and the hostile attitude of the latter toward the Muscovites' activities becomes quite understandable. Even the successes of Muscovites in the development of applied topics (the theory of hydraulic shock by N.E. Zhukovsky or his achievements in the field of aero- and hydrodynamics) did not become the basis for their approving assessments—in St. Petersburg (see, for example, Steklov Vladimir 1991, p. 25), it was customary to look down on the results of "Moscow celebrities". This inherently ideological divergence[4] became the basis to establish confrontational relations between mathematicians of the two capitals, which led to mutual alienation and various conflict situations, which were often dealt with the Moscow Mathematical Society. One of these conflicts was the controversy regarding the well-known results of S.V. Kovalevskaya concerning the integration of the equations of motion of a rigid body around a fixed point; for her paper in 1888, she was awarded the prize of the French Academy of Sciences. A.A. Markov found gaps in her demonstrations and sharply attacked her. The Muscovites P.A. Nekrasov and G.G. Appelrot defended her work and tried to fill these gaps.[5]

The dramatic period of Russian history started with the outbreak of the First World War. The revolutionary events of 1917, which were transformed into a fratricidal Civil War, had serious effects on the scientific community and did not moderate the sharpness of hostility regarding the relations between the mathematicians of the two cities. And although in 1918, after the Bolshevik government moved to Moscow, Petrograd (as St. Petersburg became known since August 1914) lost the privilege to be the capital, the mathematicians of the northern capital, which still housed the Academy of Sciences (recently no longer Imperial), looked their Muscovites colleagues with their usual disregard. Of course, the Muscovites were offended by such a kind of attitude. Moreover, now they represented the capital and, most importantly, they had something to be proud of—they were not at all the "poor provincials": the results of N.N. Luzin (1883–1950) as well as these of his students brought Muscovites to the forefront of modern mathematics!

And when the news reached them that the mathematicians of the northern capital were going to solemnly celebrate the 100th anniversary of the birth of Pafnuty Lvovich Chebyshev, the founder of the St. Petersburg School, which, as

---

[4] Which should be considered in the context of the general cultural confrontation between St. Petersburg and Moscow (Demidov 2009).

[5] It was not possible to do this by the means which the mathematicians of that time disposed. It was only at the end of the twentieth century that it became possible to restore the order. V.V. Kozlov (see Kozlov Valerii 2000) played an outstanding role in the study of this issue.

then believed, fell on May 26,[6] they decided to participate in the anniversary celebrations in order to pay tribute to the graduate of Moscow University, who made there the first steps in science.[7]

In the early morning, a noisy campaign of people, mostly young, poured into the square in front of the Nikolaevsky railway station[8] in Petrograd. These were the mathematicians who arrived from Moscow, most of them were Professor Nikolai Nikolaevich Luzin's students. What was exactly the date and day of the week it is difficult to establish today. We can only say that it was at the beginning of June, as L.A. Lyusternik (1899–1981) (Demidov 2021; Lyusternik Lazar' 1965) attributed this event, and, as we have already said, May 26 was then the birthday of Pafnuty Lvovich Chebyshev: the Muscovites arrived just on time for the celebrations that were held by their Petrograd colleagues to commemorate the 100th anniversary of the great mathematician. As L.A. Lyusternik recalled: "the university professors N.N. Luzin, V.A. Kostitsyn, S.P. Finikov with their wives, the old guard of Luzitania—V.V. Stepanov, [...] V.N. Veniaminov, P.S. Uryson, the "Ivanovites"[9] A.I. Nekrasov—Rector of the Ivanovo Polytechnic Institute, D.E. Menshov, A.Ya. Khinchin, A.N. Vlasov, the graduate students of the 1st and 2nd Moscow State Universities—S.D. Rossinsky, V.S. Bogomolova, A.Yu. Zelenskaya, S.S. Kovner and N.K. Bari, who had just graduated from the university and «left» there, 8–9 senior students, among them Yulia Rozhanskaya, "Tatulya" (Tatiana Yulievna) Aikhenvald, Bela Pevzner, Mitya Perepyolkin, Kolya Nyuberg and others, including the author himself"—L.A. Lyusternik. The noisy campaign, «hand in hand, moved along the pavement of Nevsky Prospekt» (on which there was practically no transport—the city just beginning to recover after the events of the great revolution) toward the Palace Square. The details about this "landing" we draw today from two sources—from the above-mentioned memoirs of Lazar Aronovich Lyusternik (Lyusternik Lazar' 1965) and the recently published memoirs of Vladimir Aleksandrovich Kostitsyn.[10] Both of them described the events

---

[6] As it was later established by V.E. Prudnikov, who discovered the corresponding date in the parish register of the Church of the Transfiguration of the Lord, in the village of Spas-Prognanye, Kaluga province, Chebyshev was born on May 4 (16), 1821 (see Prudnikov Vasilii 1976).

[7] Chebyshev studied at Moscow University in 1837–41 and, after his graduation, started "to prepare for a professorship." He went to St. Petersburg in 1847 after defending in 1846 his master's thesis "An Experience in Elementary Analysis of Probability Theory" (Moscow, 1845). His tutor at Moscow University was professor N.D. Brushman (1796–1866); for his mentor, he retained deep respect and gratitude until the end of his days (Prudnikov Vasilii 1976).

[8] The currently Moscow railway station.

[9] In the cruel time of the Civil War, they fled from hunger and cold in Ivanovo-Voznesensk.

[10] Kostitsyn V.A. (1883–1963) was a man of a completely extraordinary fate. Son of a gymnasium teacher in the city of Efremov, Tula province, in 1902–1907 he studied at the Mathematical Department of the Faculty of Physics and Mathematics of Moscow University, where his supervisor was D.F. Egorov. He acquainted friendship with N.N. Luzin. While still a student, he became a Bolshevik and actively took part in the revolution of 1905—he led the university combat squad. After the defeat of the revolution, he served a year and a half in prison and was forced to emigrate to France. He lived in Paris, where he graduated from the Sorbonne and took his first steps in mathematics.

from memory, already being elderly,[11] and some episodes (often important) conveyed in different ways, and sometimes they even enter in mutual contradiction. So, Kostitsyn calls P.S. Aleksandrov (and even counts him among the speakers at the meeting of the Petrograd Mathematical Society—more on that below), while Lyusternik writes that Pavel Sergeevich fell ill and did not go to Petrograd. However, there is nothing surprising in such inconsistency—very often the memory fails. In addition, in 1921, both narrators belonged to completely different circles of the Moscow mathematical community—Kostitsyn to the professorship (he was at that time one of the most influential figures in the Soviet scientific administration and went, in particular, to settle some problems concerning the management of the Pulkovo Observatory), and Lyusternik should be ranked among the very "lower classes" on its hierarchical ladder—he was just a student. So their position and, accordingly, their awareness, and therefore their points of view, were completely different. However, we should trust Lazar Aronovich regarding the issue of Alexandrov's presence in the "landing": He shared his memoirs at a solemn meeting dedicated to the centenary of the Moscow Mathematical Society (Lyusternik Lazar' 1965), in the presence of Alexandrov himself, while Kostitsyn wrote on this event settled in Paris, without having no way to verify the reported information. In other matters, we should rather trust Kostitsyn, who lived "at the top". So, Lyusternik writes (Lyusternik Lazar' 1965, p. 33): "The invitation to take part in the anniversary, of course, was sent to Moscow University—alma mater of P.L. Chebyshev—and to the Moscow Mathematical Society", of which he was a founding member. Kostitsyn reports (Kostitsyn Vladimir 2017, p. 126): "about 40 mathematicians from Moscow, young and not so young … had to leave for Petrograd without invitation, because the Academy of Sciences, faithful to its traditions, did not invite anybody from Moscow". And therefore, when the Muscovites had already reached the Academy of Sciences, Petrograd masters "were frightened by such an extensive delegation and showed neither hospitality nor correctness: under

In Paris, he made a close acquaintance with V.I. Lenin. In 1916 he returned to Russia and actively participated to the events of the war with Germany, the revolution, the Civil War, and the organization of scientific and educational institutions. As he got a fairly high position in the scientific and the educational administration of the Soviet state, V.A. Kostitsyn did a lot to restore the normal life of the Russian physical and mathematical community. In particular, he contributed to the organization of the Research Institute of Mathematics and Mechanics of Moscow University, as well as to the establishment of publishing activities (for example, the release of an updated "Mathematical Sbornik"). In 1927, he went on a scientific mission to France, from where he never returned to the USSR. The main achievements of Kostitsyn relate to the calculus of variations, mathematical physics, astrophysics, geophysics, as well as mathematical biology (Kostitzin V.A. *Biologie mathématique.* Préface de V. Volterra. Paris: Armand Colin. 1937, etc.), where his results are recognized as classical. Recently, his memoirs and diaries (Kostitsyn Vladimir 2017) are published in two volumes, which became a valuable source on the history of science of the twentieth century.

[11] Records by V.A. Kostitsyn date back to May 1950, the memoirs of L.A. Lyusternik in 1961, when he "for some time for health reasons" was "torn off from his usual activities" (Lyusternik Lazar' 1965, p. 21).

the pretext that the honoring program had already been established, the representatives of Moscow were not given the possibility to present their address" (Kostitsyn Vladimir 2017, p. 127).

This approach is not particularly surprising. As we said at the beginning of our text, the Moscow and St. Petrograd mathematicians were in continuous state of confrontation for almost forty years, often turning into open clashes. We have already spoken on the disputes between Muscovites and Petersburgers regarding S.V. Kovalevskaya's results. Let us recall the debate over the results of V.G. Imshenetsky 1887–1891 concerning his methods to find rational integrals of linear differential equations, whose coefficients as well as the free term are entire rational functions (see Yushkevich Adol'f 1968, p. 433). The Muscovites (K.A. Andreev, P.A. Nekrasov and others) supported the St. Petersburg academician, who fiercely was criticized by his St. Petersburg colleagues (A.A. Markov, A.N. Korkin, K.A. Posse). These debates ended with a meeting of the Moscow Mathematical Society in June 1892, after which Imshenetsky died in a hotel of Moscow. As we have already said, in St. Petersburg and Moscow, mathematical communities were developed in a completely different spirit, which ultimately determined the views on mathematics and theoretical preferences that accentuated the differences in their research topics.[12]

The First World War began in 1914, the revolutions broke out in 1917, and the followed Civil War did not bring peace to their relations. All these events were hard for the science as well as for the education. The lack of a normal supply of food and fuel has put the urban population on the edge of survival. Large cities had a particularly hard time, especially Petrograd and Moscow, which became the capital when the authorities of the state moved in 1918. In 1921, the Civil War ended and a peaceful life gradually began to improve. The Institute of Physics and Mathematics of the Academy of Sciences started to work in Petrograd, with V.A. Steklov as director (it was from this Institute the Steklov Mathematical Institute, one of the leading mathematical institutes of the twentieth century, emerged in 1934), and already worked underway at Moscow University in order to organize the Research Institute of Mathematics and Mechanics,[13] inaugurated in 1922.

Such was the situation in Petrograd and Moscow in June 1921, the days when Petrograd mathematicians decided to celebrate the centenary of the birth of the great Russian mathematician P.L. Chebyshev.

[12] This "cold war" of mathematicians in the two capitals left its mark on the life of the entire Russian mathematical community, since the professorship of Russian universities largely consisted of graduates of the capital's universities.

[13] Lyusternik recalled (Lyusternik 1965, p. 25): "I will add: both in the carriage (that is, on the road from Moscow to Petrograd—S.D.), and in the student hostel (where the youth from the Moscow "landing" were settled; professors were placed in the House of Scientists—S.D.), the news was discussed—the upcoming opening in Moscow at the Moscow State University of the Institute of Mathematics and Mechanics. One must think that this news was also discussed at the hotel of the House of Scientists, where V.A. Kostitsyn stayed with his wife". Kostitsyn was one of the founders and leaders of the new institute.

The great meeting of the Academy of Sciences dedicated to this date (it was assembled a day after the Muscovites arrived in Petrograd) opened with A.V. Vasiliev's address, who, writes Kostitsyn (Kostitsyn Vladimir 2017, p. 128), «told the biography of Chebyshev and gave a very good overview of his activities. The academician Markov was supposed to speak about Chebyshev's works on number theory, who, apparently, is never able to speak on the topic and must always show his opposition to the established power. And now he spoke not so much about Chebyshev, but about the good era in which Chebyshev lived, while now "every milliterate comrade can come here, declare that I understand nothing in number theory, and kick me out". And one should have heard the tone in which this hateful word "comrade" was uttered».

Further, here we will already give the tribune to Lyusternik (Lyusternik Lazar' 1965, p. 25),[14] «V.A. Steklov read the report "Theory and Practice in the Works of P.L. Chebyshev", which was published two years later[15]».

The Muscovites had the opportunity to give their lectures in the next day at the meetings of the Petrograd Mathematical Society, perfectly in accordance with the anniversary of Chebyshev, which were held in the university building. From the Petersburgers, as Lyusternik writes (Lyusternik Lazar' 1965, p. 25), Ya.V. Uspensky, B.A. Venkov, G.M. Fikhtengolts and others spoke, from Muscovites—N.N. Luzin, A.I. Nekrasov, P.S. Uryson. And also, we must add, V.A. Kostitsyn. From his memoirs, we learn about this (see Kostitsyn Vladimir 2017, p. 127): "Nekrasov, Alexandrov,[16] Uryson and I reported from us".

Lyusternik either forgot about Kostitsyn's address, as Kostitsyn did about Luzin's talk, or preferred not to make any mention on the emigrant's report (Kostitsyn emigrated to France in 1928, however, by that time he was no longer alive). From Lyusternik we know the following about the content of the Moscow's addresses (about all except Kostitsyn's) (Lyusternik Lazar' 1965, p. 26): «A.I. Nekrasov spoke about his work on the theory of nonlinear integral equations connected to the theory of wave propagation. He developed the method in order to solve them, which was called the Nekrasov's method. Luzin constructed an example of an analytic function tending to infinity everywhere on the boundary of the circle of convergence; to do this, it was necessary to build on a sequence of circles tending to the boundary. Of course, the Luzitanians "cheered" for their patron, who

[14] This information is absent in Kostitsyn's memoirs.

[15] With some abbreviations, the text of Steklov's report was reproduced in the first volume of Uspekhi matematicheskikh nauk for 1946—in the 2 (12) issue (pp. 4–11), dedicated to the 150th anniversary of P.L. Chebyshev.

[16] We have already said that Kostitsyn's memory was not correct—he was convicted that Aleksandrov participated in the "landing". So, describing the walk of Muscovites through night Petrograd (and it was June—the time of white nights), Kostitsyn even writes (Kostitsyn Vladimir 2017, p. 127): «... Pavel Samuilovich Uryson did not notice anything, except for the beauty spilled around, and his alter ego, Pavel Sergeevich Aleksandrov, looked with hatred at the "dear teacher"». For us, such a "mistake" should serve as a serious warning: It is necessary to be critical of the information delivered by memory.

was supposed to show the “arrogant Petersburgers” a class in mathematics. When Luzin finished, Pavel Uryson ran up to him and exclaimed: “Nikolai Nikolaevich, you beat them with your circles! ...”. P.S. Uryson’s report was consecrated to some topological properties of flat continuums. Neither we nor the speaker himself understood that this was the first sign of a new mathematical school—the topological one».

Lyusternik concluded this passage with the following words: “This is how the first mathematical conference in Soviet times took place, the first meeting of mathematicians from two major scientific centers”. We will add the following to these words: This is how the first attempt by Muscovites to break the ice in relations with Petersburgers ended without visible external success. One could not talk about any serious breakthrough in these relations (although the Muscovites made any effort to improve them—this is how N.N. Luzin paid a visit to A.A. Markov, who then was settled in Peterhof (Lyusternik Lazar’ 1965, p. 27), if in 1926, that is, a good five years after the event, the academician Ya.V. Uspensky in a letter to academician A.N. Krylov gave such an assessment to the work of N.N. Luzin and his school (I quote from Ermolaeva Natal’ya 1989, p. 193): «Regarding Luzin, I know that he is a good specialist in his field (set theory and related Cantor-Lebesgue rubbish), a brilliant professor who created a school for his students in Moscow and with his influence abolished real mathematics in Moscow».

The next decade introduced a whole series of changes in the relationship between mathematicians of the two mathematical schools, which developed all the more successfully since so much began to depend on the capital, which Moscow has now become. However, an external factor put an end to this confrontation, the intervention of an outside force—the sovereign will of the "father of peoples" I.V. Stalin. In the reforms he carried out, the future building of Soviet science was conceived in the form of a pyramid, controlled from the top, in which the “headquarters of Soviet science” are located—in accordance with his ideas, the reorganized Academy of Sciences of the USSR, which should always be with him “at hand”, i.e., in Moscow. In accordance with this, the Presidium of the Academy and 14 leading institutes should been transferred to Moscow, which was carried out in 1934. So, in 1934 the V.A. Steklov Mathematical Institute moved to Moscow and the representatives of the two great schools were forced to live and work together. Their amazing synthesis arose, strengthened by scientists who came to the capital from other centers of mathematical thought—a synthesis that laid the foundation in order to form the Soviet mathematical school—one of the determinants regarding the development of world mathematical thought in the second half of the twentieth century.

And the "landing" of the mathematicians from Moscow, who arrived in the early June morning of 1921 at the Nikolaevsky railway station in Petrograd, became the first step for the construction of this renowned mathematical school, although none of the participants in this event could even think about it then. Here is how L.A. Lyusternik wrote about it (Lyusternik Lazar’ 1965, p. 28): “The first meeting after the revolution of the two main mathematical schools took place and it showed that despite all the difficulties and losses, mathematical science in our country is alive.

Moreover, it sprouts young sprouts. And, for the most part, we, boys and girls in science, found ourselves at the head of the great river of Soviet mathematical science. We were, in a sense, touched by its origin. Of course, obviously at that moment we didn't realize it, but we felt something like that".

References

Demidov, Sergey (1988). Der philosophische Kontext der Herausbildung der Moskauer funktiontheoretischen Schule. Schriftenreihe für Geschichte der Naturwiss., Techn.und Medizin (NTM). Bd. 25. S. 25–31.

Demidov, Sergey (2009). Saint-Pétersbourg et Moscou, deux capitals. Eds. C. Bartocci, P. Odifreddi. La mathématique. V. 1. Paris: CNRS Édition. P. 683–703.

Demidov, Sergey (2021). Демидов С.С. К 200-летию со дня рождения Пафнутия Львовича Чебышева // Вестник Московского Университета. Серия 1: Математика. Механика. № 3. С. 3–12. (In Russian)

Ermolaeva, Natal'ya (1989). Ермолаева Н.С. Новые материалы к биографии Н.Н. Лузина // Историко-математические исследования. Вып. 31. С. 191–203. (In Russian)

Kostitsyn Vladimir (2017). Костицын В.А. «Моё утраченное счастье ...»: Воспоминания, дневники. Т. 1. М.: Новое литературное обозрение. (In Russian)

Kozlov, Valerii (2000). Sofya Kovalevskaya: a mathematician and a person // Russian Mathematical Surveys. Vol. 55. Issue 6. P. 1175–1192.

Lyusternik Lazar' (1965). Люстерник Л.А. Выступление на юбилейном заседании Московского математического общества. Успехи математических наук. Т. 20. Вып. 3(123). С. 21–30. (In Russian)

Lyusternik Lazar' (1967). Люстерник Л.А. Молодость Московской математической школы / / Успехи математических наук. Т. 22. Вып. 2(134). С. 189–239. (In Russian)

Nekrasov Pavel (1904). Некрасов П.А. Московская философско-математическая школа и её основатели. Математический сборник. Т. 25. Вып. 1. С. 1–249. (In Russian)

Prudnikov Vasilii (1976) Прудников В.Е. Пафнутий Львович Чебышев (1821–1894). М.: Наука. (In Russian)

Steklov Vladimir (1991). Стеклов В.А. Переписка с отечественными математиками. Воспоминания / Отв. ред. В.С. Владимиров. Л.: Наука. (In Russian)

Yushkevich Adol'f (1968). Юшкевич А.П. История математики в России до 1917 года. М.: Наука. (In Russian)

Die Vorgeschichte und die Gründung des mathematisch-physikalischen Seminars an der Universität Göttingen (1850)

Menso Folkerts

In der ersten Hälfte des 19. Jahrhunderts erfuhren die deutschen Universitäten radikale Veränderungen: Zwischen 1803 und 1818 wurden zahlreiche Universitäten aufgelöst und andere neu gegründet, und eine neue „Ideologie" der höheren Bildung wurde formuliert und verbreitet. Die Bildung sollte dazu beitragen, alle geistigen Fähigkeiten des Individuums zu entwickeln und zu entfalten. Nach den Vorstellungen des Neuhumanismus sollte die Bildung durch das Studium der griechischen Kultur gefördert werden, weil diese Kultur das vollkommenste Beispiel der menschlichen Harmonie und Kreativität darstelle. Aus der Sicht des philosophischen Idealismus sollte die Ausbildung dazu führen, die Einheit des Wissens zu erfassen und die rationale Entfaltung aller Erkenntnis zu begreifen. Trotz unterschiedlicher theoretischer Begründungen des neuen Universitätsbegriffs stimmten die verschiedenen Richtungen darin überein, dass man das alte Nützlichkeitsprinzip der Universitäten aufgeben müsse. Vielmehr sei es Aufgabe der Universität, die Studenten zu selbständiger Tätigkeit zu erziehen und in ihnen die Idee der Wissenschaft, die sie im Innern trügen, zu erwecken.[1]

Der gegebene Ort an den Universitäten, um diese Fähigkeiten zu erlangen, war die philosophische Fakultät. Die Reformer verlangten, dass das intensive Studium der klassischen Philologie, der Philosophie und der Naturwissenschaften dem Fachstudium in den höheren Fakultäten vorangehen oder es begleiten müsse. Schon vor 1800 war an der Universität Halle ein Seminar für das Hauptfach der künftigen Gymnasiallehrer, die Klassische Philologie, eingerichtet worden.[2] Nach

M. Folkerts (✉)
Ludwig-Maximilians-Universität München, Munich, Germany
e-mail: m.folkerts@lrz.uni-muenchen.de

[1] Zu dieser Entwicklung siehe Turner (1971, 1987a) (vor allem S.225f.) und Turner (1987b).
[2] Siehe Turner (1987b), S.146.

A. Verdugo Rohrer and J. Zender (eds.), *History of Mathematics and Its Contexts*, Trends in the History of Science, https://doi.org/10.1007/978-3-031-60828-5_9

diesem Vorbild entstanden ab 1810 in Berlin und an anderen preußischen Universitäten Seminare für die Klassische Philologie und auch für das Fach Geschichte.[3]

Seminare für die Mathematik und für die naturwissenschaftlichen Fächer wurden erst später gegründet. Das erste derartige Seminar wurde für Physik im Jahre 1834 von Karl Gustav Jacobi und Franz Neumann in Königsberg eingerichtet. Es diente aber primär zur Förderung des wissenschaftlichen Nachwuchses und nicht zur Vorbereitung künftiger Gymnasiallehrer.[4]

Das grundlegende Werk zur institutionellen Geschichte der Mathematik in Deutschland im 19. Jahrhundert (Lorey 1910) behandelt die mathematisch-physikalischen Seminare nur am Rande. Erst in den letzten Jahrzehnten ist dieses Thema in den Fokus der mathematikhistorischen Forschung getreten, und vor allem Gert Schubring hat über diese Seminare zahlreiche Arbeiten veröffentlicht, die auf umfangreichen Quellenstudien beruhen.[5]

Auf Anregung des Mathematikers Ludwig Adolf Sohncke und des Physikers Ludwig Friedrich Kämtz wurde im Jahre 1839 an der Universität Halle erstmals ein mathematisch-naturwissenschaftliches Seminar gegründet, das die pädagogische Komponente in den Vordergrund rückte. Im Bericht aus dem Jahr 1849 wird festgehalten, dass in den vorangegangenen vier Jahren 54 Studenten an diesem Seminar teilnahmen.[6]

Das Seminar in Halle wurde zum Vorbild für ein ähnliches Seminar, das 1850 an der Universität Göttingen entstand. Über die Vorgeschichte und die Gründung dieses Seminars gibt es bisher noch keine zusammenfassende Darstellung.[7]

Die Anregung für ein solches Seminar ging von dem Göttinger Mathematiker Moritz Stern aus. Aus einem Brief von Stern an Theodor Braun, den zuständigen Vertreter im Universitäts-Kuratorium, wissen wir, dass Stern in einem Gespräch mit Braun diesen Wunsch geäußert und Braun ihn ermuntert hatte, Vorschläge zu machen. Stern konkretisierte seine Vorstellungen in einem Schreiben an Braun vom 10. Oktober 1849.[8] Er verwies auf die Statuten der Seminare in Königsberg und Halle. Sie seien aber „unpraktisch“, denn nach ihnen sollten besonders ausgezeichnete Abhandlungen in einer Zeitschrift publiziert werden, was aber niemals geschehen sei. Seinem Brief fügte Stern auch den Entwurf von Statuten für das künftiges Seminar bei, und er ging auch auf die Motive ein, die ihn

[3] Siehe Turner (1987b), S.145.

[4] Grundlegend hierzu: Olesko (1991).

[5] Übersichtsartikel: Schubring (1983, 1990, 1991, 2000); Spezialstudien zu Bonn (Schubring, 1985, 1989, 2004) und zu Münster (Schubring 1985a).

[6] Langhammer (1981), S.252. Langhammer informiert ausführlich über die Gründung des Seminars, seine Statuten und seine Aktivitäten.

[7] Für diesen Aufsatz konnte ich auch die Aufzeichnungen von Gert Schubring aus dem Universitätsarchiv Göttingen verwenden.

[8] Göttingen, Universitätsarchiv (im Folgenden: UA Göttingen), Kur. 7448, f.331r–332r.

geleitet hatten.[9] In Hinblick auf die Anzahl der Direktoren legte Stern sich nicht fest. Er schreibt[10]:

> In Beziehung auf die Mitglieder des Direktoriums habe ich den Entwurf sehr allgemein gehalten. Nöthigenfalls müßte wohl darauf gerechnet werden, daß das Institut nur von zwei Direktoren, einem für die mathematische und einem für die physikalische Abtheilung geleitet würde. Nach der Beschaffenheit der hiesigen Lehrkräfte würden aber die Arbeiten jeder Abtheilung sehr bequem unter zwei Direktoren vertheilt werden können. Für die physicalische Abtheilung würde ich also die Herren Professoren Weber und Listing nennen, für die mathematische Herrn Professor Ulrich und mich selbst, da ich doch einmal die Nennung meines Namens nicht füglich umgehen kann.

Das Kuratorium wandte sich zunächst an Carl Friedrich Gauß und bat ihn am 24. Oktober 1849 um eine Stellungnahme.[11] Gauß sollte sich vor allem dazu äußern, ob vier Direktoren (Wilhelm Weber, Benedikt Listing, Georg Karl Justus Ulrich, Moritz Stern) eingesetzt werden sollten oder nur ein Direktor für jede Abteilung, nämlich Weber und Stern. Der Brief war mit dem Zusatz „Vertraulich" gekennzeichnet, jedoch erlaubte das Kuratorium, Weber über die Anfrage zu informieren und sich mit ihm zu beraten.

Gauß antwortete am 6. November.[12] Er begrüßte generell die Einrichtung eines Seminars, um künftige Gymnasiallehrer auf ihren Unterricht in Mathematik und in den Naturwissenschaften vorzubereiten. An einigen Punkten in Sterns Entwurf übte er Kritik: Er schlug vor, die besten Seminarteilnehmer zwar zu belohnen, aber man solle keine Rangordnung vornehmen. Vielmehr sollten in jedem Semester maximal vier hervorragende Teilnehmer ausgezeichnet werden. Die Studenten sollten nicht nur Vorträge halten, sondern auch schriftliche Aufgaben lösen und Aufsätze verfassen; dadurch könnte die Leistung von allen Direktoren geprüft werden, und sie könnte auch für spätere Zeugnisse verwendet werden. Bei der Frage, ob zwei oder vier Direktoren ernannt werden sollten, müsse man klären, ob jeder Professor am Seminar als „Direktor" bezeichnet werden solle oder nur die Person, die für die Geschäftsführung zuständig sei. Gauß kritisiert, dass in Sterns Entwurf nicht die Funktion angegeben sei, die einem Direktor zugeteilt wird. Wenn „Direktor" dasselbe bedeute wie „Vorstand" im Statut des Seminars in Halle, so müssten nach Sterns Entwurf nur Weber und Stern als Direktoren wirken, und Ulrich und Listing wären überflüssig. Gauß äußert entschieden, dass Ulrich als langjähriges Mitglied der Prüfungskommission die Mängel in der Ausbildung der künftigen Mathematiklehrer besonders gut kenne. Falls für jede Sektion ein Direktor zuständig wäre und Stern diese Funktion für die mathematische Sektion übernehmen würde, so würde nach Gauß' Worten „eine Unterordnung des Prof. Ulrich unter jenen mir den

[9] UA Göttingen, Kur. 7448, f.333r–341v. Sterns Entwurf und seine Motivation sind im Anhang 1 wiedergegeben.

[10] UA Göttingen, Kur. 7448, f.331v–332r.

[11] SUB Göttingen, Cod. Ms. Sternwarte 2, 125.

[12] Hannover, Niedersächsisches Landesarchiv, Kleine Erwerbungen A 70, Nr. 1. Der Brief ist im Anhang 2 wiedergegeben.

Verhältnissen in allen Beziehungen nicht angemessen scheinen". Gauß plädiert dafür, dass für die Leitung der gemeinschaftlichen Angelegenheiten nur ein Direktor bestimmt wird, der turnusgemäß wechseln sollte. Er schlägt vor, dass das künftige Statut von allen für das Seminar vorgesehenen Professoren beraten wird. Gauß empfiehlt, dass Weber diese Besprechung einberuft, und fügt hinzu, er habe dieses Vorgehen schon mit Weber besprochen.

Auch wenn Gauß' Stellungnahme in den Akten nicht weiter erwähnt wird, haben seine Vorschläge die Diskussion maßgeblich beeinflusst. Zunächst schickte das Kuratorium Sterns Entwurf an das Oberschulkollegium mit der Bitte, sich über den Nutzen des geplanten Seminars für die Ausbildung der Mathematik- und Physiklehrer an den höheren Schulen zu äußern.[13] In der Antwort[14] betonte das Kollegium die Bedeutung einer solchen Institution für die Vorbereitung auf das Lehramt, und es empfahl, dass alle Mitglieder des Seminars auch an den praktischen Übungen zur „beschreibenden Naturlehre" teilnehmen sollten.[15]

Am 12. Dezember gab das Kuratorium in einem Schreiben an Ulrich, Weber, Listing und Stern seine Entscheidung bekannt.[16] Da die angeschriebenen Professoren „als Theilnehmer an der Leitung des Instituts zunächst in Betracht kommen würden", sollten sie den Entwurf in allen Teilen prüfen, die wünschenswerten Änderungen angeben und sich über ihre Bereitwilligkeit aussprechen, an dem Institut teilzunehmen. Hinzugefügt wurde:

> Wir bemerken dabei nur, daß ein festes Directorium in der Hand eines Einzigen Uns den Vorzug zu haben scheint; Wir wollen jedoch hierdurch, falls die Herren anderer Ansicht sein sollten, eine Darlegung Ihrer Gründe nicht ausschließen. Schließlich überlassen Wir den Herren, sofern Sie solches für angemessen halten, noch andere dortige Professoren zu den Berathungen über den Gegenstand zuzuziehen.

Am 19. Januar 1850 wurde das Kuratorium über das Ergebnis der Beratungen informiert.[17] Die angeschriebenen Professoren erklärten sich bereit, die Leitung des geplanten Seminars zu übernehmen. Sie fügten eine geänderte Fassung des Entwurfs von Stern bei.

Einer der vier Direktoren sollte die Geschäfte führen. In der mathematischen Abteilung sollten sich die Mitglieder vor allem im Vortragen üben, in der physikalischen Abteilung im Experimentieren. Ein älteres Mitglied der physikalischen Abteilung sollte an der Leitung beteiligt werden. Die Empfehlung, noch andere Professoren einzubeziehen, wurde aufgegriffen. August Grisebach, Wolfgang Frhr. Sartorius von Waltershausen und Friedrich Rudolf Leuckart waren

---

[13] Brief vom 11.11.1849 (UA Göttingen, Kur. 7448, f.348rv).

[14] Datiert: 5.12.1849 (UA Göttingen, Kur. 7448, f.349r–350r). Unterschrieben wurde der Brief von Heinrich Friedrich Theodor Kohlrausch (1780–1867), ab 1830 Oberschulrat und Generalinspektor der gelehrten Schulen Hannovers.

[15] Brief vom 5.12.1849 (UA Göttingen, Kur. 7448, f.349r–350r).

[16] UA Göttingen, Kur. 7448, f.351r (Entwurf mit einigen Korrekturen und endgültige Fassung).

[17] UA Göttingen, Kur. 7448, f.353r–356v.

bereit, die Mitglieder des Seminars zu ihren Lehrveranstaltungen zur Botanik, Mineralogie und Geologie zuzulassen oder für sie spezielle Übungen anzubieten.

Das Kuratorium genehmigte am 11. Februar 1850 die Vorschlage.[18] Ulrich und Stern wurden zu Direktoren der mathematischen, Weber und Listing zu Direktoren der physikalischen Abteilung ernannt. Weber wurde für das erste Jahr mit der Geschäftsführung beauftragt. Mit dem Sommersemester 1850 wurde das Seminar eröffnet.

Die Gründung und die Tätigkeit des Mathematisch-physikalischen Seminars wird, allerdings nur kurz, von W. Lorey behandelt.[19] Es existiert auch eine ausführlichere, bislang ungedruckte Darstellung über das Seminar. Sie wurde 1914 von Felix Klein und Constantin Carathéodory verfasst.[20] Klein war damals vom Kurator der Universität aufgefordert worden, einen Fragebogen über die Einrichtung und Verwaltung der Universitätsinstitute auszufüllen, und hatte in diesem Zusammenhang gemeinsam mit Carathéodory einen Bericht erstellt, der die Gründung des Mathematisch-physikalischen Seminars und die Sammlung mathematischer Modelle behandelt. Nach kurzen Bemerkungen über die Gründung des Seminars folgen Abschnitte über die Direktoren, die innere Einrichtung und das Lesezimmer. Besonders ausführlich werden die Geschichte und der gegenwärtige Zustand der Sammlung mathematischer Instrumente und Modelle dargestellt.

Über die weitere Entwicklung des mathematisch-physikalischen Seminars gibt es keine zusammenhängende Darstellung.[21] Am Seminar wirkten auch Bernhard Riemann (im Jahre 1851) und Ernst Schering (von 1853 bis 1861) mit; sie erhielten jeweils 20 Taler für die Einübung der neu eingetretenen Mitglieder. Weber und Ulrich schieden 1871 aus dem Vorstand aus, und Alfred Clebsch wurde 1872 volles Vorstandsmitglied. 1874 waren Stern, Lazarus Fuchs und Schering in der mathematischen Abteilung tätig. 1886 wurde Felix Klein zum Mitdirektor des Seminars ernannt.

---

[18] UA Göttingen, Kur. 7448, f.364r–365.

[19] Lorey (1916, S. 118f).

[20] „Universität Göttingen – Philosophische Fakultät“, maschinenschriftlich, 9 S. (Original: SUB Göttingen, Nachlass F. Klein, 2 E 13–16; Durchschlag: UA Göttingen, Math.-Nat. Fak. 68). Im Original ist handschriftlich notiert, dass der Bericht von Carathéodory und Klein zusammengestellt wurde.

[21] Archivalien in UA Göttingen, Kur. 7448. Dort gibt es u.a. Jahresberichte und Jahresrechnungen, Mitgliederverzeichnisse und Zeugnisse.

Anhang 1: Sterns Entwurf und seine Motivation[22]

Entwurf der Statuten eines mathematisch-physikalischen Seminars.

§1.

Das mathematisch-physikalische Seminar bezweckt zunächst die Ausbildung von Lehrern für den mathematischen und physikalischen Unterricht an Gymnasien und anderen höheren Lehranstalten. Zugleich soll es den Studirenden, welche sich vorzugsweise der Mathematik und Physik widmen, die Gelegenheit bieten, sich mit solchen Theilen dieser Wissenschaften bekannt zu machen, welche in den gewöhnlichen academischen Vorträgen kurz oder gar nicht behandelt werden. Auch soll dieses Institut überhaupt zur Hebung des Studiums der mathematisch-physikalischen Wissenschaften beitragen.

§2.

Das mathematisch-physikalische Seminar besteht aus zwei Abtheilungen, einer mathematischen und einer physicalischen. Jede Abtheilung steht unter der Leitung besonderer Direktoren, welche das Königl. Univ.-Curatorium aus den academischen Lehrern der Mathematik und Physik ernennt.

§3.

Für die Arbeiten jeder Abtheilung sind wöchentlich zwei Stunden bestimmt. In der einen Stunde werden von einem der Direktoren Vorträge gehalten und zwar in der mathematischen Abtheilung über einen mathematischen, in der physikalischen über einen physikalischen Gegenstand. Die Direktoren sind verpflichtet hierzu vorzugsweise solche Gegenstände zu wählen, welche gewöhnlich nicht in den academischen Vorlesungen behandelt werden. Auch steht es ihnen frei, mit diesen Vorlesungen praktische Uebungen zu verbinden.

Die zweite Stunde ist ausschließlich praktischen Uebungen gewidmet. Diese bestehen in Vorträgen, welche die Seminaristen abwechselnd halten, und zwar in der mathematischen Abtheilung über einen Gegenstand aus der reinen oder angewandten Mathematik, in der physikalischen über einen Gegenstand aus der Experimental- oder mathematischen Physik.

Während eines solchen Vortrages, der in der Regel nicht über die Dauer einer Stunde ausgedehnt werden soll, ist es sowohl dem Direktor als den übrigen Mitgliedern gestattet, Einwürfe oder sonstige dahin gehörende Bemerkungen zu machen. Die Themata, auf welche sich die Vorträge beziehen, werden von den Direktoren angegeben oder, unter deren Genehmigung, von den Mitgliedern frei gewählt. Daneben können die Direktoren auch Aufgaben stellen, welche sämmtliche Mitglieder zu lösen haben. Die Lösungen werden in den Uebungsstunden besprochen. In der physikalischen Abtheilung sollen die Mitglieder auch noch im Gebrauche physikalischer Apparate geübt werden.

[22] Quelle: UA Göttingen, Kur. 7448, f.333r–341v.

§4.

Mitglied des Seminars kann jeder förmlich immatriculirte Student werden, welcher schon seit zwei Semester an einer Universität studirt hat. Die Mitglieder sind verpflichtet an allen Stunden Theil zu nehmen. Die Aufzunehmenden haben sich bei einem der Direktoren zu melden. Es steht den Direktoren frei, Zuhörer zu ihren Vorträgen im Seminar zuzulassen.

§5.

Die Direktoren haben gegen Ende jedes Semesters in einer gemeinsamen Berathung diejenigen vier Mitglieder des Seminars zu bestimmen, welche sich in diesem Semester am meisten ausgezeichnet haben und dieselben dem Königl. Universitäts-Curatorium zu bezeichnen. Jedes dieser Mitglieder erhält aus der Universitätscasse eine Gratification von 15 rth. Auch haben die Direktoren jährlich einen gemeinschaftlichen Bericht über den Zustand des Seminars an das Universitätscuratorium abzustatten.

§6.

Die Direktoren des Seminars sollen in jedem Semester, neben ihren sonstigen Vorlesungen auch ihre jedesmalige Thätigkeit am Seminare in den Lektionscataloge ankündigen.

Motive zu dem anliegenden Statutenentwurfe.

Der Nutzen eines mathematisch-physikalischen Seminars, als einer Bildungsanstalt für solche die es sich zum Lebensberufe machen, die Wissenschaft der Mathematik und Physik fortzupflangen und zu erweitern, bedarf kaum eines weiteren Nachweises. Ein solches Institut findet sein fertiges Vorbild in den philologischen Seminaren, deren Einfluß auf die Entwickelung der humanistischen Bildung unbezweifelt ist, und die deshalb als ein integrirender Theil jeder Universität angesehen werden. Die philologischen Seminare entstanden in der bis in unsere Zeit hineinragenden Bildungsepoche, in welcher Kenntniß des Alterthums fast als die ausschließliche Unterlage jeder wissenschaftlichen Erkenntniß angesehen wurde. Je lauter aber die sogenannte realistische Richtung ihr Recht verlangt, je größer für alle Gebildeten das Bedürfniß wird, die Grundlagen kennen zu lernen, auf welchen die so mächtig in unsere Zustände eingreifenden mechanischen und physikalischen Entdeckungen und Erfindungen beruhen, desto nothwendiger wird es auch, daß den zukünftigen Lehrern der Mathematik und Physik ein akademisches Institut geboten werde, welches sich ihre weitere Ausbildung zum besonderen Zweck macht.

Man könnte vielleicht, von dem angedeuteten Gesichtspunkte ausgehend, die Beschränkung des Seminars auf Mathematik und Physik für einen Mangel halten und verlangen, daß dasselbe die Naturwissenschaften in weiterem Sinne umfasse. Die Universität Halle besitzt wirklich ein Seminar für Mathematik und die gesammten Naturwissenschaften. So wünschenswerth es nun auch seyn mag, daß jeder Zweig der Naturwissenschaften in besonderen Instituten gepflegt werde, so wird doch deren Vereinigung mit dem mathematisch-physikalischen Seminare den Zweck des letzteren weit mehr hemmen als fördern.

Soll nemlich diese Vereinigung blos äußerlich bestehen, so daß die Mitglieder einer Abtheilung dieses Seminars nicht zugleich Mitglieder einer anderen seyn müssen, so hat sie gar keinen wissenschaftlichen Zweck und hemmt nur den Geschäftsgang. Dann wird es gewiß bei Weitem vorzuziehen seyn, wenn jede Abtheilung sich als besonderes Seminar constituirt. Sollen dagegen die Mitglieder verpflichtet seyn, sich an *allen* Arbeiten des Seminars zu betheiligen, so möchte einer solchen Anforderung nur von Wenigen genügt werden können, da dies nicht nur sehr umfassende Kenntnisse, sondern auch mehr Muße voraussetzt, als den meisten Studirenden neben ihren sonstigen Arbeiten übrig bleibt.

Mit besonderer Rücksicht auf unsere hiesigen Verhältnisse, wird hierbei noch zu beachten seyn, daß die Mitglieder dieses Seminars, insofern sie den Lehrerberuf im Auge haben, auch häufig Mitglieder des philologischen und pädagogischen Seminars seyn werden, wodurch bereits ein beträchtlicher Theil ihrer Zeit absorbirt wird. Auch ist schon für die specielle Ausbildung in einzelnen Zweigen der Naturwissenschaft durch andere Institute gesorgt, so für die Chemie durch das chemische Laboratorium, für die Physiologie durch das physiologische Institut.

Dagegen ist eine getrennte Ausbildung des Mathematikers und des Physikers nicht mehr denkbar. Eine physikalische gründliche Bildung setzt von selbst die Mathematik voraus. Der Mathematiker aber, obgleich er als solcher der Physik nicht bedarf, wird dennoch ohne Einsicht in dieselbe höchst einseitig bleiben, da gerade die Physik die schönsten Anwendungen der Mathematik bietet; so wie die moderne Ausbildung der Mathematik auch hauptsächlich durch physikalische Probleme veranlaßt worden ist.

Aus diesen Gründen bezeichnet der Entwurf das Seminar als ein mathematisch-physikalisches.

Was die Aufnahme in das Seminar betrifft, so hat es passend erschienen, hierbei eine möglichst große Freiheit walten zu lassen. Aus diesem Grunde enthält der Entwurf keine Beschränkung in Beziehung auf das Fachstudium des Aufzunehmenden, wie dies bei anderen ähnlichen Instituten der Fall ist. So z.B. verlangen die Statuten des Königsberger mathematisch physicalischen Seminars, daß der Aufzunehmende Mathematik und Physik studirt. Eine ähnliche, wenn auch weniger enge, Beschränkung enthalten die Statuten des schon früher erwähnten Seminars zu Halle. Sollen Beschränkungen dieser Art dazu dienen den Zudrang Unberufener zu verhüten, so ist zu bemerken, daß ein solcher, nach den Bestimmungen des Entwurfs, wohl nicht zu fürchten ist. Einen pekuniären Vortheil haben nach §4 nur die ausgezeichnetsten Mitglieder zu erwarten; es darf daher bei jedem, welcher sich zur Aufnahme meldet, eine entschiedene Neigung zur Mathematik und Physik vorausgesetzt werden. Die ernstliche Beschäftigung mit diesen Wissenschaften ist bei Studirenden, welche sich nicht ausgesprochen denselben widmen, eine so seltene Erscheinung, daß sie gewiß Aufmunterung und nicht Zurückweisung verdient. Auch ist zu bedenken, wie mancherlei Verhältnisse einen jungen Mann veranlassen können, bei den entschiedensten

Anlagen für ein gewisses Studium, dennoch ein anderes zu ergreifen. Ein mathematisch-physikalisches Seminar, welches einen Studenten der Theologie wie *Keppler*, oder einen Studenten der Medicin wie *Olbers*, statutenmäßig zurückweisen wollte, würde sich dadurch gewiß kein Verdienst um die Wissenschaft erwerben.

Von der schon entwickelten Ansicht ausgehend, daß sich nicht leicht Unberufene zur Aufnahme melden werden, verlangen auch die Statuten nicht, daß der Aufzunehmende sich einer wissenschaftlichen Prüfung unterwerfe, die doch für den Examinator wie für den Examinanden immer manches Lästige hat. Wer seine Fähigkeiten überschätzend in das Seminar eintritt, wird sich bald genug von selbst zurück ziehen. Um jedoch eine gewisse Garantie für die Reife der Mitglieder zu haben, ist die Bestimmung getroffen, daß der Eintretende bereits ein Jahr studirt haben muß. Vielleicht wäre es aber besser auch noch diese Beschränkung fallen zu lassen.

Durch die Bestimmung, daß die Direktoren auch Zuhörer zu ihren *Vorträgen* zulassen dürfen, kann das Seminar seinen Nutzen noch auf weitere Kreise ausdehnen. Dagegen möchte es rathsam seyn von den *Uebungsstunden* die Zuhörer gänzlich auszuschließen.

Es bedarf keiner besonderen Ausführung, daß eine Anerkennung ausgezeichneter Leistungen der Seminaristen zur Hebung des Instituts beitragen und am passendsten in einer Geldunterstützung bestehen wird. Nach §5 würden jährlich für diesen Zweck 120 rth zu verwenden seyn, welche Summe als ein Minimum anzusehen seyn dürfte. Es wird von dem Ermessen des Universitäts-Curatorium abhängen, ob nicht eine größere und unter einer größeren Zahl zu vertheilende Summe zu diesem Zwecke zu bestimmen sey.

Anhang 2: Gauß' Brief an das Universitäts-Kuratorium, Göttingen, 6.11.1849[23]

An Königliches Universitäts Curatorium
Die Errichtung eines physisch-mathematischen Seminars betreffend.

Indem ich die den rubricirten Gegenstand betreffenden Papiere hieneben zurücksende, beehre ich mich, meine Ansicht über denselben in Folgendem gehorsamst vorzulegen.

Von einer Einrichtung, welche speciell die Beförderung einer gründlichern und umfassendern Vorbildung künftiger Gymnasiallehrer für Mathematik und Naturwissenschaften bezweckt, ist man unstreitig berechtigt, nützliche Wirkungen für den Gymnasialunterricht in diesen Fächern zu erwarten.

In dem von dem Professor Stern eingereichten Entwurf eines Statuts habe ich mehreres gefunden, dessen Zweckmäßigkeit mir zweifelhaft scheint. Wenn z.B.

[23] Quelle: Niedersächsisches Landesarchiv Hannover, Kleine Erwerbungen A 70, Nr. 1.

darin festgesetzt wird, daß denjenigen vier Seminaristen, deren Leistungen in einem halben Jahre die ausgezeichnetsten gewesen sind, jedem eine Gratification von 15 rth verabreicht werden soll, so weiß ich nicht, ob man, besonders in den ersten Jahren, so bestimmt voraussetzen kann, daß bei einer Rangirung der Seminaristen nach dem Grade der Auszeichnung ihrer Leistungen, allemahl dieser Grad z. B. bei dem vierten, wenn er auch relativ seinen Hintermännern vorsteht, auch absolut bedeutend genug sein wird, um dafür eine Remuneration zu geben. Mir deucht, daß es zweckmäßiger sei, diese Zahl 4 (oder welche andere man sonst wählen möge) nicht als nothwendig, sondern nur als facultativ hinzustellen, so daß für die ausgezeichnetsten Leistungen bestimmte Gratificationen in Aussicht gestellt würden, deren Anzahl in jedem halben Jahre jedoch nicht über vier hinausgehen soll.

Die eignen Arbeiten der Seminaristen sind in dem Entwurfe nur auf das Abhalten mündlicher Vorträge beschränkt. Ohne diesen Übungen ihre große Nützlichkeit abzusprechen, würde ich doch für angemessen halten, die eigne Geistesthätigkeit auch durch schriftlich zu lösende Aufgaben und andere über aufgegebene Materien auszuarbeitende schriftliche Aufsätze zu wecken und zu üben. Diese schriftlichen Arbeiten würden auch eine sichrere und keinem Verdacht einer Partheilichkeit ausgesetzte Grundlage für die Auswahl der zu remunerirenden Seminaristen geben, da der mündliche Vortrag immer nur von einem der Directoren angehört ist, die schriftlichen Arbeiten aber von der Gesammtheit alle Directoren geprüft werden können, auch, wenn im Archiv aufbewahrt, zuweilen noch in späterer Zeit als Anhaltspunkte für auszustellende Zeugnisse dienen können.

Was die Frage betrifft, ob nicht zwei Directoren zu ernennen sein, anstatt der von dem Pr[ofessor] Stern vorgeschlagenen vier, so bin ich nicht völlig gewiß, ob ich die Bedeutung dieser Frage richtig auffasse. Im Sternschen Vorschlage verstehe ich die Bedeutung des Worts Director so, daß damit *jeder* an der Anstalt als Lehrer theilnehmende Professor bezeichnet wird, oder daß dieser Name ganz dasselbe, und nur dasselbe bedeuten soll, wie Vorstand in dem Hallischen Statut, während von der Function, die in diesem Statut einem Director zugetheilt wird, in dem Sternschen Aufsatze keine Rede ist.

Ist bei der Frage des Rescripts die Benennung Director aber so gemeint, so wäre der Fragepunkt folglich der, ob nicht von der Theilnahme der Professoren Ulrich und Listing ganz abgesehen werden solle, so daß die Prof[essoren] Weber und Stern an der Anstalt wirksam würden. Meiner Ansicht nach würde

1. in allgemeiner Rücksicht, zu erwägen sein, daß wenn nur zwei Professoren die Träger der Anstalt sind, jedem doppelt so viel Arbeit, mithin ein sehr bedeutender Zeitaufwand, angesonnen werden müste.
2. in specieller Beziehung auf den mathematischen Theil, würde ich gerade *vorzugsweise* die Theilnahme des Prof[essors] Ulrich für wünschenswerth halten (seine Geneigtheit vorausgesetzt), da derselbe als vieljähriges Mitglied der Prüfungs-Commission besser als jeder andere die am häufigsten vorkommenden Mangelhaftigkeiten in der Vorbildung der Candidaten auf

deren Abhülfe durch die Anstalt hingewirkt werden soll, kennen zu lernen Gelegenheit gehabt hat.

Sollte hingegen die Frage so gemeint sein, daß zwar Prof. Ulrich und Listing Theilnehmer sein, aber für jede Abtheilung ein Director für die Oberleitung der Geschäfte bestellt werden sollte, so würde die Ernennung des Prof. Stern zu dieser Function in der mathematischen Section, also eine Unterordnung des Prof. Ulrich unter jenen mir den Verhältnissen in allen Beziehungen nicht angemessen scheinen. Überhaupt scheint mir für die Leitung der gemeinschaftlichen Angelegenheiten die Bestellung Eines Directors oder Dirigenten, wie man ihn nennen mag, zureichend, und könnte dieß Amt nach einem bestimmten Turnus unter sämmtlichen theilnehmenden Lehrern wechseln.

Da wenigstens bei der ersten Gründung einer solchen Anstalt die Theilnahme an derselben auf einem freiwilligen Zusammentreten, und das Gedeihen auf der Voraussetzung beruhet, daß alle Theilnehmer sich den Geschäften mit Liebe unterziehen, so scheint es mir nothwendig, daß das Statut nicht aus einseitiger Vorlage, sondern lediglich aus der gemeinschaftlichen Berathung aller künftigen Theilnehmer entstanden sein muß, dessen Resultat dann dem königl[ichen] Universitäts Curatorium zur Prüfung und Genehmigung zu unterbreiten sein würde. Ich würde daher anheim geben, daß etwa der Prof. Weber autorisirt würde, ein solches Zusammentreten zu gemeinschaftlicher Berathung einzuleiten.

Die Ansichten des Prof. Weber sind im Wesentlichen mit meinen oben entwickelten übereinstimmend, und hat er auf meinen Wunsch dieselben in einem kleinen Aufsatz schriftlich ausgesprochen, welchen ich mich beehre hier beizufügen.[24]

Gehorsamst
Gauß

Göttingen den 6 November.
1849

Anhang 3: Endgültige Statuten[25]

Nach Verordnung des hohen Universitäts-Curatoriums vom 11. Februar d. J. wird mit nächstem Sommerhalbjahre eine neue Lehranstalt an unserer Universität eröffnet werden, nämlich ein mathematisch-physikalisches Seminar, zu Förderung des Studiums der Mathematik und Physik überhaupt und zur Ausbildung von Lehrern für diese Fächer an höheren Lehranstalten insbesondere. Viele Theile der Mathematik und Physik konnten bisher in den gewöhnlichen akademischen Vorträgen nur kurz oder gar nicht behandelt werden, weil nach der Einrichtung dieser Vorträge mit jedem Semester ein solcher Wechsel der Zuhörer eintrat, daß nur

[24] Dieser Aufsatz ist vermutlich nicht erhalten.

[25] Quelle: *Nachrichten von der G. A. Universität und der Königl. Gesellschaft der Wissenschaften zu Göttingen*, No.6. März 11. 1850. In: Nachrichten 1850, S.73–75.

selten auf die in früheren Semestern gelegte Grundlage später fortgebauet werden konnte. Das mathematisch-physikalische Seminar soll nun alle diejenigen, welche sich dem Studium dieser Wissenschaften vorzugsweise widmen, dauernder fesseln und dadurch die Gelegenheit zu einem zusammenhängenden planmäßigen Lehrcursus darbieten. Es wird dadurch zugleich ein Weg gebahnt, die zum wahren Verständniß dieser Fächer nothwendige Selbstthätigkeit der Studirenden zu wecken und zweckmäßig zu leiten, indem sie theils zu eignen Vorträgen, theils zu Behandlung mathematischer und physikalischer Aufgaben angehalten werden, wie es für die Ausbildung zu einem künftigen Lehrer dringendes Bedürfniß ist. Eine Ausdehnung dieser Anstalt über die gesammte Naturwissenschaft ist nicht für nöthig erachtet worden, weil für mehrere Fächer schon seit längerer Zeit durch besondere Anstalten gesorgt ist, namentlich für das Fach der Chemie und der Physiologie durch eigene Laboratorien für die Studirenden im chemischen und im physiologischen Institute unserer Universität, welche allen Bedürfnissen in diesen Fächern vollständig genügen. Außerdem würde die Verschmelzung zu heterogener Studien das Band gelockert haben, welches die Theilnehmer am Seminar verbinden soll, und welches eine gemeinschaftliche Theilnahme an allen Vorträgen und Arbeiten im Seminar voraussetzt, wie sie nur bei dem innigen Zusammenhange der mathematischen und physikalischen Studien wohl bestehen kann. Jedoch ließ sich damit – um einem vom Oberschulcollegium hervorgehobenen praktischen Bedürfnisse vieler Lehranstalten zu entsprechen, an welchen die Lehrer der Mathematik und Physik den Unterricht in der beschreibenden Naturlehre zu versehen haben – ohne Nachtheil die Einrichtung verbinden, daß den Mitgliedern des Seminars auch Gelegenheit zur praktischen Ausbildung in der beschreibenden Naturlehre gegeben würde. Uebrigens bietet das neuerdings erweiterte physikalische Institut unserer Universität den Raum zu einem physikalischen Laboratorium dar, worin die Theilnehmer des Seminars zu eigenen Arbeiten Gelegenheit erhalten werden, eine Gelegenheit, welche überall fehlt, wo die physikalischen Institute noch auf einen bloß für das Bedürfniß der Vorlesungen eingerichteten Apparat beschränkt sind. Die Erweiterung des gedachten Instituts durch ein Laboratorium für die Studirenden, welches für die Physik eben so wichtig wie für die Chemie ist, stützt sich aber wesentlich auf das Bestehen des Seminars, welches die Sicherheit gewährt, daß immer eine größere Anzahl Studirender von dieser für sie getroffenen Einrichtung Gebrauch macht und wirklichen Nutzen daraus zieht.

Die Einrichtung des Seminars und die zu seinem Gedeihen von dem hohen Universitäts-Curatorium gewährten Mittel ersieht man aus den folgenden Statuten. Die Professoren Ulrich und Stern sind zu Directoren der mathematischen, die Professoren Weber und Listing zu Directoren der physikalischen Abtheilung ernannt worden; die Professoren Grisebach und Sartorius von Waltershausen und Doctor Leuckart haben übernommen, für die praktische Ausbildung in der beschreibenden Naturlehre Sorge zu tragen.

Statuten des mathematisch-physikalischen Seminars zu Göttingen.

Für das mathematisch-physikalische Seminar zu Göttingen werden folgende Bestimmungen aufgestellt:

§. 1.

Das Seminar bezweckt zunächst die Ausbildung von Lehrern für den mathematischen und physikalischen Unterricht an höheren Lehranstalten.

Zugleich soll es den Studirenden Gelegenheit bieten, sich mit solchen Theilen der Mathematik und Physik bekannt zu machen, welche in den gewöhnlichen akademischen Vorträgen kurz oder gar nicht behandelt werden. Auch soll die Anstalt überhaupt zur Hebung des Studiums der mathematisch-physikalischen Wissenschaften beitragen.

§. 2.

Das Seminar besteht aus zwei Abtheilungen, einer mathematischen und einer physikalischen. Jede Abtheilung wird von Directoren geleitet, welche das Universitäts-Curatorium aus den akademischen Lehrern der Mathematik und Physik ernennt.

Die Directoren beider Abtheilungen bilden den Vorstand des Seminars.

Einer derselben übernimmt – und zwar vorerst abwechselnd jedesmal für ein Jahr – die Geschäftsordnung für die ganze Anstalt.

§. 3.

Für die Arbeiten der mathematischen Abtheilung sind wöchentlich zwei Stunden bestimmt. In der einen Stunde hält einer der Directoren einen Vortrag über mathematische Gegenstände, und zwar vorzugsweise über solche, welche in den regelmäßigen akademischen Vorträgen gewöhnlich nicht behandelt werden. Mit diesen Vorträgen können auch praktische Uebungen verbunden werden.

Die zweite Stunde ist ausschließlich praktischen Uebungen gewidmet. Diese bestehen in Vorträgen über einen Gegenstand der reinen oder angewandten Mathematik, welche die Seminaristen abwechselnd halten. In der Regel darf ein solcher Vortrag nicht mehr als eine Stunde in Anspruch nehmen. Während desselben ist es sowohl dem Director als den Mitgliedern des Seminars gestattet, Einwürfe oder sonstige dahin gehörende Bemerkungen zu machen.

Die Gegenstände der Vorträge werden von den Directoren angegeben, oder mit deren Genehmigung von den Mitgliedern gewählt.

Daneben können die Directoren auch Aufgaben stellen, welche sämmtliche Mitglieder zu lösen haben. Die Lösungen werden in den Uebungsstunden besprochen.

§. 4.

Für die Arbeiten der physikalischen Abtheilung, deren Local das physikalische Laboratorium ist, werden wöchentlich zwei bis vier Stunden bestimmt, von denen je zwei zusammen gelegt werden können.

Die Arbeiten bestehen in Vorträgen der Directoren über einzelne Gegenstände der theoretischen Physik, verbunden mit Uebungen in Beobachtungen und Messungen.

Ferner haben die Mitglieder über den Inhalt einzelner Abhandlungen und Aufsätze aus akademischen Schriften und physikalischen Zeitschriften zu berichten, welche von dem Director als Leitfaden zu den physikalischen Uebungen ausgewählt werden.

Jährlich wird eine Demonstration der physikalischen Instrumenten sammlung gehalten, womit die Anleitung im Gebrauche der einzelnen Instrumente zu verbinden ist.

Die Mitglieder haben über alle Beobachtungen und Versuche, an welchen sie Theil nehmen, Protokoll zu führen, und über die dazu geeigneten Beobachtungen und Versuche selbständige Ausarbeitungen zu liefern.

§. 5.

Mitglied des Seminars kann jeder immatriculirte Student werden, welcher schon seit zwei Semestern an einer Universität studirt, oder auf einer polytechnischen Schule hinreichende Vorbildung erlangt hat.

Die Mitglieder sind verpflichtet an allen Stunden Theil zu nehmen.

Die Aufzunehmenden haben sich bei dem geschäftsführenden Director zu melden.

Es steht den Directoren frei Zuhörer zu ihren Vorträgen im Seminar zuzulassen.

Mitglieder, welche den Forderungen der Statuten nicht entsprechen, können durch einen Beschluß des Vorstandes von der Theilnahme am Seminare ausgeschlossen werden.

§. 6.

Es wird eine Summe von halbjährlich 80 rth ausgesetzt, aus welcher gegen das Ende jedes Halbjahrs denjenigen Mitgliedern, welche sich in demselben am meisten ausgezeichnet haben, Stipendien zu bewilligen sind. Es sollen höchstens vier Stipendien bewilligt werden.

Die Directoren haben nach gemeinsamer Berathung dem Universitäts-Curatorium Vorschläge darüber zu machen.

Ein älteres Mitglied der physikalischen Abtheilung, welches von den Directoren der physikalischen Abtheilung dem Universitäts-Curatorium vorgeschlagen wird, erhält eine halbjährige Vergütung von 20 rth für die Einübung der neu eintretenden Mitglieder, ohne hierdurch seinen etwaigen Anspruch auf ein Stipendium zu verlieren.

Außerdem werden, soweit nöthig, die bei den physikalischen Arbeiten entstehenden Kosten ersetzt werden.

Für diese und andere Ausgaben erhält das Seminar die jährliche Summe von 60 rth.

Diese darf nicht überschritten werden; die Ersparnisse eines Jahres sind jedoch auf das folgende zu übertragen, um größere Bedürfnisse bestreiten und ausgedehntere physikalische Arbeiten ausführen zu können.

§. 7.

Jeder Director führt ein Verzeichniß über die laufenden Arbeiten, in welchem zugleich die größeren Arbeiten der Mitglieder anzugeben und zu beurtheilen sind.

Dasselbe bildet die Grundlage eines gemeinschaftlichen Berichts über den Zustand des Seminars, welchen die Directoren jährlich dem Universitäts-Curatorium vorzulegen haben.

§. 8.

Um dem Bedürfnisse der Lehranstalten zu begegnen, soll den Mitgliedern des Seminars zugleich Gelegenheit gegeben werden, sich in der beschreibenden Naturlehre praktisch auszubilden.

§. 9.

Die Directoren des Seminars und die Lehrer, welche den Unterricht in der beschreibenden Naturlehre übernehmen, haben in jedem Halbjahre neben ihren sonstigen Vorlesungen, auch ihre jedesmalige Thätigkeit am Seminare anzukündigen.

Hannover, den 11. Februar 1850.

Königlich Hannoversches Universitäts-Curatorium.

Braun.

References

Langhammer, Walter 1981. Some Aspects of the Development of Mathematics at the University of Halle-Wittenberg in the Early 19th Century. In: *Epistemological and Social Problems of the Sciences in the Early Nineteenth Century*, hrsg. v. H. N. Jahnke und M. Otte, 235–254. Dordrecht: Reidel.

Lorey, Wilhelm 1916. *Das Studium der Mathematik an den deutschen Universitäten seit Anfang des 19. Jahrhunderts.* Berlin und Leipzig: B. G. Teubner.

Nachrichten 1850. *Nachrichten von der Georg-Augusts-Universität und der Königl. Gesellschaft der Wissenschaften zu Göttingen vom Jahre 1850*, Nr. 1–17. Göttingen.

Olesko, Kathryn M. 1991. *Physics as a Calling: Discipline and Practice in the Königsberg Seminar for Physics.* Ithaca, NY: Cornell University Press.

Schubring, Gert 1983. *Seminar – Institut – Fakultät: Die Entwicklung der Ausbildungsformen und ihrer Institutionen in der Mathematik.* Bielefeld: Interdiziplinäres Zentrum für Hochschuldidaktik der Universität Bielefeld, Heft 1/1983.

Schubring, Gert 1985. Die Entwicklung des Mathematischen Seminars der Universität Bonn 1864-1929. *Jahresbericht der Deutschen Mathematiker-Vereinigung* 87: 139–163.

Schubring, Gert 1985a. Das Mathematische Seminar der Universität Münster, 1831/1875 bis 1951. *Sudhoffs Archiv* 69: 154–191.

Schubring, Gert 1989. The Rise and Decline of the Bonn Natural Sciences Seminar. *Osiris* 5: 57–93.

Schubring, Gert 1990. Zur strukturellen Entwicklung der Mathematik an den deutschen Hochschulen 1800–1945. In: *Mathematische Institute in Deutschland 1800–1945*, hrsg. v. W. Scharlau, 264–278. Braunschweig / Wiesbaden 1990: Deutsche Mathematiker-Vereinigung, Friedr. Vieweg & Sohn.

Schubring, Gert 1991. Spezialschulmodell versus Universitätsmodell – die Institutionalisierung von Forschung. In: *‚Einsamkeit und Freiheit' neu besichtigt. Universitätsreformen und Disziplinenbildung in Preußen als Modell für Wissenschaftspolitik im Europa des 19. Jahrhunderts. Proceedings of the Symposium of the XVIIIth International Congress of History of Science at Hamburg-Munich, 1–9 August 1989*, hrsg. v. G. Schubring, 276–326. Stuttgart: Steiner.

Schubring, Gert 2000. Kabinett – Seminar – Institut. Raum und Rahmen eines forschenden Lernens. *Berichte zur Wissenschaftsgeschichte* 23: 269–285.

Schubring, Gert 2004. Das Bonner naturwissenschaftliche Seminar (1825–1887). Eine Fallstudie zur Disziplinendifferenzierung. *Acta Historica Leopoldina* 43: 133–148.

Turner, R. Steven 1971. The Growth of Professorial Research in Prussia, 1818 to 1848 – Causes and Context. *Historical Studies in the Physical Sciences* 3: 137–182.

Turner, R. Steven 1987a. IV. Universitäten. In: *Handbuch der deutschen Bildungsgeschichte. Band III. 1800–1870. Von der Neuordnung Deutschlands bis zur Gründung des Deutschen Reiches*, hrsg. v. K.-E. Jeismann u. P. Lundgreen, 221–249. München: Beck.

Turner, R. Steven 1987b. The Great Transition and the Social Patterns of German Science. *Minerva* 25: 56–76.

D'Alembert: Mathematics and the Enlightenment

Christian Gilain

First, we must remember what is meant by mathematics in the eighteenth century. The article "Mathématique, ou Mathématiques" (Mathematics) in the *Encyclopédie,* edited by Diderot and d'Alembert, indicates:

> Mathematics are divided into two classes; the first, which we call pure Mathematics, considers properties of abstract magnitude [...]. The second class is called mixed Mathematics; its object of study is the properties of concrete magnitude insofar as it is measurable or calculable. (t. X, 1765, pp. 188-189)[1]

Pure mathematics include geometry, arithmetic, algebra, and infinitesimal analysis. Mixed mathematics include mechanics, astronomy, optics, etc., in general, all mathematized natural sciences. This classification given by the *Encyclopédie* is current at the time, even if in the second half of the eighteenth century the expression "physico-mathematical sciences" is often used instead of "mixed mathematics".

In this text, I present some of d'Alembert's contributions to the field of pure mathematics and some remarks about his vision of mathematics in general, taking into account recent researches that lead to changes in the previous historiographical narrative.

C. Gilain (✉)
Sorbonne Université, Paris, France
e-mail: christian.gilain@imj-prg.fr

[1] Original text: "Les Mathématiques se divisent en deux classes; la première, qu'on appelle Mathématiques pures, considère les propriétés de la grandeur d'une manière abstraite [...]. La seconde classe s'appelle Mathématiques mixtes; elle a pour objet les propriétés de la grandeur concrète, en tant qu'elle est mesurable ou calculable" (*Encyclopédie*, t. X, 1765, p. 188–189).

A. Verdugo Rohrer and J. Zender (eds.), *History of Mathematics and Its Contexts*, Trends in the History of Science, https://doi.org/10.1007/978-3-031-60828-5_10

Integration of Rational Differentials, Theory of Imaginary Quantities and the Fundamental Theorem of Algebra

In 1702, in the framework of the development of the new infinitesimal calculus, Leibniz studies the integration of rational functions. He introduces the method of partial fractions and poses the following problem: can all these integrals be expressed by using only the quadratures of the hyperbola and of the circle (that is by using only elementary transcendental functions)?

Leibniz shows that the answer to this problem of integral calculus depends on the following algebraic question: Can any real polynomial be written as a product of real factors of 1st or 2nd degree? But his answer is negative... He uses as argument the "counter-example" of the polynomial $x^4 + a^4$. His mistake results from the fact that Leibniz thinks that there are imaginary quantities such as $\sqrt[4]{-1}$ which cannot be written in the form $p+q\sqrt{-1}$, with p and q both real. In other words, for Leibniz there are "imaginary" quantities which are not "complex" numbers. Even if mathematicians saw fairly quickly that this "counter-example" was a mistake, the general problem remained open for almost half a century...

D'Alembert comes to this problem in the framework of his research into integral calculus. In his memoir "Researches into integral calculus"[2] sent to the Berlin Academy in 1746, he proofs two results:

1. The set of the expressions of the form $A + B\sqrt{-1}$ (with A and B both real) is closed under explicit algebraic or elementary transcendental operations.
2. Any polynomial with real coefficients has at least one root of the form $A + B\sqrt{-1}$, with A and B both real.

At first, in 1745, d'Alembert had thought that he could immediately deduce the second result from the first. But afterward he clearly indicated that in the absence of knowledge of the form of the roots of an algebraic equation of any degree, *if we want to be rigorous*, we have to distinguish the case of implicit algebraic functions from the case of usual explicit ones. He therefore looked for a *direct* proof of the second result.

This second result is what is called the *fundamental theorem of algebra* (FTA), also called the d'Alembert–Gauss theorem. Here is the structure of the proof of the FTA as given by d'Alembert in 1746.

Let $P(x) = x^m + ax^{m-1} + \cdots + fx + g$ be any real polynomial.

1. D'Alembert considers the plane curve expressed by the equation

$$G(x, y) = x^m + ax^{m-1} + \cdots + fx + y = 0.$$

[2] "Recherches sur le calcul intégral" (d'Alembert 1748).

2. According to Newton x can be expanded in a fractional power series in the neighborhood of $y = 0$[3]: $x = Ay^{\frac{n}{p}} + By^{\frac{r}{s}} + \text{etc.}$

D'Alembert deduces that for y near 0, positive or negative, we obtain
$x = p + q\sqrt{-1}$, with p and q both real (due to the closure of these expressions under explicit algebraic operations).

3. To get from this *local* result to a *global* one he uses a reduction to absurdity. He considers the set (say E) of all y which give a root

$x = p + q\sqrt{-1}$ (with p and q both real) of $G(x, y) = 0$. D'Alembert, supposing that this set is bounded and applying the local result to the bounds, arrives at a contradiction and deduces that the result can be extended to any real value. (That is, $E = \boldsymbol{R}$.)

4. Then taking $y = g$, he proves the existence of at least one x of the form $p + q\sqrt{-1}$, with p and q both real, which is a root of the polynomial

$$G(x, g) = P(x).$$

Noting that the conjugated expression $p - q\sqrt{-1}$ is then also a root of the polynomial P, d'Alembert goes on to deduce the FTA in the real form: Every real polynomial can be written as a product of real factors of the first or second degree.

In the article "Equation" of the *Encyclopédie*, d'Alembert puts this theorem at the foundation of the theory of algebraic equations and explicitly qualifies it as a "fundamental proposition of Algebra"[4] (t. V, 1755, p. 848). D'Alembert's analytical demonstration gives the FTA the status of an *existence* theorem: There are technical gaps (for example about the local convergence of the series) but no logical error.[5] Finally, in 1769 d'Alembert extends his theory to include polynomials with coefficients of the form $a + b\sqrt{-1}$, thus completing his results of closure of the set of "complex" numbers.

In contrast, the algebraic demonstrations of the FTA by his contemporaries (Euler, Lagrange) admitted a priori the possibility of splitting any polynomial into factors of the first degree. That is, they assumed the existence of polynomial roots of an unknown nature but with which they calculated as with real numbers. In 1799, this appeared to Gauss as a vicious circle. (In fact, at the beginning of the twentieth century, these demonstrations could be rehabilitated once the general existence of the splitting field of a polynomial had been proved.)

[3] In this case, the coefficients A, B, etc. are real because they are obtained from the coefficients a, b, etc. by rational operations.

[4] Original text: "proposition fondamentale de l'Algèbre".

[5] For more details, see (d'Alembert *OC* I/4a).

The Algebraization of the Integration of Irrational Differentials: Toward Elliptic Integrals

D'Alembert's work on the topic of the integration of algebraic differentials comes in a direct line from Maclaurin's *Treatise of fluxions* (1742). In this treatise, Maclaurin presented his conception of the integral calculus, based on a classification of integrals into categories of increasing complexity, according to the degree of algebraic curves used to express them and of the quadrature-rectification hierarchy.

In his memoir "Researches into integral calculus", after his study of integrals of rational differentials, d'Alembert pursues this research program by considering the integrals of irrational differentials expressed by means of conic arcs. But whereas Maclaurin used geometrical methods, drawing many figures, d'Alembert draws none and develops *purely algebraic methods.*[6]

In particular, he shows that the following integrals can be expressed by means of elliptic or hyperbolic arcs:

- $\int \frac{\mathrm{d}x}{\sqrt{P}}$, with P a polynomial of 3rd degree;
- $\int \frac{\mathrm{d}x}{\sqrt{P}}$, with P a polynomial of 4th degree;
- $\int \frac{x\mathrm{d}x}{\sqrt{P}}$, with P a polynomial of 3rd degree.

Euler was impressed by d'Alembert's results on the reduction of integrals to conic arcs. He wrote to d'Alembert from Berlin in December 1746:

> But what I particularly liked about your paper, is the reduction of several integral formulas to the rectification of the ellipse and the hyperbola. I had already thought about this, but I was never able to reduce the formula
>
> $$\frac{\mathrm{d}x}{\sqrt{\alpha + \beta x + \gamma x^2 + \delta x^3 + \varepsilon x^4}}$$
>
> and I regard your solution as a master-piece. (d'Alembert *OC* V/2, pp. 57-58)[7]

Later, this line of research on integrals of irrational differentials had been also developed by Euler who showed that it was now possible to introduce *new transcendental* (non-elementary) *functions* into Analysis, and to calculate with them. These integrals came into the class of "elliptic integrals", of the form $\int \frac{R\mathrm{d}x}{\sqrt{P}}$ where

[6] For more details, see (d'Alembert *OC* I/4a).

[7] Original text: "Mais ce qui m'a plu surtout dans votre pièce, c'est la réduction de plusieurs formules intégrales à la rectification de l'ellipse et de l'hyperbole: matière à laquelle j'avais déjà pensé, mais je n'ai jamais pu venir à bout de la formule [...] et je regarde votre résolution comme un chef d'œuvre de votre pénétration."

R is a rational function and P is a polynomial of degree 3 or 4, class studied systematically by Legendre from the end of the eighteenth century.

Study of Systems of Differential Equations: Theory of Linear Systems with Constant Coefficients

Here, to understand the origin of d'Alembert's theory, it is necessary to start from his work on mechanics. In 1743, in his *Traité de dynamique* (Treatise of Dynamics), d'Alembert studies the small oscillations of the compound pendulum (i.e. of a string fixed at one point and loaded by discrete weights) and shows that the local movement of bodies is expressed by a system of second-order linear differential equations with constant coefficients.

Thus, in the case of the double pendulum with simplified hypotheses, d'Alembert (1743, art. 101) arrives at a system which he writes in this form

$$-\mathrm{d}^2x = (2x - y).\frac{2\mathrm{d}t^2}{T^2},$$

$$-\mathrm{d}^2y = (2y - 2x).\frac{2\mathrm{d}t^2}{T^2}.$$

In these formulas, t represents the time, x and y the small displacements of the two bodies relatively to the vertical (T is a constant).

For the integration of such a system, d'Alembert's method consists of introducing constant *multipliers* α and ν for both equations, respectively. These multipliers are chosen in such a way that by adding the new equations, one can obtain a single ordinary differential equation

$$-\mathrm{d}^2u = Ku.\frac{2\mathrm{d}t^2}{T^2}, \text{ with } \quad u = \alpha x + \nu y.$$

This is a second-order equation with constant coefficients.

But, in 1743, d'Alembert fails to find an explicit analytical solution of the equation.

Later, in his pure mathematics memoir "Further researches into integral calculus",[8] he goes back to the question by developing a *general theory* of systems of linear differential equations with constant coefficients. He orders this new theory as follows:

1. The systems of *first-order* linear differential equations are placed in the position of *fundamental object* of the theory;

[8] "Suite des recherches sur le calcul integral" (d'Alembert 1750, 4th part, problem V).

2. The other linear differential systems, whatever their order, are reduced to the preceding type.

Thus, first of all, he studies a differential system of two equations of the first order

$$\mathrm{d}x + (Cx + Dy)\mathrm{d}t = 0,$$

$$\mathrm{d}y + (Kx + Ly)\mathrm{d}t = 0,$$

(with *C*, *D*, *K*, *L* constant).

Here, d'Alembert uses only one multiplier, ν, and he shows that the system can be reduced to a single equation of the first order

$$\mathrm{d}u + (C + K\nu)u\mathrm{d}t = 0, \text{ with } \quad u = x + \nu y,$$

if the coefficient ν verifies a characteristic equation of degree 2

$$K\nu^2 + (C - L)\nu - D = 0.$$

Taking the roots p and p' (real or imaginary) of this latter equation, he easily deduces the solutions u and u' of the first order differential equation. And, finally, he obtains x and y in the form of linear combinations of two exponential functions

$$x = ae^{-qt} + be^{-q't} \text{ and } y = ce^{-qt} + de^{-q't}$$

(with $q = C + Kp$, $q' = C + Kp'$, and *a*, *b*, *c*, *d* constant).

The method and the result extend to systems of n first order linear differential equations with constant coefficients, the characteristic equation being then of degree n.

Then, d'Alembert generalizes his study to systems which include p differential linear equations of order q thanks to a new idea: The possibility of transforming such a system in an equivalent one of the first order by introducing new variables, the successive differential coefficients (except one) of the unknowns *x*, *y*, etc. Thus, he takes:

$$\frac{\mathrm{d}^{q-1}x}{\mathrm{d}t^{q-1}} = x', \frac{\mathrm{d}^{q-2}x}{\mathrm{d}t^{q-2}} = x'', \text{ etc.}, \frac{\mathrm{d}^{q-1}y}{\mathrm{d}t^{q-1}} = y', \frac{\mathrm{d}^{q-2}y}{\mathrm{d}t^{q-2}} = y'', \text{ etc.}$$

and obtains an equivalent system with $p \times q$ *first order* differential equations, with the unknowns x, x', x'', etc., y, y', y'', etc., a system to which he can apply his preceding theory.

D'Alembert's method, consisting of giving a central place to systems of first-order linear differential equations and reducing all differential equations of any order to these systems, was original at the time.[9] It was taken up and generalized to include nonlinear differential equations by Cauchy circa 1820, in his analysis course at the *Ecole polytechnique*, course in which he reorganizes the whole theory of differential equations.[10]

The Theory of Partial Differential Equations: Toward a New Branch of Analysis

Determining the beginning of a theory in relation to what is only its prehistory is a complex historiographical question. In his book *Families of Curves and the Origins of Partial Differentiation,* S. Engelsman has proposed a reasonable criterion for the identification of the beginning of the theory of partial differential equations (PDE) by conjugating three conditions:

> a) partial differential equations must occur explicitly, b) the integration problem for partial differential equations must be stated explicitly, and c) methods must be available to solve the integration problem, at least for a certain class of partial differential equations. (Engelsman 1984, p. 150).

In spite of the interest of the works done on partial differentiation in the 1730s (notably that of Euler on modular equations), Engelsman shows that all these conditions had not been fulfilled. In contrast, these conditions are fulfilled by d'Alembert in his memoir on vibrating strings, "Researches on the curve formed by a stretched string set into vibration",[11] sent to the Berlin Academy in 1746.[12]

1. D'Alembert establishes that the (normalized) equation of the movement of the string is: $\alpha = \beta$, where $\alpha = \frac{d^2y}{dt^2}$ and $\beta = \frac{d^2y}{ds^2}$ (notations of the author) are the second-order partial derivatives of *y*, with $y(t, s)$ the ordinate at the instant *t* of the point M of the curve of abscissa *s*. This is a linear second-order PDE, now called the *wave equation.*
2. He looks for the general solution $y(t, s)$ of this equation, to show that there are an infinity of solutions other than that of the sinusoidal curve, which is already known.
3. For this, he uses a method reducing the integration of the PDE of the second order to that of a system of two exact differential forms of order 1.

[9] For more details, see (d'Alembert *OC* I/4a).
[10] See (Gilain 1989).
[11] "Recherches sur la courbe que forme une corde tendue mise en vibration" (d'Alembert 1749).
[12] See (d'Alembert *OC* V/2).

With this method, d'Alembert finds the solution of the PDE under this general form

$$y = \psi(t + s) + \Gamma(t - s),$$

where ψ and Γ are "arbitrary" functions.

Next he looks for the solution which corresponds to what we call today "initial conditions" and "boundary conditions" given by the mechanical problem.

While acknowledging d'Alembert's priority in solving this problem, Euler developed a more general conception of the nature of the permissible initial curve of the string, and so of the "arbitrary" functions that can enter into the solutions of the PDE. As we know, a long debate went on between them and with other scholars (Daniel Bernoulli, Lagrange) on this point.[13]

If d'Alembert thus appears as the initiator of the theory of the PDE in the 1740s, it is only in the 1760s that the study of these equations is developed in the framework of pure analysis, without reference to any physical context. D'Alembert devotes a whole paper to this subject in volume 4 of his *Opuscules mathématiques* published in 1768.[14] Euler (1770), going further, presents a synthesis of the state of the theory of the PDE in the 3rd volume of his great treatise on integral calculus. It was during these years that the study of the partial differential equations became a new branch of Analysis.

The Foundations of Infinitesimal Calculus and the Theory of Limits

Already in his *Traité de dynamique* of 1743 (art. 37), d'Alembert emphasized that, although it has the advantage of making proofs shorter, the method of infinitesimals is *not rigorous*. He develops this point of view in the *Encyclopédie*. In the article "Différentiel", he indicates that the differential calculus does not assume the existence of infinitely small quantities but consists only in determining algebraically the limit of a ratio.[15] He places himself in the continuity of Newton and Maclaurin, but in the article "Fluxion" of the *Encyclopédie* (t. VI, 1756), he criticizes the use made by British scholars of the notion of movement to found the new calculus. For d'Alembert this is an idea foreign to the subject.

He reaffirms this conception of the foundations of the calculus in the article "Limite" of the *Encyclopédie* (t. IX, 1765, p. 542), and in a whole chapter[16] of his *Eclaircissemens sur différens endroits des Elémens de Philosophie* (Clarifications of different parts of the Elements of Philosophy) published in 1767.

---

[13] See for instance (Demidov 2008).

[14] See (Jouve 2008) and (Guilbaud and Jouve 2009).

[15] Original text: "ce calcul [différentiel] ne consiste qu'à déterminer algébriquement la limite d'un rapport dont on a déjà l'expression en lignes" (*Encyclopédie*, t. IV, p. 986).

[16] "Sur les Principes Métaphysiques du calcul infinitésimal" (d'Alembert 1767, § XIV).

D'Alembert's position in favor of a theory of limits as the foundation of the infinitesimal calculus met with a real echo. From the 1770s on, several authors adopted this point of view, often referring explicitly to d'Alembert.[17]

In 1805, it is the method of limits which is taught at the *Ecole polytechnique* by Lacroix, Cauchy's teacher of analysis. From 1817 on, Cauchy, in his course at the same institution, reconstructs the whole differential and integral calculus on the basis of a more precise concept of the limit.[18] Thus d'Alembert appears as the initiator of a *tradition* which Cauchy continues, giving it however a whole other dimension.

Pure Mathematics and Mixed Mathematics in the Work of D'Alembert

The previous topics enlighten us on the relationships between pure mathematics and mixed mathematics in d'Alembert's work. Indisputably, d'Alembert wrote more papers or books on mixed mathematics (solid mechanics, fluid mechanics, celestial mechanics) than on pure mathematics. However, as we have seen, d'Alembert made important contributions to pure analysis, notably to the integral calculus, domain which is central in mathematics at the time, as much for its theoretical importance as for the richness of its applications to various physico-mathematical sciences then in full development.

Whereas some of d'Alembert's works are to be situated directly in a framework of pure mathematics (like the themes treated in the 1st, 2nd and 5th sections above), other parts of his work originated in the search for solutions to mechanical problems (like the themes of the 3rd and 4th sections). However, in these latter cases, d'Alembert, at a certain moment, develops a study of pure analysis with its own logic and leading onto theories which can afterward be applied to mechanics or astronomy.

Mathematics and the Enlightenment

Recent studies have been done on d'Alembert's mathematical work, particularly in the context of the edition of his *Complete Works.*[19] They improve our knowledge of his various contributions to the mathematical sciences and of his influence on other mathematicians. However, we still find much prejudice against d'Alembert.

---

[17] On the reception of d'Alembert's conception and the subsequent development of the concept of limit, see Schubring (2005, chap. III and following).

[18] See (Gilain 1989).

[19] *Œuvres complètes de D'Alembert*, series I to V, 8 volumes published (2002–2022), Paris: CNRS Editions. See more information on the website <http://dalembert.academie-sciences.fr>. See also the Special Issue of *Centaurus*: "Tercentenary of Jean Le Rond D'Alembert's birth (1717–1783). A review of the latest research", vol. 59/4 (2017).

One current historiographical trend supports the thesis that eighteenth-century mathematicians are characterized by an utilitarian position which considers mathematics as a simple tool for the empirical sciences and do not concern themselves with the rigor of concepts and reasoning. This thesis was developed, for example, in the Focus of an issue of the *Isis* Journal:

> Eighteenth-century mathematicians were usually content to reach correct and useful results that would aid in the understanding of the natural world. [...] The usefulness of the result was deemed proof enough that the method must be essentially correct. Mathematical rigor was deemed the province of unimaginative pedants. (Alexander 2006, p. 720)

Moreover, the authors affirm that there is a direct link between this supposed scientific position and the philosophy of the French encyclopedists: "enlightened mathematics was not rigorous. Strict mathematical rigor was a negative value in the mid-eighteenth-century world of the *Encyclopédie* because it was artificial and rigid." (Richards 2006, p. 703).

The work of d'Alembert who was at the same time a mathematician, an encyclopedist and a philosopher of the Enlightenment, contradicts this thesis.[20]

We may add that d'Alembert himself presented his conception of the role of mathematicians and of the utility of pure mathematics, in the article "Géomètre" (Geometer) of the *Encylopédie.* (At the time, "géomètre" is the word often used for a mathematician and "géométrie" for pure mathematics). He refers to the "preface" to the History of the Paris Academy of Sciences, where Fontenelle presented the new academy and wrote:

> It seems first of all that if one wanted to limit mathematics to its most useful applications, one would have to cultivate it only in as far as it has an immediate and perceptible relationship to the arts, and leaves aside all the rest as a vain theory. But this idea would be wrong. […] All parts of mathematics are founded on geometry and to go further, on algebra itself. (Fontenelle 1702, pp. vij-viij)[21]

[20] This agrees with Schubring's opinion: "It is a widespread opinion to think of eighteenth-century mathematics as unconcerned with the foundations and as interested only in the further development of analysis. As we have seen with the concepts of negative numbers as well as with the infinitely small quantities, the mathematicians were, in contrast, very anxious to clarify basic concepts." (2005, p. 285). See also (Gilain 2018).

[21] Original text: "Il semble d'abord que si l'on voulait renfermer les Mathématiques dans ce qu'elles ont d'utile, il faudrait ne les cultiver qu'autant qu'elles ont un rapport immédiat & sensible aux Arts, & laisser tout le reste comme une vaine théorie. Mais cette idée serait bien fausse. […] toutes les parties des Mathématiques, sont fondées sur la Géométrie, & pour aller jusqu'au bout, sur l'Algèbre même.".

At the Paris Academy of Sciences, this was reflected institutionally by the existence of three classes of mathematical sciences: Geometry, Mechanics and Astronomy. D'Alembert successively has belonged to these three classes; on his work as an academician of sciences, see (d'Alembert *OC* III/11).

However, d'Alembert adds another element in favor of the general usefulness of pure mathematics, with the following *philosophical and political* argument:

> But independently of the physical and palpable uses of geometry, here we shall envisage another side of its advantages, a side to which we have perhaps not yet paid enough attention: It is the usefulness of the study of geometry in preparing, almost imperceptibly, the way towards a philosophical spirit [...]. It is perhaps the only means of gradually shaking the yoke of oppression and profound ignorance under which certain countries of Europe still groan. [...] Soon, the study of geometry will lead to that of mechanics which [...] will lead to the study of true physics; and finally true physics will lead to true Philosophy, which, by the prompt and general light which it will spread, will soon be more powerful than all the efforts of superstition [...] (*Encyclopédie*, t. VII, 1757, pp. 628–629)[22]

It is a remarkable plea for the cultural and humanist value of mathematics. However, on the emancipatory powers of mathematics, we may think that unfortunately d'Alembert was a little too optimistic...

References

Abbreviations

HAB: *Histoire de l'Académie des sciences et belles-lettres* (Berlin).

OC: *Œuvres complètes de d'Alembert*, Paris: CNRS Editions (2002 -).

Alexander, Amir R. (2006). Tragic Mathematics: Romantic Narratives and the Refounding of Mathematics in the Early Nineteenth Century. *Isis* 97/4: 714-726.

d'Alembert, Jean le Rond. (1743). *Traité de dynamique.* Paris: David l'aîné.

d'Alembert, Jean le Rond. (1748). Recherches sur le calcul intégral. *HAB* II: 182–224; *OC* I/4a, 99–164.

d'Alembert, Jean le Rond. (1749). Recherches sur la courbe que forme une corde tendüe mise en vibration. *HAB* III: 214–219.

d'Alembert, Jean le Rond. (1750). Suite des recherches sur le calcul intégral. *HAB* IV: 249–291; *OC* I/4a, 175–245.

d'Alembert, Jean le Rond. (1767). Eclaircissemens sur différens endroits des Elémens de Philosophie. In *Mélanges de littérature, d'histoire et de philosophie*, t. 5, 3–272, Amsterdam: Zacharie Chatelain & Fils.

d'Alembert, Jean le Rond. *OC* I/4a. *Œuvres complètes de d'Alembert. Textes de mathématiques pures (1745–1752)*, s. I, vol. 4a, ed. C. Gilain, Paris: CNRS Editions, 2007.

[22] Original text: "Mais indépendamment des usages physiques & palpables de la Géométrie, nous envisagerons ici ses avantages sous une autre face, à laquelle on n'a peut-être pas fait encore assez d'attention: c'est l'utilité dont cette étude peut être pour préparer comme insensiblement les voies à l'esprit philosophique, & pour disposer toute une nation à recevoir la lumière que cet esprit peut y répandre. C'est peut-être le seul moyen de faire secouer peu à peu à certaines contrées de l'Europe, le joug de l'oppression & de l'ignorance profonde sous laquelle elles gémissent. [...] Bientôt l'étude de la Géométrie conduira à celle de la méchanique; celle-ci mènera comme d'elle-même et sans obstacle, à l'étude de la saine Physique; & enfin la saine Physique à la vraie Philosophie, qui par la lumière générale & prompte qu'elle répandra, sera bientôt plus puissante que tous les efforts de la superstition [...].".

d'Alembert, Jean le Rond. *OC* V/2. *Œuvres complètes de d'Alembert. Correspondance générale 1741–1752*, s. V, vol. 2, ed. I. Passeron, Paris: CNRS Editions, 2015.

d'Alembert, Jean le Rond. *OC* III/11. *Œuvres complètes de d'Alembert. D'Alembert académicien des sciences*, s. III, vol. 11, ed. H. Chabot, M. Jacob, and I. Passeron, Paris: CNRS Editions, 2022.

Demidov, Sergey S. (2008). D'Alembert et la notion de solution des équations différentielles aux dérivées partielles. *Bollettino di storia delle scienze matematiche*, XXVIII/2: 155–166.

Encyclopédie ou Dictionnaire raisonné des sciences, des arts et des métiers, ed. Diderot and d'Alembert, 17 vol., 1751–1765. Online publication: http://enccre.academie-sciences.fr/encyclopedie.

Engelsman, Steven B. (1984). *Families of Curves and the Origins of Partial Differentiation*, Amsterdam: North-Holland.

Euler, Leonhard. (1770). *Institutiones calculi integralis*, vol. 3, Petersbourg: Acad. Imp. Sci. ; *Opera omnia*, s. I, vol. 13, Leipzig and Berlin: Teubner, 1914.

Fontenelle, Bernard Le Bovier de. (1702). Préface. *Histoire de l'Académie royale des sciences. Année 1699*, Paris.

Gilain, Christian. (1989). Cauchy et le cours d'analyse de l'Ecole polytechnique. *Bulletin de la SABIX* 5: 3-145.

Gilain, Christian. (2018). Les Lumières dans l'historiographie des mathématiques. In *Enquête sur la construction des Lumières*, ed. F. Salaün and J.-P. Schandeler, 167–179. Ferney-Voltaire : Centre international d'étude du XVIIIe siècle.

Guilbaud, Alexandre, and Jouve, Guillaume. (2009). La résolution des équations aux dérivées partielles dans les *Opuscules mathématiques* de d'Alembert (1761-1783). *Revue d'histoire des mathématiques* 15/1: 59–122.

Jouve, Guillaume. (2008). Le rôle de d'Alembert dans les débuts d'une étude programmatique des équations aux dérivées partielles (1760–1783). *Bollettino di storia delle scienze matematiche* XXVIII/2: 167–181.

Maclaurin, Colin. (1742). *A Treatise of Fluxions*, 2 vol., Edinburgh: T.W. and W. Ruddimans.

Richards, Joan L. (2006). Historical Mathematics in the French Eighteenth Century. *Isis* 97/4: 700-713.

Schubring, Gert. (2005). *Conflicts between Generalization, Rigor, and Intuition. Numbers Concepts Underlying the Development of Analysis in 17–19th Century France and Germany.* New York: Springer.

Forgetting Past Science—Reconstructing Past Science

Oblique Perspectives on the Once Familiar, Now Half Forgotten

Jens Høyrup

1

When Gert and I frequented the *Gymnasium*, the sciences—at least the natural sciences, the humanities and social sciences present other problems to which we shall return—were supposed by most philosophers and historians of science to be by their very nature cumulative and progressive. They were in no need of recovering the forgotten, and everything was retained straightaway (except mistakes, which could be safely buried). The general public, though sometimes scared by the technological and world-shaping outcome of the applications of science, still tends to think so. For good reasons, we must admit. To the general public, the proof of the pudding is the eating, and if we measure science by, say, the ability to predict the trajectories of planets and spatial vehicles or the development of new antibiotics (or to explain the unfortunate mechanisms by which we make older ones inefficient), then science does progress.

Whether science is *cumulative* turns out to be a different question, and this is indeed where doubt set in around 1960 among philosophers of science—some of whom were already established as philosophers. Most famous of all Thomas Kuhn, a historian of science who failed to find linear progress when he explored the Copernican revolution and turned philosopher in the process of understanding *why*. He presented his explanations in *The Structure of Scientific Revolutions* from (1962) and in a paper on "The Function of Dogma in Scientific Research" presented in 1961 and published in (1963).

What general discourse took over and retained from Kuhn's argument was little more than the term "paradigm" (rarely understood exactly as Kuhn intended it, but

J. Høyrup (✉)
Roskilde University, Roskilde, Denmark
e-mail: jenshoyrup@gmail.com

A. Verdugo Rohrer and J. Zender (eds.), *History of Mathematics and Its Contexts*, Trends in the History of Science, https://doi.org/10.1007/978-3-031-60828-5_11

even his usage was rather blurred)[1] and perhaps the idea that successive paradigms are "incompatible". This was generally understood more radically than intended, emphasizing the non-cumulative character of scientific practice even more strongly than Kuhn had done himself.

Kuhn's view on the matter was summed up in the formulation (1970: 172) that scientific development was Darwinian in nature, "*from* primitive beginnings but *toward* no goal". This denial of progress toward a goal named "truth" hinged on Kuhn's identification of truth with *an ontology*, a list of constituents of the world. For instance, in the first part of the eighteenth century, Georg Stahl introduced phlogiston[2]—a material principle of combustion—as a means to bring order to the multitude of combustion and reduction phenomena; half a century later, Lavoisier pushed the same principle away in order to make place for oxygen. Similarly, in 1687 Newton introduced the gravitational force as an explanation of how the planetary system works; Einstein's theory of General Relativity replaced it by the geometrical properties of space itself. In both cases, we have steady progress in the predictive power of the discipline, but speaking of steady improvement of the ontology seems absurd.

Kuhn first meant "paradigm" to designate the "great book" (e.g., Ptolemy's *Almagest* or Newton's *Principia*)—or, in more recent times, the standard set of textbooks—on which everybody in a discipline is trained; this is where oblivion comes in. Training is indeed more than just accepting as true a set of statements—as everybody knows who has thought about how you are trained to swim, to ride a bicycle, or to drive a car. Training reaches the level of your mind which makes you (re)act without thinking about why you do so. In order to understand why you act as you do on the foundation of training requires quasi-Kantian *critique*, which is always difficult and slows down work—Newtonian physicists never bothered much about Kant's *Critique of Pure Reason.* Training, however, is not necessarily cumulative: having been trained to drive a car, you may certainly extend your training by trying to drive on ice; but beware not to rely on your training in running a motorbike when starting car training—better forget it for safety!

Now let us have a look at the dynamics Kuhn sees in scientific development "*from* primitive beginnings but *toward* no goal" (we shall permit ourselves to

---

[1] In an often cited article, Masterman (1970: 61f) states that Kuhn,

> with that quasi-poetic style of his, makes paradigm-elucidation genuinely difficult for the superficial reader. On my counting, he uses "paradigm" in no less than twenty-one different senses in his (1962), possibly more, not less. Thus he describes a paradigm:

This is followed up by 21 quotations which actually do not qualify as descriptions of what a paradigm *is* but are mostly statements about what it (sometimes) *does*. None the less: deriving what the paradigm is from such statements about what it does will not result in a sharp definition.

[2] More precisely, he elaborated ideas that had been around already in the later seventeenth century (Partington 1961: II, 666–669).

present an oblique and evidently very brief version.[3] The starting point is a field of interest: maybe (Kuhn's example, borrowed with due credit from John Heilbron's then ongoing doctoral work) electric phenomena; maybe "the situation of women". Initially, there are almost as many approaches as there are workers, nobody really understands why others are doing as they are, and there can be little progress; this is the "pre-paradigmatic phase". Then one approach seems to hold particular promise, and others begin to emulate it. Kuhn's example is Franklin's theory of electricity and his experiments; in women's studies of the early 1970s, Germaine Greer's *Female Eunuch* played a similar role. This leads to the second sense of "paradigm": *that which you take over* when being trained from a particular book or textbook tradition: a set of classificatory and explanatory concepts, methods and tricks of the trade, expectations and norms as to what the discipline should be interested in and how explanations ought to look, which handbooks and journals are important, etc. This constitutes a *discipline.* The elements of the paradigm are only in part learned from analytical explanation, in part they are genuine skills learned from non-verbalized training; and even that which is verbalized often has the character of *values* rather than rules that can be followed strictly.[4] Workers interested in the same real-life object but not sharing the paradigm in this sense do not belong to *the same* discipline in the epistemological sense, even though they may be employed at the same institute. They may still be able to learn from each other, but with the same difficulty as one learns from disciplines looking at different objects, through particular spectacles.[5]

Once a paradigm in this sense has been established, systematic and cumulative work can start. One worker's results and questions can be taken over by other workers—if needed questioned with understanding, because not only explicit statements but all their connotations, conceptual presuppositions, etc., are shared. In this way, the explanatory power of the paradigm is exploited to the full; gradually, the range of problems it has solved increases—Kuhn sees the paradigm as an instrument for "puzzle solving". In this process, the paradigm itself is sharpened as a tool (it "accommodates"). But exactly because it allows those who are well trained in the paradigm to anticipate which kind of problems it allows one to attack,[6] it helps to shelve and forget other interests which were once part of the inspiration for the field—once the paradigm of comparative linguistics had put the study of etymologies on a more secure base in the early nineteenth century,

[3] This oblique interpretation is fully unfolded in (Høyrup 2000: 277–394, *passim*).

[4] Emphasis on training and on the guidance by values rather than rules are the two main features that differentiate Kuhn's approach from others with whom he is often compared. This difference was rarely understood at the time—and in (1992) Kuhn himself had forgotten.

[5] In order to qualify this statement one should observe that "*the* paradigm" is not made of one piece: It is a nested structure. Some components of the training and scholarly norms of a linguist are shared with literary scholars, others with psychologists, still others with none of them.

[6] In this respect, the doctoral advisor is supposed to embody the paradigm, knowing in advance which problems should yield to good work, and which represent blind alleys which it would be irresponsible to ask or encourage the student to take up.

not only the methods used by the etymologist of the sixteenth and seventeenth centuries but also the questions that had occupied them came to be viewed as scholarly obscenities (as they still are today, if they have not become historical curios).

At some moment, intractable anomalies begin to abound, problems which the paradigm "promises" to be able to solve but which in fact resist understanding. Systematic work under the paradigm has exhausted its possibilities—and at the same time, it has located the points rather precisely where it fails.[7] The discipline gets into a state of mild panic, the limits of the permissible widen, and the situation begins to resemble that of the pre-paradigmatic phase—until one of the swarm of innovative proposals has so much success that it serves to establish a *new* paradigm which replaces the old one. The process repeats itself, and new collective oblivion ensues, now hitting even the concepts and norms of the previous paradigm. Those who grow up with the new paradigm and are trained in it certainly learn about the triumphs of their predecessors—but these triumphs are now told in the terms of the present cognitive framework. If confronted with original writings from before the "revolution", they tend to misunderstand what these had been speaking about and looking for. The two paradigms are, to use Kuhn's word, "incompatible", and understanding one from the standpoint of the other is as translating from one language into another; workers under the new paradigm, if not unusually attentive, tend to do so "naively", falling into the trap of "false friends" which any conscious translator knows so well.

This description is already highly abstract when we discuss the natural sciences; no wonder (notwithstanding the reviewers' amusement at the time) that Kuhn, when entering in 1978 into the thicket of historical details in a book about Max Planck and the early history of quantum mechanics, did not use the word "paradigm" a single time in its 400 pages. Within the *humanities*, Kuhn's picture may still give us some insights, but these will have to be thought through and sifted cautiously. Kuhn already pointed out that in the humanities, training is not made from textbooks but from reading of original works in which, taken together, no *one* approach normally has the monopoly (perhaps not quite as true today, but

[7] In a passage from the *Dialectics of Nature* (MEW 20, 334f), Engels already noticed this dynamic:

> Ebenso hat in der Chemie die phlogistische Theorie durch hundertjährige experimentelle Arbeit erst das Material geliefert, mit Hülfe dessen Lavoisier in dem von Priestley dargestellten Sauerstoff den reellen Gegenpol des phantastischen Phlogiston entdecken und damit die ganze phlogistische Theorie über den Haufen werfen konnte. Damit aber waren die Versuchsresultate der Phlogistik durchaus nicht beseitigt. Im Gegenteil. Sie blieben bestehn, nur ihre Formulierung wurde umgestülpt, aus der phlogistischen Sprache in die nunmehr gültige chemische Sprache übersetzt, und behielten soweit ihre Gültigkeit.

Another "Kuhnian" observation is packed into the quotation: Joseph Priestley produced oxygen, but understood it within the framework of the phlogiston theory as "dephlogisticated air"; only Lavoisier "invented" oxygen, thus engendering what developed into a new paradigm. Our ontology, the objects we see, depends on our paradigm.

still a fairly adequate picture). In particular, we may think of the proliferation of fields called "studies": "English studies", "Women's studies", "post-colonial studies", "queer studies", "science studies", etc. What these have in common is to be defined from interest in some part of the external world, not through anything like a paradigmatic approach. They are by their nature pre-paradigmatic—and since those who work in them usually already have a scholarly training, interdisciplinary. For a while, a particular approach may achieve almost paradigmatic status—for example *The Female Eunuch*—but as a rule, this approach is not eventually replaced by a new one because its possibilities have been exploited to the full, thereby creating an internal pressure for renewal and alternative; in more colorful words, it is not exploded from within. Instead, it comes under attack because any *particular* approach tends to forget and repress problems from the field which, at second thoughts or because of a new political and socio-cultural situation, seem no less important than those interesting the quasi-paradigmatic approach (for example, in the case of women's studies, questions of working and educational possibilities and legal restraints, which Greer's sexual emphasis had driven to the periphery). To say it crudely, changes of approach, interest, and emphasis in the humanities tend to depend *even less* on internal cognitive dynamics and *even more* on fashion or adaption to external demands than in the natural sciences. That does not mean that oblivion of earlier approaches does not take place in the "studies" fields. However, since they were not created by a highly technicalized discipline, the barriers for understanding them are lower. For workers in women's studies today, the difficulties in understanding Simone de Beauvoir's *Deuxième sexe* from 1949 will at most be a combination of the difficulties encountered by her readers in the year the book was written with those of reading any intellectual text from that year at 70 years' distance; such difficulties are quite different from those confronting a young geneticist who tries to grasp what things are really about in a paper from a genetics journal from 1949. There is oblivion and oblivion.

Moreover: The oblivion of the geneticist may be necessary, in so far as skills are better wielded when you do not need to reflect; when working on the connection between sequences of amino acids in DNA strings, protein production, and genes, the geneticist should not be bothered by questions and fumblings antedating the double helix and the cracking of the DNA code—these would only disturb working concepts. At least when training the next generation of workers, it is important for efficiency not to teach all the meanings a concept (etc.) has had in history—that is, everything it does not mean any longer.[8]

[8] There may of course be other reasons for such cognitive repressions, more rooted in institutional history than in epistemology. In the 1960s, it was impossible to make a thesis on "intuitionist" philosophy in the Mathematical Institute of the University of Copenhagen, since Harald Bohr—whose bust looked at everybody in the institute library—had belonged to the opposite, "formalist" school. In the Niels Bohr Institute, it was equally impossible to have a serious interest in "hidden variables", since these were at odds with Niels Bohr's "Copenhagen interpretation" of quantum mechanics.

Workers in a non-paradigmatic field need forgetfulness of preceding approaches to the same extent and for the same reason as they have to neglect other intellectual fields—one cannot be actively interested in everything without becoming superficial.

The two kinds of oblivion have a common effect: The field may forget where it came from, and thereby perhaps sever the connection to its roots in larger society. This does not necessarily happen. New roots or legitimizations may also arise or be constructed; but the invective "purely academic" illustrates the risk.

To sum up this first section: Science regarded as a depository of results remains in progress and even cumulative. No paradigm shift bereaves Watt's steam engine of its degree of efficiency, and successive paradigm formations and paradigm shifts in thermodynamics and materials science have each contributed to improve the efficiency of steam engines. But science as an *intellectual* endeavor is not. By climbing on the shoulders of predecessors (giants or otherwise), scientists transform them into stone plinths or at best pretty caryatids; past scientists are no dialogue partners.

2

Until the 1950s (often but not always in subsequent decades), *history* of science supported the accumulative-progressive view of philosophers of science. Of course, it might present personal history, with or without integration into social, institutional, and cultural context and with or without integration with technology (mostly without, it must be said). As a history of ideas, however, what it depicted was the caryatids of contemporary science—what a Kepler or a Newton had done was presented in modern formulae. If (for example, and indeed true for both) they had developed their thinking on the partial basis of ancient harmonics (musical theory), that was at most mentioned as a curiosity, not explored; but mostly it was not mentioned, being irrelevant to that accumulation of still relevant results to which they had contributed.[9] Paradoxically as that may seem, historiography was thus a tool for oblivion—for the kind of oblivion inherent in the practice of the science of its time. Yet the paradox is that of any court history, by its very nature a *justification of the present rule*; and the history of science was largely written from a perspective similar to that of a court historian. The court in question was not that of political rule but that of prevailing science, and what was produced was the picture of the professional past which could please or flatter the practicing

[9] Some colleagues might protest here and point to Alexandre Koyré as an exception. To some extent at least, it is true, Koyré (a run-away theologian) rather transformed early Modern science into a plinth for contemporary *philosophy*.

scientist—"the royal road to me", in a formulation which goes back to Grattan-Guinness (2004: 15) but has since become a familiar jibe among historians of science of a new breed.[10]

This new breed came, if not to dominate the field then at least to have some importance from the 1970s onward. Its hermeneutics is not that of Gadamer—his concept of *Horizontverschmelzung* ("fusion of horizons") from *Wahrheit und Methode* (according to which the present conceptual world is a kind of fulfillment of the historical development and hence ultimately the key in which the past should be read) is indeed a justification of the traditional historiography of science. Instead, it tries to reconstruct the scientific concepts of the past from the web of texts (scientific, but not only scientific) from a particular epoch—to some extent following the Wittgensteinian principle "don't ask for the meaning, ask for the use", at least trying to formulate past meanings on the basis of past use, and trying to avoid importing present-day meanings without being aware of it. We may call it "anamnestic", since it tries to recover memory of the past which had been repressed.

At times, those who make "anamnestic" history of science are as aware as other humanists and as philosophers of history that what we can achieve is only a *reconstruction*, an abstraction from which many aspects of the past have disappeared, and which may even be mistaken in some of the aspects it retains. They are also conscious that we can only reconstruct, that is, explain the past, by making use of our own language and concepts: not naively, not by using the concepts of actual scientific practice at face value, not by one-to-one mapping between past and present concepts, but so to say network-to-network—but still rooting our explanations (and the questions we ask of the past) in the world we live in. Others seem to be blissfully ignorant of these paradoxes of the historiographical endeavor (or at least write as if they were) and believe that they can tell the truth about the past "on its own conditions", or to find out *wie es eigentlich gewesen*, in the all-encompassing sense which is often attributed to Ranke though it was not his[11]—being just as convinced to be right as the workers under the paradigm in one of the hard sciences and just as unable to understand other approaches than their own.

Many factors have gone into the rise of this trend. One (but only one) is the emergence of a professionalized environment. The pertinence and quality of the work of a historian of science are now judged in the first instance by other historians of science, not by active scientists. There is a second instance, namely the one where the relevance of paid positions in the history of science is considered; here, practicing scientists—or rather, formerly practicing scientists who

---

[10] Grattan-Guinness goes so far as to deny that what we are speaking of here should be *history*, classifying it instead as "heritage". In that case, however, one may ask whether (for example) Livy wrote history.

[11] When formulating the phrase in the preface to his first book in 1824 (von Ranke 1824: vi), he merely contrasted moralizing historiography to the simpler task of getting *some* facts right (*nur Geschichten, nicht die Geschichte*).

have climbed the administrative ladder—may have their say. But this *is* a second instance and a less powerful influence on the research style of the single worker than the professional peers.

3

Given that (say) chemists have limited influence on how historians of science approach the history of their field, we may ask what they get out of the historians' attempt to overcome their professional oblivion. A curing anamnesis? Anamnesis curing for what?

This seemingly simple question—like almost any seemingly simple question—is a crystal with many faces. We shall reflect upon a few of them.

In the daily work of the practicing scientist, the answer is that history (of any kind) means little. Normal science is not trite routine; as lived by the participant of course it involves much routine, but the moments of getting new insight, new ideas, are felt to be what is important and distinguishes the undertaking. However, these insights, etc., are produced within the framework of the prevalent practice and mode of thought ("paradigm"), and here neither the projection of these on the past nor views of the past which might cast doubt on their absolute validity is of any help.

When a crisis approaches, one might believe historical insight to have a better possibility to intervene. Not only is the amount of innovation greater[12]; it is also less constrained by the prevalent paradigm. Kuhn claimed indeed that one of the things that characterize this situation is increased willingness to take up past ideas. New ideas may certainly be inspired from anywhere, also by distorted or less distorted versions of past theoretical structures. But they might as well come from a fairy tale which impressed one as a child, or from experience in the kitchen. Newton's falling apple is a lie, but nothing would prevent it from having been true; Einstein, when writing his first paper on the Special Theory of Relativity in 1905, *might* have been inspired from the discussion of the nature of time in Aristotle's *Physics* IV.11—thoughts are developed in parallel for quite a while—but he was not.

Traditional history of science, like court history in general, had a different function: legitimization of the present and its power relations. Historiographic sensitivity to differences, the awareness that the past cannot be reduced to a less developed version of the present but was a world on its own, makes legitimization less straightforward. If we cannot claim that everything in the past points toward

---

[12] Even though much is still routine; if not, the new paradigm would not be created by the workers in the field—if nothing was conserved, their professional routine would not count. During the discussion at the "Symposium on Evolutionary Epistemology", Ghent 1984, Donald T. Campbell offered this beautiful aphorism "In normal science, 95% of everything is routine; during a scientific revolution, only 90% is routine".

us, then it seems less obvious that we are *on the only possible* and therefore *on the right* path.

That sensitivity to differences and acceptance of their legitimacy which characterizes "anamnestic" history of science does not convey a message that science progresses along the only possible road. This may displease scientists less than it displeases those who disguise controversial activities as scientific (and hence inescapable) progress or necessity—be it agriculture based on genetic manipulation, be it bombings calculated mathematically, and hence claimed to possess "surgical precision" because of their reliance of infallible mathematics. It may also displease those who dismiss scientific results as "mere theory" or science as a whole as "nothing but another belief system" when wanting to teach Genesis as biology or to pump the atmosphere full of carbon dioxide. "Anamnestic" history of science does not give us a picture of the scientific endeavor as a military march along a straight line; nor, on the other hand, as blind groping and random walk in all directions. Kuhn may show that Planck did not understand the implications of what he was doing when he was taking the first steps toward quantum mechanics; but Planck still provided a starting point for what has since then been adopted by every physicist, though integrated in different frameworks depending on whether they accept the "Copenhagen interpretation" or not, and without which our computers and diode lights simply would not have been made.

Working scientists thus cannot or can only accidentally take advantage of history (of any kind) in their scientific practice. Many, of course, are not interested in historical questions at all. Those who are, however, do not in general react negatively to the "anamnestic" style. They may disagree with particular interpretations, and they may have good reasons—interpretations are always under discussion, so this is a different matter than disagreement with the undertaking as such. What they get at the personal level may be a richer feeling of professional identity. What the profession gets is an opportunity for its members to reflect once more upon their role and upon the relation between what they are actually doing and what were the extra-scientific concerns from which the discipline once started—making the members more conscious of the responsibility of the discipline and the scientific enterprise in general. Anthropology and cognate disciplines may serve as an example: the appearance of Edward Said's *Orientalism* in 1978 had a strong impact on many of their workers. Recent investigations of the intertwinement between the development of nineteenth-century (French, but not only French) anthropology and the effort of the colonial powers to "divide and rule" by sharpening whatever ancient ethnic conflicts could be identified within the subjugated populations should call for similar afterthoughts and might sharpen the awareness of the profession of its use for similar purposes today. So, if history of science, while overcoming its traditional role of celebrational court history, does not become a "purely academic" concern itself which has lost sight of its original aspirations, then it may contribute to producing academics who are more responsible citizens, both of the *republic of letters* and of the *republic of everybody*.

4

Let us close the essay with some observations concerning what has occupied Gert for half a century: the history and the didactics of mathematics.

Together with medicine, mathematics is probably the academic field which has the strongest tradition for interest in its own history and which most often pays for positions for historians within its own institutes.

In contrast to medicine, which has had no difficulty in recognizing at least the major revolutions in its history and even points to these as triumphs (such as the rejection of the Galenic bodily humors and the miasma theory), mathematics has a long tradition for seeing itself as a discipline in steady progress, keeping forever what it has once brought under its dominion. In the 1970s, after the impact of Kuhn's work on scientific revolutions, this was held out as what distinguished mathematics from other fields; mathematics, thus we may interpret the view, never errs, and therefore never had to retract—it just learns more and more, and orders its knowledge better and better.

In the meantime, historians of mathematics have refuted this view—as it turns out without creating excessive scandal among mathematicians.[13] None the less, just as in the case of the physical sciences there is something alluring in the old view—we may still recognize mathematics in papyri and on clay tablets from the Bronze Age, and crack their procedures by means of (the most basic level of) the mathematics of our own days—which means that they are somehow valid in the same sense as the mathematics we practice ourselves. We shall not attempt to discuss why this is so (it is probably just as impossible to state the answer exactly as in the case of the physical sciences) nor claim that it is the full truth (it is not). What is important for our present discussion is the impact of this continuity in *teaching*.

Nobody would teach medical students the Galenic theory of humours, neither in the original nor in some updated version, and nobody would do anything similar with the phlogiston theory in school chemistry. On this account, mathematics is much more akin to literary theory and historiography. One may base an introductory university course of text analysis on the principles of Aristotle's *Rhetoric* and *Poetics*, as a colleague of mine once did; and one may even use the original texts (evidently in translation and not in ancient Greek). It is also natural to confront those who undertake the study of history with Herodotos, Thucydides, and Livy—three wonderfully contrasting historiographic models. School mathematics does not act so radically (nor would anybody base pre-university teaching of literature and history on the ancient authors), but there is still a high degree of continuity. At the age of 14, I was not only confronted with slightly updated versions of Euclidean geometry and proportion theory but also with the whole range of topics from commercial arithmetic which I have now encountered when working

[13] Gillies (1992) contains the statement that originally stirred the debate as well as a number of implicit or explicit answers.

on Italian "abbacus school" mathematics from the late Middle Ages (my school algebra, however, made use of the letter formalism created in the seventeenth century). Until the nineteenth century, reform movements in mathematics education often returned to Euclid's *Elements*, at least books I–VI, as the best basis (while the next reform movement would kick him out again—"Down with the triangles, down with Euclid", in Dieudonné's dramatic words from 1960 (Castelnuovo 1977: 43)).

The historian's traditional task at the mathematical institute has therefore not only been to write court history, to describe "the royal road to us", but also, perhaps mainly, to give some kind of inspiration to the teaching endeavor. Some didacticians have even believed in the "genetic method" (Gert knows all about it), the pedagogical version of Haeckel's law in biology (individual development recapitulates historical evolution), and thus held that children should develop mathematical conceptual structures in the same order as they have arisen historically; others were less radical, but many still thought history might be a good wall against which to throw the ball.

Literal belief in the genetic method presupposed unacknowledged oblivion, oblivion not only of the different conceptual structures by which past mathematics was carried but also lack of awareness that something has been forgotten. Anamnesis might really be needed, it seems. It probably is—but its result is somehow catastrophic. While showing that past conceptualizations were different from ours, it also makes us see how these drew on *experience* which is no longer ours, and in particular not those of modern school children—and that they were limited by lack of kinds of experience which our kids have encountered already before starting school. This shows the futility of the strict version of the genetic method and makes it clear that it is a Lichtenberg mirror "from which no apostle can stare back when a monkey looks into it".

But anamnestic history may still be useful for didactics. Firstly, it may teach didacticians something about how the construction of conceptual structures is conditioned by the experiential world of the learners. Secondly, if its results are communicated to teachers and children, it may broaden their view of what mathematics is and what it could be. Some 25 years ago, I had written a short paper on the notion of "broad lines", an idea of lines carrying an implicit breadth (permitting a measurement of areas and lengths by the same unit, just as we measure cloth in length units because the breadth is given). This turns out to have been a widespread idea in early practical geometries. Enrico Giusti, the journal editor to whom it was sent, told me afterward that he had promised to make a lecture in a middle school the day after he had received it and chose to tell about these broad lines. "The children enjoyed it. They were enthusiastic that mathematics could be thought in a different way than the teacher said".

Mathematics is certainly a field where the joy of discovery on the part of the learner is mostly supposed never to go beyond what the teacher already knows; anamnestic history may therefore be an eye-opener, a support for tolerance. It is not a priori certain that mathematics is the only field where this is the case.

References

Castelnuovo, Emma, (1977). "L'enseignement des mathématiques". *Educational Studies in Mathematics* **8**, 41–50.

Gillies, Donald (ed.), (1992). *Revolutions in Mathematics*. Oxford: Clarendon Press.

Grattan-Guinness, Ivor, (2004). "History or Heritage? A Central Question in the Historiography of Mathematics", pp. 13–31 *in* Ivor Grattan-Guinness & B. S. Yadav (eds), *History of the Mathematical Sciences*. New Delhi: Hindustan Book Agency.

Høyrup, Jens, (2000). *Human Sciences: Reappraising the Humanities through History and Philosophy*. Albany, New York: State University of New York Press.

Kuhn, Thomas S., (1962). *The Structure of Scientific Revolutions*. Chicago: University of Chicago Press.

Kuhn, Thomas S., (1963). "The Function of Dogma in Scientific Research", pp. 347–369 *in* Alistair C. Crombie (ed.), *Scientific Change*. London: Heinemann.

Kuhn, Thomas S., (1970). *The Structure of Scientific Revolutions*. 2nd ed. Chicago: University of Chicago Press.

Kuhn, Thomas S., (1992). *The Trouble with the Historical Philosophy of Science*. Cambridge, Mass.: Department of the History of Science, Harvard University.

Masterman, Margaret, (1970). "The Nature of a Paradigm", pp. 59–89 *in* Imre Lakatos, & Alan Musgrave (eds), *Criticism and the Growth of Knowledge*. Cambridge: Cambridge University.

MEW: Karl Marx & Friedrich Engels, *Werke*. Berlin: Dietz Verlag, 1956–2018.

Partington, James Riddick, (1961). *A History of Chemistry*. vol. 2. London: Macmillan, 1961.

von Ranke, Leopold, (1824). *Geschichten der romanischen und germanischen Völker von 1494 bis 1514*. Leipzig & Berlin: Reimer.

A Revolutionary in Mathematics Education (On V.R. Mrochek)

Alexander Karp

Stressing the importance of grasping the social nature and the social significance of developments in school mathematics, Gert Schubring wrote that it is important to examine school texts, school practices, and all relevant legislative and administrative measures not separately from one another, as isolated phenomena, but in their totality and through their interactions. The role of a unifying factor, in such a case, may be provided by the biography of a figure in mathematics education whose activity was clearly constrained by specific social conditions and who produced educational texts, worked with teachers, and promulgated views specifically under these conditions (Schubring 1987). Let us further note that the study of such a biography may be especially fruitful for periods of transition, periods when the political order of a country and the structure of its society underwent dramatic changes, and when views and positions were therefore revealed with particular vividness and clarity.

The present paper focuses on a country and a historical period which Schubring did not investigate, but to which these general principles naturally apply. Its subject will be one of the most prominent figures in pre-revolutionary Russian mathematics education, Vatslav Romualdovich Mrochek (1879–1937). Mrochek was one of the leaders of the pre-revolutionary Russian reform movement in mathematics education (Karp 2019), representing its most radical left wing. However, there are no studies dedicated exclusively to him, to our knowledge, apart from a paper by Baranets and Verevkin (2013), which drew attention to materials preserved in the archives of the Russian Academy of Sciences (Morozov n.d.), which we will make use of, but which is devoted mainly to Mrochek's activities outside mathematics

A. Karp (✉)
Teachers College, Columbia University, New York, USA
e-mail: apk16@columbia.edu

A. Verdugo Rohrer and J. Zender (eds.), *History of Mathematics and Its Contexts*, Trends in the History of Science, https://doi.org/10.1007/978-3-031-60828-5_12

education. The present study relies on archival materials and on the analysis of publications from Mrochek's time.

V.R. Mrochek Before the Revolution of 1917

Several questionnaire forms filled out by Mrochek already after the revolution have been preserved (Morozov n.d.; Mrochek n.d.). From these, it can be gathered that he came from the nobility, that his grandfather was a landowner, and that his father was a low-ranking officer (lieutenant), but was retired. His mother was the daughter of a retired major. His nationality is indicated as Polish. Lacking access to any additional facts, we can imagine a family of Polish gentry of modest means, living in Ukraine. Vatslav attended a gymnasium in Zhytomyr and subsequently continued his education at St. Petersburg University, graduating from its physics-mathematics faculty. According to him, he also graduated as an auditor from its historical-philological faculty at the same time, but we have not been able to find any record of this, so that it may be supposed that he attended lectures but did not necessarily take exams or fulfill other demands, whose fulfillment, however, was required for a diploma.

It is not entirely clear why Mrochek's education at the university took so long—seven years—but what is clear is that these years were not quiet ones. Like many other students of that time, Mrochek joined the Socialist Revolutionary Party, the most popular organization opposing the regime prior to the revolution. The party was seen as representing the interests of the peasantry, the most numerous estate in Russia, but it owed its fame above all to members of the intelligentsia, who were devoted to the people and believed that the best way to help the people was to follow the precepts of the so-called *Narodnik movement*, whose organization *Narodnaya Volya* (People's Will) had organized the assassination of Tsar Alexander II in 1881. The Socialist Revolutionary Party was founded in 1902 and shortly thereafter came to include a "combat group," which inaugurated its activity with the assassination of Minister of the Interior Dmitry Sipyagin in 1902. This was the party that Mrochek joined in 1904, indicating on questionnaires that he had been a member from 1904 until 1917, and that for a time in 1918, he was a member of the Party of Left Socialist Revolutionaries, which broke off from the main organization.

It should not be imagined, however, that Mrochek spent his time waiting to ambush tsarist officials. The overwhelming majority of party members confined their activity to meeting with one another, reading banned literature of one sort or another (and the amount of such literature was vast), perhaps collecting funds for prisoners, perhaps even participating in various kinds of demonstrations (which were ferociously broken up). From the already cited questionnaire, we know that in 1901 and 1902 Mrochek was arrested for participating in student strikes (and apparently spent some time in prison), and that in 1904, he was sentenced to administrative internal exile for participating in preparing a May First proclamation.

It should be borne in mind, however, that given the prohibition against any kind of activity that was in the least degree oppositional, terror against tsarist officials was seen in an entirely different manner than it might be seen from the perspective of the liberal states of today. The Russian writer Ilya Ehrenburg (1891–1967), without question a close observer of his time, depicted in a novel a member of the Russian intelligentsia, Tishin, who was quite typical (even if he did come across as a caricature): "At about the age of twenty, Alexey Spiridonovich became involved in politics, which is to say, he started composing a brief on Bogdanov's *Political Economy* and contemplating the question: was it or was it not a sin to kill a governor?" (Ehrenburg 1923).[1] Such questions were posed by very many people, and the conclusion they often came to was that it was no sin at all. There was nothing unusual in Mrochek's pre-revolutionary political biography.

In 1905, Mrochek began teaching mathematics in secondary educational institutions—gymnasia and real schools. He does not indicate precisely which institutions these were, however. He does state in his autobiography that as someone suspected of taking part in Socialist Revolutionary Party activities, he could not occupy positions at state-run schools. Mrochek was also involved in teaching various courses for teachers, as well as mathematics (technical mathematics, as he writes) at the Professors' Association Polytechnicum from 1912 until 1917 [other teachers at this institution included, for example, Stepan Timoshenko (1878–1972), subsequently a corresponding member of the Russian Academy of Sciences and a professor at the University of Michigan and Stanford (Sinkevich 2022)]—educational institutions that were not run by the state, which appeared when significant numbers of scientific staff were forced out of state-run institutions for political reasons, attracted scientists. It was not Mrochek's teaching, however, but his literary activity that brought him fame.

"The Pedagogy of Mathematics"

Probably his most famous book was *The Pedagogy of Mathematics* (Mrochek and Filippovich 1910a), published in 1910 and written in coauthorship with Filipp Filippovich (1878–1938), an individual whose biography also merits attention (Kolyagin and Savvina 2006).

Filippovich was born in Serbia, began his education there, and became an active participant in the socialist movement—so active, in fact, that in 1899 he was forced to emigrate to Russia. There, he graduated from St. Petersburg University in 1904, the same year as Mrochek. In 1905, he was arrested for revolutionary activity, but was soon freed, thanks to leniency granted by a decree of the tsar. In 1912, Filippovich returned to Serbia, where he gave himself over wholly to political activity, even serving as secretary of the Central Committee of Yugoslavia's Communist Party and as mayor of Belgrade. Later, however, he was again forced to move to

[1] All translations from Russian are by the author.

the USSR, and in 1938, like many communist immigrants to the USSR, he was arrested as a Trotskyist and executed.

The Pedagogy of Mathematics was written jointly, and it is practically impossible to distinguish which author wrote what. Kolyagin and Savvina (2006) claim that the parts on geometry were written by Filippovich, not Mrochek, but they offer no proof—their claim is likely based on the fact that Filippovich wrote about geometry without Mrochek as well, and that these chapters also have less historical material, which Mrochek liked to insert everywhere. In our view, however, it is hardly possible to claim this as a certainty, and in what follows we will take the view that both Mrochek and Filippovich can be considered coauthors of everything in the book.

It should be said that, in keeping with their overall views, Kolyagin and Savvina (2006) scold the authors somewhat for their "kowtowing to the West": "Unfortunately, Mrochek and Filippovich clearly overestimated the role of foreign methodological trends. They write too enthusiastically about Americans," and further: "Blind adherence to American or European models, as is well known, cannot lead to positive results either in the sphere of education or in any other sphere in the life of the Russian people." Even so, after rebuking Mrochek and Filippovich, they manage to find mitigating factors: there are numerous references to works by Russian educators, and the erring authors do not fail to "bow down before the authority" of the famous Russian educator Ushinsky (p. 47). All this is not new: Lankov (1951) openly called Mrochek "a troubadour of foreign influences," which Kolyagin and Savvina (2006) nonetheless consider a gross exaggeration (p. 47).

The point of the book, however, came from the fact that Mrochek and Filippovich, who evidently had good knowledge of German and French (Mrochek wrote about this on the questionnaire, indicating that he could read English, too), saw their task as consisting in telling Russians about how Russia had fallen behind the newest methodological developments. In their introduction, they state outright that the era following 1905 (the year of the First Russian Revolution) made the need for a radical uprooting of the old system of schooling obvious, furthering "the peaceful conquest of civilized countries by experimental psychology and experimental pedagogy," and further: "The work of the International Mathematics Commission of 1908, elected by the Fourth International Congress of Mathematicians (Rome, April 1908), shows that mathematicians of all countries have recognized, finally, the ungainliness of existing school education and have undertaken decisive reforms" (pp. IV-V). The orientation toward "civilized countries" and the international reform movement, understood in the most radical fashion, was in fact the book's main distinguishing characteristic. Indeed, the book was written on the wave of responses to the work of the International Commission, about the importance of which the authors of this book wrote in another brochure, published in the same year (Mrochek and Filippovich 1910b), which will be discussed below.

Autocracy's sycophants, Mrochek and Filippovich wrote, believe precisely that "Public education must be opposed at all costs. If Russia has had any success in resisting the spirit of the times, this has been due partly to people's illiteracy." Recognizing, however, that gymnasia could not be done away with, the forces of autocracy decided instead "to protect [the people] from any inclination to think, to kill independence, and to inure them to acting mechanically. The best means for this turned out to be... Latin and Greek!" (p. 50). Mathematics was dealt a blow as well: "Reduced to 28 h out of a total of 226 h, constrained to nourish itself on the textbooks of Malinin, Davydov, and Kiselev,[2] banished to the fringes of the school, [mathematics] soon broke with traditions and turned into a weapon for making people stupid. Mathematics was learned by those who wanted to learn it; usually, a student's success in ancient languages made his teachers turn a blind eye to his total neglect of even elementary mathematics" (p. 51).

This sad picture formed part of a history energetically delineated by the authors—a history of mathematics education, a history of education, and history in general. The first two chapters of the first part are devoted to the evolution of the pedagogy of mathematics. They are followed by chapters about the visual method and the laboratory method, about psychology and pedagogy, and lastly, about the main principles of the pedagogy of mathematics.

The reader learns that there are three developmental periods. The first is from ages 2–3 to 6–7; about this period, little is said. The second is from ages 8–9 to 13; it coincides with the beginning of classes in school. We are informed that during this period, teaching should consist mainly of exercises and building mechanical skills. Finally, the third period is from age 13 for girls and 14 for boys, all the way up to 23–24. During this time, the body develops as well as the soul; half-children are impressionable, and this is the most difficult moment for education. What is to be done, then? The authors answer with a quotation from the American psychologist Stanley Hall, stressing that drilling should be abandoned, and that education should be based on a feeling of freedom and interest—students should be taught only that which arouses their strong interest (p. 97). How this is to be done in practice, however, is left unexplained; instead, the authors once again repeat that schools do everything in a way that is completely wrong.

Some general principles can nonetheless be discerned in the first part of the book. Abstract reasoning comes under fire: "the role of pure reasoning is utterly negligible; on its own, it is incapable of conveying understanding" (p. 102). The authors explain that "Biologically, the student is a sensory-motor apparatus, with the nervous system in charge. The sensory part governs perceptions; the spinal cord and the brain govern their processing; the motor centers govern reproduction and generalization" (p. 103). They draw the conclusion that "reasoning is mastery of muscular efforts." After knocking down reasoning from its throne, as they put it, and after noting that the process of abstraction (we would say "abstract reasoning") takes place more slowly in childhood and is not amenable to any accelerating techniques, the authors come to the practical conclusion that "it is necessary radically to change the method of teaching, bringing general educational manual labor

[2] The principal mathematics textbooks used in gymnasia. Kiselev's textbooks were also used much later, until the mid-1970s.

into the foreground" (p. 104). From this, they go on to extol the laboratory method in the teaching of mathematics.

The second part is devoted to concrete sections of the course in mathematics, from arithmetic and elementary geometry to quadratic equations. It must be said at once that what we are presented with here are mainly certain general historical and theoretical considerations. The connection with manual labor and with the laboratory method is by no means always apparent.

For example, the chapter on "The Foundations of the Elementary Course in Algebra" opens with the following general observation: "While we in Russia still adhere to Euler's point of view and continue to 'teach science,' in Western Europe and America they have been able to move on to a living, modern, and applied approach to constructing the course" (p. 287). But then, instead of telling readers how to construct a living and applied course, the authors write at length about the history of algebra, after which they identify three approaches to teaching algebra. The first approach is oriented around transformations, the second is oriented around equations, and the third is oriented around functional dependency. The authors cite textbooks that follow each of these approaches, but themselves incline mainly toward the third, as befits adherents of the international reform movement.

The chapter on visual geometry can be more easily connected with the general psychological-pedagogical considerations articulated earlier, although even here the authors' recommendations possess a quite general character, that is, they mainly write about the need for such a connection. First, unsurprisingly, we are told that the old order of teaching geometry must be radically transformed and that the course in geometry must begin in the earliest grades and "assemble raw materials into facts and concepts," as well as "give precision to knowledge acquired through everyday experience." Then, the authors argue that teaching should be grounded in observation rather than reasoning, concretization rather than abstraction, technical drawing rather than cartoony sketches, and the construction of models and accurate measurements rather than rote memorization from the textbook (p. 184). They come to the conclusion that the course.

(1) must be constructed on the basis of child psychology and interwoven with manual work—the construction of models,
(2) must develop spatial conceptualization,
(3) must develop the idea of functional dependency,
(4) must offer material for a deductive course in geometry and mathematics and prepare students for the necessity of proving geometric truths,
(5) must provide a store of geometric knowledge for practical needs.

The fourth requirement seems somewhat unexpected, coming as it does after the critiquing and dethroning of the abstract, but even prior to this the abstract had never been excluded and prohibited altogether.

The authors then go on to discuss the content of the course—various shapes in space and in the plane, the volumes and areas of surfaces, and also symmetry.

They enter into various details, although they are occasionally guilty of a certain degree of mathematical imprecision.

The writing of a second volume of the book was planned, in which other topics from the school course were to be discussed, but no such volume was ever written.

Mrochek's Other Works

The Pedagogy of Mathematics was not Mrochek's first book: already in 1908, he published *Linear Trigonometry and Elementary Goniometric Functions*, which was subsequently reissued in 1913 (Mrochek 1913)—this is the edition we will refer to.

The book opens with an expansive historical overview, but subsequently the course is quite traditional in style; we would merely note that it includes many sections that are not studied today even in educational institutions of the most advanced type, and which, indeed, were not part of the curriculum of *real schools* (*Realschule*) even then, which is reproduced in the book by the author. There are other differences as well, for example, almost at the very beginning of the textbook—and what we have before us is precisely a textbook, with a theoretical exposition, proofs of theorems, and numerous problems—today's readers encounter a peculiar argument demonstrating that the sides of a right triangle are not proportional to its angles. This is shown by examining a triangle with angles 30^0, 60^0, 90^0, which stand in a relation of 1:2:3 to one another, while the sides in such a triangle stand in a relation of $1 : \sqrt{3} : 2$. Not satisfied with this example, the author also examines a triangle with angles 45^0, 45^0, 90^0, where, likewise, no proportionality is found. After this, he concludes that "the assumption of proportionality must be regarded as definitively rejected" (p. 6).

And this is how vectors are introduced:

> the magnitudes with which Geometry operates are absolute magnitudes, like arithmetic magnitudes; they are referred to with the English term "scalar." Most of the magnitudes encountered in nature, however, have not only a certain size, but also a certain direction, for example, time (past and future), the space traversed by a moving body, temperature (above and below zero), and so on. The same is also true of many analytic magnitudes, such as force, velocity, and acceleration (in mechanics), and so on. All of these magnitudes are called "vectors." Goniometric functions are vectors as well. (p. 103)

Today, this manner of exposition seems imprecise and even inaccurate. We would venture to note, however, that at the beginning of the twentieth century, Mrochek was no flagrant exception—people used the same language in writing as they did in normal conversation, without considering it necessary to adopt a special mathematical style.

A word should be said about the problems. There are many of them. They include applied, or as we would call them today, *real-world problems*. But there are also pages of exercises devoted to transformations or solving trigonometric equations that remained standard for decades.

As we might conclude, the textbook was not a flop financially—otherwise, no new edition would have been prepared—but to speak of it as a commercial success would hardly be appropriate.

Another relatively large composition by Mrochek was published in 1912 in the journal *Obnovlenie shkoly* ("School Renewal") and was titled "Arithmetic in Its Present and Past." About half of it is devoted precisely to the past, that is, to a historical sketch of the teaching of arithmetic and the studying of arithmetic (although the author paints his picture with broad strokes and tells about much else along the way). But its second half is devoted to modern times, and here the author does not hold back his polemical temperament, denouncing the monotonous routine of teaching in Russia (as is clear from its title, this was exactly what the journal envisioned as its main task). Two directions can be distinguished in his criticism.

The first is grouping problems into different types and teaching how to solve them in connection with their type: problems about pools that are filled in different ways, problems in which one runner catches up with another, and so on. Mrochek writes passionately and at length about the stultifying influence of this kind of instruction and tells the story of a girl who, after solving a quite difficult, but standard problem belonging to one of the mechanically memorized types, was unable to determine how much money one would get in exchange for one and a half dozen eggs that cost 12 kopecks per dozen, and who protested that egg problems had not been covered in class.

The second issue that Mrochek addresses is the relation between arithmetic and algebra, which was discussed at the time and later as well, but which was especially urgent at the time because schoolchildren who had already studied algebra were not allowed to use it to solve problems that were specifically supposed to be solved arithmetically. On this count, however, a certain amount of casuistry entered scholarly discussions: for example, writers discussed what constitutes a purely arithmetic problem and what differentiates it from an algebraic problem. Mrochek, of course, was in favor of making use of algebra and condemned those who opposed this, quoting at great length from the French writer Charles-Ange Laisant (1841–1920). Then he again turns to history, explaining how the system criticized by him had taken shape historically, and finally concludes triumphantly: "The old schooling, the old recipes have been banished once and for all! Ours is the age of the freely developing, creative personality, the age of labor in society and for society!" (p. 65).

Below, we will discuss still other contributions by Mrochek. Here, we would just mention that he published many translations of foreign mathematics educators (the same Laisant, for example) as well as various compositions on nonmathematical topics, such as "The Panama Canal" (in 1914) or "Bridges Past and Present" (in 1915). It may be supposed that these writings were also motivated by his need to make a living, although he was undoubtedly interested in technology and over twenty years later had plans to write again about canals (Baranets and Verevkin 2013).

Reviews of Mrochek's Works

It must be said that Mrochek's works were by no means universally praised. We will confine ourselves here to offering several quotations from V.V. Bobynin (1849–1919), considered the most prominent historian of mathematics of the pre-revolutionary period (Ovchinnikov 2011; Rybnikov 1950), regarding Mrochek's *Linear Trigonometry*, or more precisely, its historical introduction, and we will also comment on two reviews of *The Pedagogy of Mathematics.*

Bobynin (1913) begins as follows:

> Due to the poor dissemination of the study of the History of Mathematics in Russia, a deplorable habit has developed in Russian mathematical literature of treating the facts and particulars of the History of Mathematics in a very unceremonious manner. Passing off the products of one's own fancy as the results of historical research in mathematics, asserting the nonexistence of whole branches of the history of mathematics simply because they are unknown to the author, and other phenomena of similar nature, are not unusual in Russian mathematical literature. Among Russian mathematician authors who are guilty of such unceremoniousness in their approach to the History of Mathematics, perhaps the one who goes furthest in this direction is Mr. Mrochek (p. 305).

This is followed by concrete examples: for example, not understanding too well that the word "academy" did not have the same meaning in antiquity that it acquired later, Mrochek naively relates that the first academy in human history was established in Alexandria, under whose aegis were then established a museum and a library; while in reality the name given to an association of scientists was "museum" (not "academy"), and it was the "museum" that was established under the aegis of the library, rather than the other way around.

Bobynin concludes in the same vigorous vein:

> In conclusion, it merely remains to be said that, despite the extreme poverty of his historical mathematical knowledge, Mr. Mrochek permits himself to pronounce condemnations and categorical judgments in an authoritative tone, with the appearance of an expert... (p. 307).

To us, Bobynin's passion may seem somewhat excessive (possibly because over the years the unceremonious treatment of the particulars and facts of history has noticeably grown), but we cannot fail to note it.

The review of *The Pedagogy of Mathematics* signed by K.L. (L.K. in the table of contents: from the discussion that subsequently followed, it is clear that the author was the well-known Russian mathematics educator Konstantin Lebedintsev 1878–1925) focuses on the nonhistorical. The reviewer begins by noting that the book under review is a work of compilation based on foreign sources, which in itself would not have been a problem had the authors been able to offer an apt and concise exposition of what it is that the supporters of reforms are striving for, and of what they have already achieved. But this, in the opinion of K.L., is not what the authors have given us:

> The cardinal principles of contemporary methodology – the need to present all sections of mathematics in a concrete, inductive manner and to systematize the studied material at the end of the course, the need for a close connection between school mathematics and real life, the need for as full a development as possible of the students' own creativity – drown in a whole sea of quotations, names, and polemical outbursts (K.L. 1910, p. 210).

The reviewer goes on to offer examples of exaggerations, inaccuracies, and contradictions. For instance, he writes about the obvious exaggeration of the achievements of psychology, which is alleged already to offer the possibility of constructing an entire system of pedagogy on the basis of experimental psychology. He calls attention to the fact that in one place Mrochek and Filippovich assert that the only correct pedagogical methodology is the laboratory method, while elsewhere they write that there can be no single correct methodology. Certain methodological propositions advanced by Mrochek and Filippovich are also criticized.

The review concludes with the assertion that although the book contains many opinions that will be shared by any supporter of reforms, these opinions did not originate with the book's authors, and therefore, it would be better to become acquainted with them from primary sources. K.L. praises the bibliography provided in the book, although here, too, he comments with irony on the inclusion of Mrochek's book on trigonometry in a section on "books of special interest."

After the review, however, there is also a note by the editor Veniamin Kagan (1869–1953). Agreeing with almost all of K.L.'s critical remarks and even adding that excessive enthusiasm for experimental psychology is harmful, and that the tone of the book's authors is too light and self-assured, he nonetheless deems it possible to recommend the book to his readers because it will acquaint them with various trends in thinking that are prevalent in the West and have been thoroughly investigated by the authors.

The discussion, however, did not end there. Mrochek and Filippovich (1911) replied with an expansive refutation of this review, and of Kagan's remarks as well, writing about the development of child psychology, and arguing that tradition alone cannot be relied on for guidance. Further, they reject some of the charges leveled against them, remarking that they never said what they are accused of saying, and likewise refuting certain specific methodological criticisms (for example, when discussing whether instances of division in which the same nomination is preserved should be distinguished from instances in which the nomination is not preserved—"123 nuts were divided equally among 3 boys: how many nuts did each boy get?" and "Each boy was given 3 nuts, and there were 123 nuts all together: how many boys got nuts?"—Mrochek and Filippovich believe that there is no difference, since ultimately it is abstract numbers that are divided in both cases).

The shrewd editor added a comment to Mrochek and Filippovich's note, indicating that he was decidedly not in agreement with K.L., who had asserted that the entire contents of the book was familiar to Russian readers already, but that, on the other hand, many arguments contained in the the authors' response to the review failed to convince him.

Lebedintsev himself (1911) (now writing under his full name) was not at a loss for words. Beginning with the observation that Mrochek and Filippovich make it seem as if he himself were coming out against the reforms, which was a characterization he could not agree with, he takes up several pages to show that they did in fact say what he had accused them of saying, and without sparing words analyzes their every remark, explaining, for example, that the two instances of division are in fact different—not mathematically, but psychologically, and that this is an objective reality. He ends by accusing Mrochek and Filippovich of boundless arrogance, with which they practically openly pronounce themselves to be the leaders and standard-bearers of reform in Russia, while in reality their book fails to achieve not only this extraordinary goal, but also any far more modest goal that it might have achieved had its authors approached their subject with greater diligence.

Nor did Dmitry Sintsov, possibly the most active member of the Russian subcommission of the International Commission on Mathematical Instruction, remain on the sidelines. Referring to the heated back-and-forth described above, and remarking that many undiscussed topics had been left to other reviewers, Sintsov (1913) comments wryly that the authors of *The Pedagogy of Mathematics* place not a little worth on their book; that the book's readers can expect to gather a great deal of interesting information from it, such as the fact that in the early sixteenth century "the technology of instruments and clocks achieved a marvelous degree of perfection in Nuremberg," or that "the previously existing route through Aden and the Red Sea closed after the conquest of Egypt by the Mohammedans," so that the authors could not be accused of narrow specialization, and finally, that the authors make it seem as if the international reform movement in mathematics education had grown out of the Russian liberation movement, which may give rise to some perplexity on the part of the readers. He then proceeds to a detailed analysis of different chapters, identifying outright mistakes and inconsistencies, both historical and mathematical. He concludes by remarking that Mrochek and Filippovich had a great number of very good books and pronouncements at their disposal, but that they "poured everything into their book, supplementing it with their own additions and elucidations, which are, as we have seen, not always successful. Therefore, I would not consider it possible to recommend this book or, all the more so, to acknowledge it under the banner of the supporters of reform, among whom I have the right, as it seems to me, to count myself" (p. 76).

Mrochek's Work on Preparing and Conducting the Congress of Mathematics Teachers

It may be supposed, however, that Mrochek's self-assured tone, which so irritated Bobynin and Sintsov, had a different effect on teachers at large. In addition, although it cannot be gainsaid that the international movement for reforms had somehow developed without the participation of the Russian revolutionary movement, the conditions in Russia were such that someone who loudly shouted "Down with it!" was perceived in a positive light by a substantial part of the population, even if the subject was teaching arithmetic. (We might recall that the production of Ibsen's play "An Enemy of the People" at the Moscow Art Theater in 1900 almost led to a revolutionary demonstration merely because its protagonist, Dr. Stockmann, speaks out in protest, which was perceived by the audience as a revolutionary summons, although of course there was no basis for this in Ibsen's play (Stanislavskii 1980).

When the International Commission on Mathematical Instruction was created in 1908, followed by the creation in 1909 of its Russian subcommission, which pushed for an analysis of the Russian situation and for the writing of summary reports to bring the Russian situation to the awareness of the international community, broad discussions sprang up in Russia. In their pamphlet about reforms, Mrochek and Filippovich (1910b) wrote as follows about responses to the creation of the International Commission:

> This movement swept Russia as well. The Second Russian National Congress on Pedagogical Psychology passed a resolution in the spirit of the reforms sketched out here; similarly wide-ranging resolutions were passed by the First Russian National Congress of Teachers of City Schools. Lastly, the Mathematics Department of the Pedagogical Museum in St. Petersburg, which had already been intensively working on reforming school mathematics since 1907, in January 1910 embarked upon a series of reports on topics proposed by the International Mathematics Commission (p. 202).

Mrochek and Filippovich, as individuals who were extremely active and knew foreign languages, became prominent figures at the Mathematics Department of the Pedagogical Museum. The Pedagogical Museum had been created back in the years of the great reforms of Alexander II, when military educational institutions and the publishing establishments associated with them were centers of a pedagogical, and to some extent political, renewal (Barskov 1914). Subsequently, it was precisely here that all kinds of exhibits, conferences, and congresses were organized. Zakhar Maksheyev, an active member of the Russian subcommission of the International Commission on Mathematical Instruction and the author of one of the reports prepared by the commission (Karp 2018), became the museum's director in 1906. He organized discussions of questions posed by the Commission, in which Mrochek took active part over a number of years. We might mention, for example, the already cited issue of *Matematicheskoe obrazovanie*, which describes the meetings of the mathematics department in 1912–1913 (Zasedanie 1913) and Mrochek's participation in the discussion (along with A.P. Kiselev and P.A. Nekrasov, individuals with altogether different views).

Mrochek participated in various congresses and meetings—for example, in the All-Zemstvo Congress on Public Education, about which he wrote in another already cited journal, *Obnovleniye shkoly* (Mrochek 1911). But, of course, the most important event in which he participated was the Congress of Mathematics Teachers, which took place at the end of 1911 in St. Petersburg at the Pedagogical Museum. Mrochek served as secretary on the congress's organizational committee (along with Filippovich and D.V. Levitus). Mrochek and Filippovich also became the compilers of at least the first volume of the Congress's proceedings (under the editorship of Maksheyev). Moreover, at the congress, Maksheyev said: "Two of the members approached me with the request to organize this Congress. These individuals, who first raised the idea of a Congress and whose names I consider it necessary to mention, were V.R. Mrochek and F.V. Filippovich" (Trudy 1913, v. 1, p. 572).

At the Congress's general assembly, Mrochek presented a report on "Experimental Problems in Mathematics Pedagogy." This was a carefully assembled overview of various foreign studies, in the course of which Mrochek, unsurprisingly, extolled manual labor and even drew the conclusion that "the time to act is before the age of 15; afterward, there will be enough time for thinking" (p. 93). The report ended with the words: "contemporary mathematics has now given the answer to Martin Ohm and his ilk: yes, rigorously scientific mathematics is inaccessible to children!" (p. 95). The mathematician Martin Ohm[3] (1792–1872), the brother of the universally known physicist, claimed to be a supporter and even the creator of a scientific approach to school mathematics: he is quoted by Mrochek as saying that students should be taught scientific geometry, not visual geometry, and therefore was chosen by Mrochek as the personification of evil. It is not clear how exactly mathematics can reveal what is and what is not accessible to children (here, of course, one might suspect a misprint—perhaps it was psychology that Mrochek had in mind). It may be supposed that responses to the report were mixed, but the discussions surrounding the reports were almost entirely left out of the volume. Only the presentation of the well-known psychologist A.P. Nechayev (1870–1948) was printed: he spoke about the achievements of experimental psychology as well as about the limits of applying new findings in physiology, pointing out that the famous studies of I.P. Pavlov were, after all, about dogs, not people.

Mrochek spoke numerous other times as well—on the school course in calculus (the report, in this case, was delivered by Filippovich), on education in higher educational institutions, and on behalf of the Polish mathematics circle in Warsaw. Then, during the session on educational literature, he delivered a presentation: "Overview of the Russian Literature on the Methodology of Arithmetic."

The ensuing discussion, in this instance, was published (although, probably, only in part), and we can see that a vehement argument took place between Mrochek and the well-known textbook author D.L. Volkovsky (1869–1934). The dispute was summed up by another prominent member of the Russian subcommission on mathematics education, General M.G. Popruzhenko (1854–1917):

> In summing up the discussions of the reports on the methodologies of arithmetic, regrettably I must note the excessive vehemence introduced into the exchanges and the unfoundedness of certain conclusions.... Thus, for example, the classification of methodologies proposed by Mr. Mrochek gives rise to various doubts regarding the psychological principles that he has taken as its foundation, and in any case, does not lead to the conclusion that the latest

[3] Schubring wrote about Martin Ohm; see, for example, the already cited Schubring (1987), which contains further references.

> methodologies are "more scientific" or the best there are, overshadowing all previous ones. (Trudy 1913, v.2, p. 134).

Excessive vehemence also characterized discussions of K.F. Lebedintsev's report on fractions in the course in arithmetic. Mrochek began as follows: "One doesn't know if one should be bemused or outraged when listening to such reports. A question that has a hundred years behind it, that is solvable and has long been solved in practice, is here presented as some kind of Gordian knot, and its would-be cutting is presented as a pedagogical discovery of America" (Trudy 1913, v.2, p. 229). He went on to quote from his book *The Pedagogy of Mathematics*. Lebedintsev responded, although his response was far more reserved.

We have no accounts of Mrochek's presentations at the Second Congress of Mathematics Teachers or even any indication that he participated in it. It appears that he did not. But in any case, taking into account the fact that Mrochek also wrote about the outcomes of the First Congress, deriding the opponents of reforms and lauding their supporters (Mrochek 1912); that however much other textbook authors argued with him, his book was also on the market; that he not only wrote about the importance of using models, but himself developed special rods that could be used to make models, which were sold and advertised; we can conclude that he was unquestionably one of the most visible figures in mathematics education before the revolution.

Mrochek After the Revolution of 1917. Conflict with Lunacharsky

The October Revolution of 1917, which brought the Bolsheviks to power, destroyed the old system of education, as had been planned, and the regulations governing it. A new system began to develop under the control, at first, of the State Commission on Education, and somewhat later, of the People's Commissariat for Education and the People Commissar for Education, Anatoly Lunacharsky (1875–1933). Inevitably, a great deal of confusion and inconsistencies arose, if only because the government soon moved to Moscow, while Lunacharsky still remained in St. Petersburg for some time; even more important was the fact that at first the new government did not have very many of its own people in the field of education.

Under these circumstances, Mrochek became an influential figure. One thing was certain: he could in no way be seen as belonging to the old monarchic teaching establishment, or even to the moderate liberal one. In the questionnaire form already cited above, he indicates that in 1917 he became a member of the Left Socialist Revolutionaries Party, which from November 1917 to June 1918 formed a coalition with the Bolsheviks. He took part in the creation of a union of internationalist teachers, but most importantly, following a resolution enacted on January 31, 1918, Mrochek became the director of the Central Pedagogical Museum, which replaced the Pedagogical Museum of Military Educational Institutions. The resolution specified that, in addition to Mrochek, the directorship of the museum would include two old Bolsheviks (F.V. Lengnik and P.N. Lepeshinsky). Mrochek also took part in the meetings of the State Commission (Rogovaya 2012, p. 185). And at the January 8, 1918, meeting, he became a member of the agency in charge of organizing courses and congresses (p. 135).

In the same autobiography (Morozov n.d.), Mrochek indicates that until the summer of 1918 he had been a member of the directorship (*kollegia*) of the People's Commissariat for Education; but this seems to have been a projection of a later reality onto the past on his part, as during the first year of the Soviet government, there was no directorship in the form in which it appeared later, and the functions of the various agencies, including the State Commission, kept changing as well. But Mrochek was definitely present at its meetings, and, for example, at the March 1, 1918, meeting, he was elected to serve on the Commission for the Evacuation of Children, while already at the March 2 meeting, he refused to serve on this commission, because "he considered this undertaking impracticable and unfeasible" (p. 206); in the event, Lunacharsky began this meeting with the declaration that since there was going to be a peace agreement with Germany, the evacuation was no longer urgent.

Mrochek, however, was not destined to sit among those in charge for long. We have the transcript of a meeting of the Regional Commission on Education from May 15, 1918, chaired again by Lunacharsky, that plainly reveals what happened (Morozov n.d.).

The Central Pedagogical Museum was given a building in the very center of the city, which began to be used for housing various individuals (residential housing was in short supply). The facilities manager had permitted a certain Alexandrov somewhat to extend his residency in the building. In this, he had been acting on instructions from comrade Pavel Bezskalko (1887–1920), an old acquaintance of Lunacharsky's from his emigre days (Lunacharsky 1923), who was considered one of the founders of proletarian literature, and who at that point occupied some kind of post in the education system. Mrochek categorically objected. The facilities manager, acting on behalf of the staff, then arrested Mrochek. Mrochek summoned an armed unit for help. Comrade Bezskalko, arriving on the scene, freed Mrochek, but Mrochek was not content to leave well enough alone and began insisting that he and the Museum were independent of the Regional Commission. Moreover, when the commander of the armed unit naturally asked all participants in the dispute to present their documents, and it turned out that Bezskalko did not have his on him, Mrochek ordered that he and all outside parties be removed. In response to which, Comrade Bezskalko called Mrochek a scoundrel.

The commission concluded that neither side was right. Noting that Comrade Bezskalko had been somewhat harsh, the commission nonetheless reasoned that Mrochek's fault was greater, since he had oppressed junior staff, had had their premises searched, and had had them evicted, thereby oppressing the proletariat (although this led to a theoretical discussion about whether the footmen and doormen of a St. Petersburg house could be considered the proletariat). But the main point was that Mrochek had claimed independence, and this was one thing that no one was prepared to tolerate. Lunacharsky remarked that Mrochek was a very difficult person and very difficult to deal with. In consequence, the commission passed a resolution stating that Mrochek had overstepped his authority, and that given his failure to observe boundaries, the Regional Commission was forced to conclude that it could no longer work with him.

The matter passed to the State Commission and, evidently, to the Petrograd municipal authorities (the cited transcript mentions that Comrade Zinoviev himself had taken an interest in the matter: at the time, Zinoviev was the head of the Petrograd government—the chairman of the Council of People's Commissars of the Petrograd Labor Commune, and so on). At a meeting on June 8, 1918, the State Commission passed a resolution to approve the resolution passed by the Commissariat of the Petrograd Labor Commune to relieve Comrade Mrochek of all duties (Rogovaya 2012, p. 546).

Subsequent Years and Resignation from the Communist Party

In July 1918, the so-called uprising of the Left Socialist Revolutionaries took place. In August, Mrochek left that party and joined the Communist Party, which now remained as the sole ruling party. In subsequent years, he worked in various places: in spring 1920, he was sent to work for the Vyborg district of Petrograd (St. Petersburg) Committee of the Russian Communist Party; prior to this, he had worked as instructor of political education for the First Artillery Command Courses, led discussion groups dedicated to studying social and economic issues, taught courses in social-political science at the Artillery Academy, and so on. At the First Artillery Command Courses, he also taught some mathematics (Mrochek n.d., d. 552531/1), and he worked on methodology at the Technical Pedagogical Institute.

Things did not stay quiet for long, however. By 1923, charges were brought against him for failure to attend party meetings and for promoting Socialist Revolutionary ideology (one can only guess to what extent these charges were connected with the trial of the Socialist Revolutionaries that took place in Moscow in 1922). In particular, Mrochek was accused of saying that in the phrase "workers-peasants' government," the emphasis should be placed on the word "peasants" (in other words, that like the Socialist Revolutionaries, he considered the peasantry to be more important than the working class); that fascism was small-scale enterprise, which came after imperialism; that the class that made history was democracy, etc. (Mrochek n.d., d. 552531/2). It was noted in his personal file that he was a good agitator, lecturer, and instructor and possessed encyclopedic knowledge, but was "too vain and puts his personal wellbeing above everything else" (Mrochek n.d., d. 552531/5, p. 6). The proposal was made to dismiss him from lecturing.

On August 17, 1923, the Conflict Commission of the Central District reached the conclusion that the charges of promoting Socialist Revolutionary ideology that had been brought against Mrochek were justified and that, in addition to this and to his failure to attend party meetings, Mrochek's involvement in teaching was driven by self-interest and ulterior motives. It was stated that he was an opportunist with a Socialist Revolutionary bent, that there was nothing Marxist in him, and that he was also fond of squabbling. But it was noted, too, that he was "a brilliant person in terms of his theoretical knowledge." It was decided to "expel him conditionally for one year, in view of his theoretical background and practical experience" (Mrochek n.d., d. 552531/2).

Mrochek replied with a long letter, which he sent to the Regional (Gubernia) Commission. In response to the charges about his ideological position, he writes with passion (the ellipses in the quote below are negligible):

> Yes, I am a member of the intelligentsia and for a long time I was a Socialist Revolutionary, but in October 1917, I came to the Military-Revolutionary Committee and was given a number of assignments, and there were no apprehensions about my political stance. When in 1918, the Proletkult elected me as head of the Department of socialist education, there were no apprehensions about my political stance. When in the fall Yudenich was approaching, instead of lecturing I was appointed assistant to the chief of staff of the defense, and there were no apprehensions about my political stance. When during the days of Kronstadt in 1919 I was sent to conduct political conversations in an interservice cadet unit, there were no apprehensions about my political stance. Why, then, after four and a half years of working in the ranks of the Russian Communist Party has it suddenly been discovered that I have a 'political stance'?" (Mrochek n.d., d. 552531/3, p. 4)

And further:

> I do not deny the party's right to demand of a lecturer that he presents this or that essential topic in a particular manner; but to allow the District Conflict Commission the right to be an arbiter in scientific questions and to decide arguments about how many classes there are according to Marx, or whether the peasants are a class, by a majority of the votes??? My mistake on August 17 was to argue with people who do not know the scientific formulations of Marxist questions, and who in the best case have only leafed through Marx and Engels – no wonder they became upset when I remarked that the Communist Manifesto was a historic agitational document, not a theoretical treatise....

In conclusion, he requests that his case be reconsidered, without allowing for the settling of petty scores and using Marxism as an alibi, because "the party must not squander old revolutionaries." And after his name is cleared, he further requests that he be considered as having resigned from the party of his own will.

On the letter are notes made by its reader, to the effect that Mrochek's case was handled fairly, that a person with a political stance must be assessed from all sides, and that those who were settling scores with Mrochek exposed themselves on their own and were punished. Be that as may, Mrochek was now no longer a party member.

Collaboration with N.A. Morozov

As Mrochek recounts in his autobiography, from 1920 on he served as professor of mathematics methodology and dean at the Technical Pedagogical Institute, later called the Higher Pedagogical Courses, while also lecturing elsewhere, including the Herzen Pedagogical Institute. From 1930 on, he was a staff member at the Lesgaft Scientific Institute, "working on a number of historical studies on the assignment of N.A. Morozov" (Morozov n.d.).

Nikolai Morozov (1854–1946) belonged to the most famous generation of the Narodniks and was prosecuted in court already in the "pre-terrorist" trial (the so-called trial of the 193 of 1877–1878), but then reconsidered his views and became a kind of theoretician of terror. He was arrested even before the assassination of the tsar in 1881, but sentenced to indefinite detention, which he served in Russia's worst prisons, the Peter and Paul Fortress and the Shlisselburg Fortress, remaining there from 1882 until his liberation, which was granted to him following the revolution of 1905. In total, at one time or another (he was arrested again later), he spent about 30 years in prison (Tvardovskaya 2003).

While in prison, he read and wrote. His main interests were astronomy and natural science. It is said that when he left prison, he brought out with him the manuscripts of over twenty books. His most famous assertions are that all contemporary accepted chronology is simply wrong, and that, for example, there was no antiquity, and that many of the same events and people are referred to in history multiple times and under different times. Today's encyclopedias charitably describe his writings as the work of a dilettante, but the image of a martyr for truth who makes great discoveries even in prison could not but have great appeal. The Soviet government, which even initially did not have much love for old revolutionary heroes, and by 1935 had liquidated the association of political prisoners, and shortly thereafter many of its members as well, nonetheless made Morozov into a cult figure, although in his views he was by no means a communist. He was made an honorary academician and the director of the Lesgaft Scientific Research Institute.

To be sure, Morozov himself did not make any trouble. It is telling that the transcript of the Regional Commission on Education meeting cited above bears the following note in Morozov's handwriting (in the opinion of Baranets and Verevkin 2013): "The reason I abstained from interceding for Mrochek, despite his enormous erudition"—Morozov did not need a conflict with the authorities (it is hardly possible to agree with Baranets and Verevkin 2013 that the reason was Mrochek's difficult character, which was documented in this transcript—Morozov was already acquainted with Mrochek's character from their personal interactions).

It was under Morozov that Mrochek worked for the remainder of his life. As far as can be judged, he did not argue with Morozov's views on chronology. The topics on which he worked under him were by no means limited to mathematics education. Baranets and Verevkin (2013) cite Mrochek's letter to Morozov from 1935, in which he writes, for example: "My plans are very broad and may be summed up as follows: to provide a conspectus of developments in the history of religions, states, technology, and science, based on the eight volumes of *Christ* [Morozov's composition] and a number of supplementary materials" (p. 34). Prior to this, in the same letter, he writes. "I have been told that your review of the Armenian chronicles has produced a confirmation of your main thesis: Mongols = Crusaders. Thus, the proofs accumulate, one after the other" (p. 33).

But mathematics was also a topic of his interest. Suffice it to note the outline of a course on "The History and Methodology of Exact Science," likewise cited in Baranets and Verevkin (2013). The outline was composed in 1928, but clearly already in connection with Morozov's work. In it, we find a section on "Problems," which contains the following subsections: "1. The origins of our school problems. 2. The history of 'types' and 'rules'. 3. The content of problems in historical evolution. 4. The classification of problems in methodology" (p. 31), etc., all of which makes us recall Mrochek's older compositions. Mrochek's texts on using the slide rule were also published.

Arrest and Death

But political events did not allow Mrochek to fade peacefully into the past. Petrov and Roginskii (n.d.) begin their account of the so-called Polish Operation with order 00485, approved by the Politburo of the Central Committee of the Communist Party on August 8, 1937, signed by People's Commissar of the NKVD Nikolai Yezhov on August 11, 1937, and accompanied by a classified letter "On the fascist-insurrectionist, espionage, diversionary, defeatist, and terrorist activities of Polish intelligence in the USSR." They note that although formally the order was directed against spies, in actual fact practically the entire Polish population found itself under suspicion. They also note, however, that in reality, the operation had begun considerably earlier. The operation is considered to have led to the arrests of about 140,000 people, about 111,000 of whom were executed (these figures appear in Wikipedia, for example).

According to surviving documents,[4] Vatslav Mrochek was arrested specifically as a Pole on August 5, 1937. On August 25, 1937, a commission of the NKVD and the Public Prosecutor's Office of the USSR sentenced him to death under articles 58–10 and 58–11 of the Penal Code of the RSFSR. Article 58–10 was used to convict people for propaganda or agitation aimed at overthrowing, undermining, or weakening the Soviet government, especially during mass unrest or by making use of the masses' religious or ethnic prejudices. In order to be punished, however, it was not necessary to perpetrate such deeds. According to article 58–11, any kind of organizational activity aimed at preparing the aforementioned counterrevolutionary crimes was considered equivalent to them. And such organizational activity, when desired, could be found without difficulty. Mrochek was executed in Leningrad on August 27, 1937.

[4] https://visz.nlr.ru/person/show/2559#prettyPhoto.

Discussion

The reform movement in pre-revolutionary Russian mathematics education included very different kinds of people. Indeed, even Andrey Kiselev, an author of textbooks, who already for decades before the revolution had been considered the embodiment of tradition and even conservatism, and was an active member of the quite right-wing (though not ultra-nationalist) Octobrist Party, participated in meetings and events conducted by the "reformers." All the more so did the already mentioned generals Maksheyev or Popruzhenko, or the director of the Moscow Land Surveying Institute, Privy Councilor Vasily Struve (Karp 2022), adhere to the reform movement. But along with moderate reformers, there were immoderate ones, and Mrochek was one of the most conspicuous among these. Mrochek wanted a revolution, even if only in a relatively small field, and consequently, he was more of a revolutionary than a mathematician or even a mathematics educator.

After the revolution, Mrochek was described as having a difficult personality, which was possibly true (although such accusations were quite common: many were said to have been undermined by their ambition). On the other hand, his fantastic erudition might also be called into question. But what is beyond doubt is his high level of activity, which diminished only in his last years, when even he, possibly, realized that it would be better to lay low.

The views expressed by him were quite typical for the era of progressive education with its orientation toward practical usefulness and aversion toward abstract and systematic mathematics; only his passion was greater than ordinary. There is much about his story that is worthy of note, however. First, compared to later times, the tolerance displayed by the reformist generals is remarkable: Mrochek was argued with, laughed at (sometimes with irritation), but he was tolerated. Second, Mrochek himself, although he did argue for the rejection of the scientific style, wrote his trigonometry textbook in a manner that was largely traditional. And subsequently he preferred to transition away from working specifically on mathematics education altogether, and he did so, even though the post-revolutionary years were precisely a time of triumph for ideas that would seem to have been close to Mrochek's own. He, however, preferred to teach political education courses, rather than, say, put to use the special rods for assembling models that he had invented earlier, or teach a course in arithmetic based on the history of the struggle of the proletariat (while this or something like it was being done by others, for example, his future critic Lankov—Karp 2012). Mrochek can hardly be considered simply an opportunist or time-serving conformist who wrote whatever had a chance of success (his attempts to conform were simply too hapless).

During virtually the same years, in the 1920s, Soviet education found its pedagogical leader in Pavel Blonsky (1884–1941), who had also belonged to the Socialist Revolutionaries earlier and had then gone over to the Bolsheviks, and who had also unquestionably imbibed the ideas of progressive education (Lipkina and Martsinkovskaya 2006). The Bolsheviks simply did not have any personnel of their own, and for some time even those who were not completely like-minded had a chance for promotion—later on, to be sure, the same Blonsky was denounced as a quack in a scholar's garb; indeed, the very notion of "like-mindedness" changed meaning, and Nadezhda Krupskaya herself—the wife of Lenin—was sidelined into a respected, but powerless role. But Mrochek made attempts to meddle not in methodological matters, but in matters pertaining to the exercise of power—even if only in trifles. This was not something that anyone tolerated.

The fact that he was executed was in some sense due to a coincidence. Had he not been a Pole (or a member of any other condemned nationality), he had a chance of dying, like Blonsky, of tuberculosis or some other illness. But in any case, there was no place for him in the new system: not as a person who liked to raise objections (which is what "squabbling" meant), not as a revolutionary by temperament, not as a foreigner (notwithstanding the internationalism proclaimed by the system), but also not as a mathematics educator. Reformist ideas were rejected.

But in the history of Russian education, Mrochek remained a presence. Strange as it may seem to people of later times, when a better system of record-keeping came into a place and an effort was made not to mention anyone who had been put to death by the government, yet Mrochek's specialized compositions continued to be reissued even during the Stalinist period (for example, his text on using the slide rule), and indeed his other books were referred to as well, and not even always in a completely negative manner. Later on, the number of such references grew, and in recent decades, *The Pedagogy of Mathematics* has been brought up often. And indeed, the historian of mathematics education will see in Mrochek's books and fate much that will shed light on the vicissitudes of Russian mathematics education.

Acknowledgements The author expresses his sincere gratitude to his reviewers for their useful comments.

References

Baranets, Natalya G. and Verevkin, Andrey B. 2013. Istorik nauki Vatslav Mrochek – sud'ba v epokhu peremen [Historian of science Vatslav Mrochek – a life during a time of change]. In *Sotsiologia znaniya i filosofiya nauk*, ed. Natalya G. Baranets, 22–36. Ulyanovsk.

Barskov, Yakov. 1914. *Pedagogicheskii muzey voenno-uchebnykh zavedenii* [Pedagogical museum of military educational institutions]. St. Petersburg: Sirius.

Bobynin, Viktor V. 1913. Istoriya matematiki v knige g. Mrocheka "Pryamolineynaya trigonometriya i osnovaniya teorii goniometricheskikh funktsii" [The history of mathematics in Mr. Mrochek's book, "Linear trigonometry and the elementary theory of goniometric functions"]. *Matematicheskoe obrazovanie* 7: 305–307.

Ehrenburg Ilya G. 1923. *Neobychainye pokhozhdeniya Khulio Khurenito i ego uchenikov* [The extraordinary adventures of Julio Jurenito and his disciples]. Moscow-Petrograd: Gosudarstvennoe izdatel'stvo.

Karp, Alexander. 2012. Soviet mathematics education between 1918 and 1931: a time of radical reforms. *ZDM/International Mathematics Education 44*(4): 551–561.

Karp, Alexander. (Ed.). 2018. *Otchety rossuskoy natsional'noy podkomissii Mezhdunarodnoy Komissii po matematicheskomu obrazovaniu* [Reports of Russian National Subcommission of ICMI]. St. Petersburg: Lema.

Karp, Alexander. 2019. The Russian National Subcommission of ICMI and the mathematics education reform movement. In *National Subcommissions of ICMI and their role in the reform of mathematics education,* ed. Alexander Karp, 149–191. Springer.

Karp, Alexander. 2022. Brat znamenitogo cheloveka (o V.B.Struve). [Brother of a famous person (About V.B.Struve)]. *Matematika v shkole 4*: 59–70.

Kolyagin, Yurii M. and Savvina, Olga A. 2006. *Matematiki-pedagogi Rossii. Zabytye imena. Filipp Vasil'evich Filippovich* [Mathematics educators of Russia. Forgotten names. Filipp Vasilyevich Filippovich.]. Elets: EGU imeni I.A.Bunina.

K.L. (1910). *Retsenziya na knigu V. Mrocheka i F. Filippovicha «Pedagogika matematiki»* [Review of the book "The Pedagogy of Mathematics" by V. Mrochek and F. Filippovich]. *Vestnik opytnoy fiziki i elementarnoy matematiki 524*: 209–212.

Lankov, Alexander V. 1951. K istorii razvitiya peredovykh idey v russkoy metodike matematiki [Toward a history of leading ideas in Russian mathematics methodology]. Moscow: Uchpedgiz.

Lebedintsev, Konstantin. 1911. Otvet na zametku gg. Mrocheka i Filippovicha [Response to note by Mssrs. Mrochek and Filippovich]. *Vestnik opytnoy fiziki i elementarnoy matematiki 531*: 73–83.

Lipkina, Anna I. and Martsinkovskaya, Tatiana D. 2006. P.P. Blonsky – v poiskakh psikhologicheskikh osnov novoy shkoly [P.P. Blonsky – in search of the psychological foundations of the new schools]. In *Psikhologiya mladshego shkol'nika* by Pavel P. Blonsky, 3–23. Moscow-Voronezh: MPSI, NPO "MODEK".

Lunacharsky, Anatoly V. 1923. *Revoliutsionnye siluety* [Revolutionary silhouettes]. Moscow: "Devyatoe yanvarya".

Morozov, Nikolay A. n.d. Mrochek Vatslav R. *Archive of Russian Academy of Science*, f. 543, op. 5, d. 139.

Mrochek, Vatslav R. n.d. Central State Archive of Historical–Political Documents, f 1728, op 1, d. 552531/1–5.

Mrochek, Vatslav R. 1911. Obschezemskii S'ezd po narodnomy obrazovaniu [All-Zemstvo Congress on public education]. *Obnovlenie shkoly 2*: 71–78.

Mrochek, Vatslav R. 1912. *Itogi pervogo vserossiiskogo s'ezda prepodavateley matematiki* [Outcomes of the First Russian National Congress of mathematics teachers]. St. Petersburg.

Mrochek, Vatslav R. 1913. *Pryamolineinaya trigonometriya i nachala goniometricheskikh funktsii* [Linear trigonometry and elementary goniometric functions]. St. Petersburg-Moscow: Tovarischestvo M.O. Wolf.

Mrochek, Vatslav R., and Filipp V. Filippovich 1910a. *Pedagogika matematiki* [The Pedagogy of mathematics]. St. Petersburg: Knigoizdatel'stvo O. Bogdanovoy.

Mrochek, Vatslav R., and Filipp V. Filippovich. 1910b. *Reforma prepodavaniya matematiki* [The Reform of mathematics education]. St. Petersburg: Tipografiya Montvida.

Mrochek, Vatslav R., and Filipp V. Filippovich. 1911. Po povodu retsenzii g. K.L. na "Pedagogiku matematiki" [On Mr. K.L.'s review of "The Pedagogy of mathematics"]. *Vestnik opytnoy fiziki i elementarnoy matematiki 529*: 16–20.

Petrov, Nikita V. and Arsenii B. Roginskii. n.d. «Pol'skaya operatsiya» NKVD 1937–1938 gg. [The NKVD's "Polish operation," 1937–1938]. Memorial. http://old.memo.ru/history/polacy/00485art.htm. Accessed 21 April 2023.

Ovchinnikov, D.A. 2011. Pervyi rossiiskii istorik matematiki [The first Russian historian of mathematics]. *Matematika v shkole 4*: 62–66.

Rogovaya, Larisa A. (Ed.). 2012. *Kul'tura, nauka i obrazovanie. Okryabr' 1917–1920 g. Protokoly i postanovleniya NARKOMPROSA RSFSR. T. 1.* [Culture, science, and education. October 1917–1920. Transcripts and resolutions]. Moscow: ROSSPEN.

Rybnikov, Konstantin A. 1950. Viktor Viktorovich Bobynin. *Uspekhi matematicheskikh nauk, 5*(1), 203–210.

Schubring, Gert. 1987. On the methodology of analysing historical textbooks: Lacroix as textbook author. *For the learning of mathematics 7:* 41- 51.

Sinkevich, Galina. 2022. *Matematiki peterburgskoy storony* [The Mathematicians of the St. Petersburg Side]. St. Petersburg: SPbGASU.

Sintsov, Dmitry. 1913. Po povodu odnoy knigi [About a Book]. *Matematicheskoe obrazovanie* 2: 67–76.

Stanislavskii, Konstantin A. 1980. *Moya zhizn' v iskusstve* [My life in art]. Moscow: Iskusstvo.

Trudy 1-go vserossiiskogo s'ezda uchiteley matematiki [Proceedings of the first Russian National congress of teachers of mathematics]. Volumes 1–3. 1913. St. Petersburg: Sever.

Tvardovskaya, Valentina A. 2003. Nikolay Morozov: ot revoliutsionera-terrorista k uchenomu-evoliutsionistu [Nikolai Morozov: from terrorist revolutionary to evolutionary scientist]. *Otechestvennaya istoriya* 2: 50–72.

Zasedaniya matematicheskogo otdela Pedagogicheskogo muzeya voenno-uchebnykh zavedenii 1912—1913 г. [Meetings of the mathematics department of the Pedagogical museum of military educational institutions, 1912–1913]. 1913. *Matematicheskoe obrazovanie* 2: 91–93.

Concept of Transmission as a Bipolar Process: The Case of the Formation of a Mathematical Community in Brazil

Vinicius Mendes Couto Pereira

Concept of Transmission

The concept of "transmission" is quite used in the general historiography of science, mostly in the first half of the twentieth century, and has been understood in this context like a process used to understand how scientific ideas and practices have been taken between two cultures, one of them considered the emitting pole and the other the receptor pole. This notion has been used with the intention to study how scientific accomplishments are disseminated between different cultures.

However, according to Schubring (1999) the utilization of this traditional concept suffers from two typical deficiencies.

The first problem consists of considering that the transferred concepts remain essentially identical, even in contexts completely different. In this aspect, Schubring (1999) emphasizes that this approach tends to neglect that usually a concept is immersed both in a concept field and a cultural field, which means that a concept, once transmitted, no longer remains identical to the original, having been transformed in the process.

The second problem of the traditional practice, pointed by Schubring, is that it conceives the historical developments of a scientific discipline in terms of some "centrisms," which means, privileging certain cultures as more elevated and civilized, being represented like more important centers.

It is taken to knowledge that the traditional approach of this concept tends to disregard all the possible social and cultural influences that could step in the process of transmission, being able to induce the thought that the development of mathematical knowledge is built in a linear and cumulative way.

V. M. C. Pereira (✉)
Universidade Federal Fluminense, Niterói, Brazil
e-mail: viniciusmendes@id.uff.br

A. Verdugo Rohrer and J. Zender (eds.), *History of Mathematics and Its Contexts*, Trends in the History of Science, https://doi.org/10.1007/978-3-031-60828-5_13

From those reasons, Schubring (1999) proposes an approach that transcends the traditional notion of transmission, considering the strong connection between private concepts or conceptual fields and a specific culture, understanding that there are no purely endogenous developments of a certain culture and that is difficult to conceive a fertile development without transmission between cultures.

That way, based on the premise that passive reception doesn't exist, that is, that transmitted knowledge is transformed by receiving social and cultural groups according to your own value sets. Schubring (1999) understands the transmission process as a bipolar process: one pole being the transmitted knowledge and the other its transformation according to the "cultural identity" of the receivers.

In addition, Schubring (2002) elaborates structural patterns which permit comparing transmissions of mathematics from different European metropolitan countries to the emerging educational systems at the periphery in the Americas, and the eventual transformation of some of these systems to new, even metropolitan centers. Based on these considerations, Pereira (2017) lists significant elements that may be present on the transmission processes:

- the relevance of the role of teaching in establishing the transmission process;
- studies of models referring to scientific institutions and to their training functions, instituted in the receiving countries, having as reference models already established in the metropoles;
- presence of translations of textbooks produced in the metropolises;
- investigation of the transmission elements that may have come from the transfer of mathematicians from the metropolises.

In this article, we intend to show how this concept, in the sense given by Schubring (1999), has been used in researches to understand how the particular concepts and conceptual fields were transmitted and received and how a mathematical community was emerged in Brazil.

Brazil's Case

Until 1808, when the Portuguese royal family came to Brazil and Rio de Janeiro became the capital of the Kingdom of Portugal, the Portuguese had been unwilling to establish a genuine teaching institution in Brazil. The first institution that taught superior mathematics, the Royal Military Academy, a school for military engineers—modeled after the *École Polytechnique*, was founded in 1810, and an intensive activity of translating and publishing modern French books began. As Schubring (2002) reminds us, the first generation of professors at this school were almost all graduates from the reformed Coimbra University and mainly from its mathematics faculty.

Therefore, considering this context and that, in this first moment, the role played by the incipient math community in Brazil was receiving mathematical production from European countries, Pereira (2017) poses two research questions.

In this way, Pereira (2017) intends to study which way a Brazilian mathematics community emerges, looking to investigate how a country that is firstly only a receptor of other countries' production, becomes independent in the sense of being able to establish its own mathematical production.

On the other hand, given the impossibility of studying all the mathematical produced in this period, we seek to understand specifically how the process of transmitting the basic concepts of Analysis took place since this was consolidated as a key theory in the nineteenth century, as pointed out by Schubring (2005): "*We will be able to conclude that an international community at least of top mathematicians formed for the first time during the last third of the nineteenth century, a community which intensely advanced research on the basis of shared conceptual notions of analysis.*"

Thus, on the one hand, Pereira (2017) presents a specialized study, on a well-defined conceptual field in mathematics, but, on the other hand, a longitudinal study: the establishment of a Brazilian community with its own production.

Formation Stages of a Mathematical Community in Brazil

Seeking answers to the research questions formulated, Pereira (2017) was able to identify a way for the emergence of a mathematical community in Brazil, the reception until it owns production.

The arrival of the Portuguese royal family in 1808 was decisive to the beginning of the scientific process in Brazil, since the scientific activities, until this point, were hampered by the Portuguese crown. Inside this new reality, the requests coming from the needs of established the royal court opened a new movement dynamic in Brazilian culture.

In this context, was founded the Royal Military Academy, through the law of December 4, 1810, signed by the regent prince, with the objective of giving, in the colony, a complete course in mathematical sciences and military sciences in their entirety, both in tactics and in fortification and artillery.

At first was offered a course for the future "Engineer and Artillery Officers" lasting seven years, consisting of one a "Mathematical Course" lasting four years and a "Military Course" lasting three years.

In particular, the "Mathematical Course" offered by the Royal Military Academy represented the start of mathematical teaching at a superior level in Brazil, inaugurating a long tradition of teaching higher mathematics completely restricted in the context of training engineers, lasting until the first decades of the twentieth century.

The Royal Charter that established the Academy specified a series of regulations regarding its structure, such as the programs and textbooks that should be taught in each chair and the professors who should teach each of the courses.

The choice of instituted textbooks suggests the intention that the works were in line with the scientific progress of the time since the works mentioned had authors such as Euler, Bezout, Monge, Legendre, Lacroix, Laplace, Francoeuer, Prony, Delembre, Lacaille, Delandre, Hauy, and Brisson.

The first Calculus textbook adopted in Brazil was *Traité Élémentaire de Calcul Differentiel et de Calcul Integral* by the French man Sylvestre François Lacroix.

During this time, several works adopted in the classes were translated into Portuguese. In the case of Calculus, for example, the work written by Lacroix was translated into Portuguese by the second-year teacher, Francisco Cordeiro da Silva Torres, under the title: *"Elementary Treatise on Differential Calculus and Integral Calculus by Mr Lacroix"* and, as happened in other translations, it was explained that the translation was done *"by order of his royal highness, translated into Portuguese for use by the students of the Royal Military Academy of this Court."*

In addition, as Pereira (2017) points out, there are signs of a certain stability in the curriculum until 1850. In particular, Lacroix's work also remained a reference work for Calculus classes at the Royal Military Academy for a long time.

After that, the Royal Military Academy passed through different structural reforms, changing its name a few times, for different historical reasons. Particularly, in 1858, the Central School was founded and in 1874 the Polytechnic School of Rio de Janeiro was founded, and for the first time, the teaching of engineering was completely disconnected from military teaching.

Still in the context of the Royal Military Academy and its developments, the title of Doctor of Mathematical Sciences was instituted, through the decree of March 10, 1842.

In this way, candidates had to be fully approved in all disciplines and present a dissertation on any point of mathematical science that had been taught in the last three years, with no requirement of originality in the concepts addressed. The first doctoral dissertation was presented in 1847.

In this context, the influence of the positivist philosophy of Auguste Comte must be highlighted, especially in the Polytechnic School of Rio de Janeiro, between the 1870s and 1920s.

The influence of positivist philosophy on the Polytechnic School of Rio de Janeiro marked the transition between the first steps of formation of a mathematical community in Brazil, in accordance with Pereira (2017).

Therefore, in this terms, the first step is characterized as the period from the foundation of the Royal Military Academy until the beginning of the positivist influence on the Polytechnic School of Rio de Janeiro.

The first positivist manifestations emerged at the Escola Central in the 1850s, but still in a very timid manner. After that, as already emphasized, the Polytechnic School of Rio de Janeiro consolidated itself as a stage for the performance of positivism, with the main actors being both the performance of significant professors and the various works inspired by positivist philosophy.

However, at the turn of the twentieth century, a movement of resistance to positivism began, initially having Otto de Alencar Silva (1874–1912) as the main protagonist. In 1898, Otto de Alencar dared to publish an article commenting and pointing out errors of content made by Comte in his *Synthèse Subjective.*

This publication is considered a milestone that marks the beginning of the depletion of the influence of the positivist doctrine among those who did science in the country at the time.

Otto de Alencar Silva's performance was not limited to seeking to overcome the existing positivist view, but to establishing a new culture and scientific posture in the context of the Polytechnic School of Rio de Janeiro. This fact can be evidenced when we note that Otto de Alencar exchanged correspondence with mathematicians such as Francisco Gomes Teixeira, Gaston Darboux, and Henri Poincaré.

In addition, the effort to establish their own production aligned with the production established by European mathematicians can be evidenced when we observe their effort to acquire specialized books and magazines from the USA and European countries, forming an excellent private library, by the standards of the time.

Silva (1998) reminds us that works such as Traité d'analyse and Théorie des fonctions algébriques de deux variables indépendantes, by E. Picard; Cours d'analyse mathématique, by E. Goursat; Cours d'analyse, by C. Hermite; Cours d'analyse de l'École Polytechnique, by C. Jordan; Électricité et optique, and Théorie des tourbillons, by H. Poincaré were introduced by him at the Polytechnic School of Rio de Janeiro, at the same time that he encouraged colleagues to study mathematics, even in the environment of an engineering school.

Thus, even in an ambience where engineers were trained, under the strong influence of positivism, the work started by Otto de Alencar and followed by mathematicians such as Manuel Amoroso Costa (1885–1928), Theodoro Augusto Ramos (1895–1935), and Lélio Gama (1892–1981), inaugurated a new stage in the development of a mathematical community in Brazil, marked by an attitude of self-education and the search for exchange and scientific knowledge with European mathematicians.

Furthermore, it should be noted that the generation of mathematicians forged in this third stage was fundamental for the success of the last stage, inaugurated with the foundation of the first Universities in Brazil.

The foundation of Universities in Brazil consisted of the last stage of formation of a Brazilian mathematical community, according to Pereira (2017). The first university organization is conceived through a decree that unifies existing schools and colleges, such as the Polytechnic School of Rio de Janeiro.

The environment and scientific culture forged in the third stage paved the way for the beginning and consolidation of mathematical research in Brazil, especially in Rio de Janeiro.

In 1935, one year after the foundation of the University of São Paulo, the thoughts proposed by this current of intellectuals were materialized through the creation of the University of the Federal District (UDF). The School of Sciences, linked to the new University, was under the responsibility of Lélio Gama, one of the exponents of the third stage.

This way, the work established at the School of Sciences not only became an environment in which mathematical concepts were being presented in a modern approach, in the sense of mathematics produced in European countries, but also had some publications as a product of this new moment. In particular, the work *"Séries Numéricas"* by Lélio Gama stands out, a product of the Analysis classes taught at the Federal District University and later at the National Faculty of Philosophy.

On the other hand, the founding of the University of São Paulo (USP), particularly the Faculty of Philosophy, Sciences and Letters (FFCL) should be highlighted as a decisive step in consolidating the fourth stage of development of the mathematical community in Brazil.

The University of São Paulo was founded through a decree published on January 25, 1934, initially composed of several faculties, in addition to the Polytechnic School of São Paulo (EPSP) and the Faculty of Philosophy, Sciences and Letters. We should note, in this context, USP now has two institutions offering courses involving higher mathematics, namely EPSP for training engineers and FFCL for training mathematicians and mathematics teachers.

The ideological conception surrounding the foundation of USP understood the university as a space for the formation of a new intellectual elite, which could establish a new culture. From this conception, the search for teachers in Europe took place, in order to implement this new perspective.

In this way, once the negotiations were concluded, French, Italian, and German professors were hired. The process of coming and establishing these teachers was called respectively the French Mission, Italian Mission, and German Mission in the FFCL. In particular, the Italian mission established itself in the FFCL, under the leadership of Luigi Fantappiè.

The Italian mission was present at USP between 1934 and 1940, having had a great impact on the establishment of a new mathematical culture in Brazil. Historiography about the influence of the Italian mission, under the leadership of Fantappiè, on the development of Brazilian mathematics has pointed in at least three directions.

The first one refers to the introduction of new courses in Brazil, such as group theory, continuous groups, number theory, differential forms applied to analysis, and tensor analysis.

The implementation of the Mathematical and Physical Seminar at FFCL provided an environment for the development of students, supporting them to be able to produce their own research, with the real possibility of developing unprecedented research, in partnership with their professors.

At this point, it should be noted that the seminars consisted of already successful practices in the process of professionalizing mathematics and in the consolidation of mathematical communities in European countries.

D'Ambrosio (2008) points out that the work conducted by the Italian Mission consisted of a primordial element in the modernization of Calculus teaching in Brazil, stating that Fantappiè tried to modify the Differential and Integral Calculus courses, effectively transforming them into a Mathematical Analysis course.

Thus, considering these points, a representation of the role of the Italian Mission in the development of a mathematical community in Brazil can be made explicit.

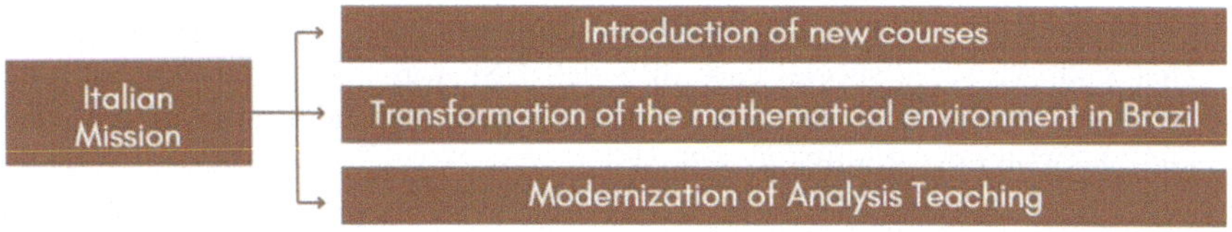

Finally, with the breakup of diplomatic relations between Brazil and Italy, the situation of Italian professors at USP became unsustainable, until in 1942, with the return of the last Italian mathematician, the Italian Mission was definitively dissolved.

In order to continue the teaching and research environment already institutionalized through the work of the Italian mission, professors were hired at levels comparable to the Italian ones, even if from a different perspective.

This way, shortly after the era of the Italian school at USP, mathematics at FFCL came to be strongly influenced by another school, the Bourbakist school.

The first boubakist to arrive in Brazil was André Weil in 1945. Afterward, several other Bourbakist professors passed through USP, such as Oscar Zariski, Jean Dieudonné, Jean Delsarte, and Alexandre Grothendieck.

According to Pereira (2017), already from the arrival of Weil, a direction toward a structuralist approach to mathematics can be noticed, making clear his concern to insert new theoretical discussions.

Pereira (2017) highlights the performance of Osmar Zariski who gave lectures on modern algebra and algebraic geometry, focusing on the study of the theory of rings and bodies, the theory of ideals, and the theory of values.

In this context, the Italian school of algebraic geometry was already considered classical, but it was criticized for understanding that it was an approach that lacked rigor. In particular, Zariski worked with the view that algebraic geometry was undergoing profound changes through the insertion of ideas and procedures from modern algebra.

André Weil also harshly criticized this traditional approach, which culminated in a strong disagreement at the World Congress of Mathematicians, in 1954, in Amsterdam, where Weil interrupted the lecture given by Francesco Severi, honorary president of the congress, pointing out a lack of rigor.

At this point, in view of the disputes about the traditional approach referring to algebraic geometry, we see USP become a meeting place for two trends in the modernization of mathematics, the traditional Italian school of algebraic geometry and the desire for a modernized approach, according to the conception defended by the boubakist school.

After the departure of Weil and Dieudonné, the hiring of Jean Delsarte and later of Alexander Grothendieck continued the insertion of the Bourbakist school at USP that would extend throughout the 1950s.

The hiring of Grothendieck is classified as the third generation of Bourbakistas at USP. Grothendieck's work is situated on a level that transcends the Bourbakist conception and much more the traditional approach of the Italian school, consisting of a new paradigm in the field of algebraic geometry.

That said, we see a special case in which USP was configured as a meeting space between two strands of mathematics that were very relevant in its time, the Italian school and the Bourbakist school.

From Reception to Its Own Production: The Importance of the Concept of Transmission as a Bipolar Process

The process of constituting a mathematical community in Brazil can be seen as a path interspersed with a network of historical facts and concepts, as brought by Pereira (2017). However, the proper understanding of the webs and components that form the network of understanding this process is a difficult task, given the complexity of all processes and historical processes involved.

Considering this, we realize that the concept of transmission as a bipolar process, in terms of Schubring (1999), becomes a powerful tool to understand how the path of formation of a Brazilian mathematical community took place, from an exclusive position of reception of structures and knowledge from European countries to the production stage of its own researches.

In fact, when we look at the various nuances of the path traced to the establishment of the mathematical community in Brazil, we realize that there are many idiosyncratic aspects, linked to the social and cultural context, which consolidated all the dynamics of building this path.

In this way, the study of this dynamic of construction, based on the concept of transmission and assuming the leading role of the reception process, becomes an absolutely enlightening tool, especially considering the idiosyncrasies that influenced the consolidation of this community.

Indeed, the path traced by Pereira (2017) presents several singularities that directly interfered in the construction of the path itself.

The first institution to teach higher mathematics in Brazil, the Academia Real Militar, was modeled after the École Polytechnique, and an intensive activity of translating and publishing the modern French books began. However, despite the inspiration of the French model, the teaching of mathematics at the Royal Military Academy developed from its specificities arising from its cultural context.

From 1847, the first doctoral dissertations were presented, which should be based on the contents taught in the institutions that succeeded the Royal Military Academy. Thus, in a way, these dissertations can be seen as a first look at the transformation of mathematical concepts that were being received.

Some dissertations already in 1848 already referenced the work of Auguste Comte, signaling the influence that the positivist philosophy would have on the science produced in the only[1] institution in the country to teach higher mathematics in this period.

In particular, with regard to mathematics, this stage began to have greater amplitude from the use of *Analytical Geometry* by Comte, playing an important role in the teaching of this discipline, not only at the Polytechnic School, but as a source of inspiration for the first Analytical Geometry books written by Brazilians.

In this way, as already described, positivism exerted a powerful influence on the production and development of mathematical knowledge, mainly between the 1870s and 1920s.

We note here that the reception of Comte's positivist philosophy took place in a completely unique way, related to the specific conditions of the country. The wide domain of the positivist view at the Polytechnic School of Rio de Janeiro directly influenced the entire way of doing mathematics, since it was the only institution that taught higher mathematics in Brazil, until then.

On the other hand, we emphasize that the process of reaction to positivism took place in a unique way, from the inauguration and dissemination of a new posture of production of mathematical knowledge. This posture was based on the search for interlocution with some of the main European mathematicians, in addition to the acquisition of specialized books and magazines in complete consonance with the mathematics produced in European research environments.

We should also point out that this reaction process took place from self-education, perhaps because it was the only possible posture in the face of the dominant context of positivist philosophy.

It should also be noted that the dynamics provided by the process of reaction to positivism generated a very favorable and conducive environment for the development of mathematical research in Brazil, in line with the mathematical production established in European research centers.

It is also remarkable that such an environment was generated in an exclusive context of engineering training.

In this way, we emphasize that this generated environment was fundamental for the development of mathematical production in Brazil, particularly in Rio de Janeiro.

On the other hand, we should also note another crucial dynamic for the development of mathematical research in the country, from the founding of USP and the establishment of the Italian Mission in Brazil.

The work of the Italian Mission also provided a conducive environment for mathematical research, but in a different movement than what happened in Rio de Janeiro.

[1] The Mines School of Ouro Preto and the Polytechnic School of São Paulo, founded in 1876 and 1893 respectively, were the other institutions that also taught higher mathematics courses in Brazil during the nineteenth century.

As already specified, the practice established by the Italian Mission contributed to the introduction of new courses, with the modernization of the teaching of Analysis and for the transformation of the mathematical environment. In addition, it is worth mentioning the establishment of usual practices in contexts where mathematical research was at a more advanced stage, such as the implementation of the Mathematical and Physical Seminars.

Furthermore, we note that even with the return of Italian mathematicians, there was a legacy through the continuation of research by assistant professors and maintenance of the environment generated by new practices.

After the end of the Italian Mission, the Bourbakist school also exerted enormous influence, transforming even more the context of mathematical research training in Brazil and influencing a new generation of Brazilian students.

Faced with the influences exercised by both the Italian School and the Bourbakist school, it is important to perceive an aspect present in the formation of the first generations of mathematicians at USP, a kind of symbiosis, of joining the characteristics arising from these two schools.

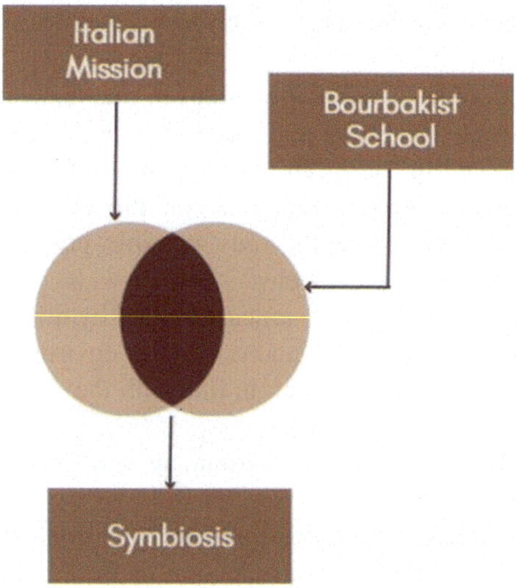

Finally, we note that the different dynamics established in Rio de Janeiro and São Paulo, from the movement of reaction to positivism and the establishment of the Italian Mission, respectively, generated environments that were fundamental for the establishment of mathematical production in Brazil.

Also, the communication between these two environments can also be seen as a symbiosis, which we are calling "*double symbiosis*" enhancing and consolidating the mathematical production environment in Brazil.

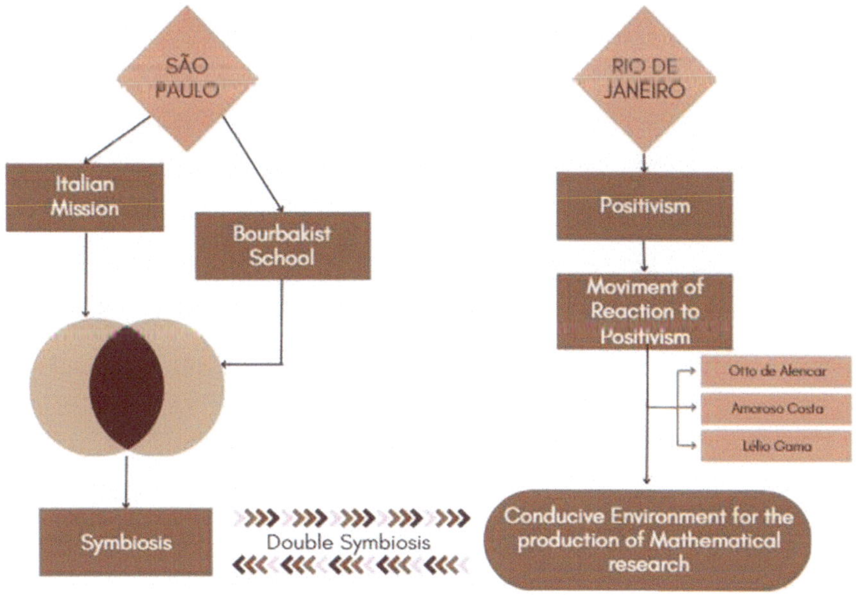

That said, we understand that the study of the formation of a mathematical community in Brazil, based on the concept of transmission as a bipolar process, has become fundamental for a better understanding of all the dynamics involved in this process.

Thus, with the reception process seen as a leading agent, we were able to understand the genesis of the formation of a mathematical community in Brazil, considering the idiosyncratic aspects that were absolutely decisive for the consolidation of the process.

Note: All images were drawn by the author.

References

D'Ambrosio, Ubiratan (2008), Uma História Concisa da Matemática no Brasil. São Paulo: Vozes.

Schubring, Gert (1999), O primeiro movimento internacional de reforma curricular em matemática e o papel da Alemanha: um estudo de caso na transmissão de conceitos. Zetetike.v.7, no.11, 29–50.

Schubring, Gert. (2002), A Framework for Comparing Transmission Process of Mathematics to the Americas. Revista Brasileira de História da Matemática, v.2, no.3, 45–63.

Schubring, Gert (2005). Conflicts between generalization, rigor, and intuition: number concepts underlying the development of analysis in 17th-19th century France and Germany. New York: Springer, 2005.

Silva, Clovis Pereira da (1998). A Contribuição de Otto de Alencar Silva para o Desenvolvimento da Ciência no Brasil. Revista da SBHC, n.19, pp 13–30

Pereira, Vinicius Mendes Couto (2017). O Desenvolvimento da Análise no Brasil – Um Caminho sobre o surgimento de uma Comunidade Matemática, PhD. Thesis University Federal of Rio de Janeiro.

Deutsche Mathematiker unter der Nazi-Diktatur

Reinhard Siegmund-Schultze

Gert Schubring zum 80. Geburtstag 2024 gewidmet, der in seinen Publikationen oftmals auf die politische Dimension der Mathematik hingewiesen hat!

Die Geschichte des „Dritten Reiches" und seiner Wissenschaft wird gewöhnlich von Auschwitz ausgehend geschrieben, und das ist gar nicht anders möglich. Horkheimer und Adornos „Dialektik der Aufklärung" wurde 1944 im amerikanischen Exil verfasst. Das Buch kann in dem Sinne gelesen werden, dass die Wissenschaft und die Technik der Moderne, und in pervertierter Form die Menschenversuche und die Militärmaschinerie der Nazis,[1] die „Fortschrittlichkeit" des Aufklärungsprojekts zunehmend in Frage stellen. Die V2-Raketen der Nazis waren die ersten, die den Weltraum hätten erreichen können. Statt dessen töteten sie gegen Ende des Krieges mehrere Tausend Zivilisten, darunter besonders viele Briten. Die V2 wurde damals im Konzentrationslager Dora im Harz-Gebirge von Häftlingen gebaut, von denen ein Drittel nicht überlebte. Die Technologie war in engem Kontakt mit dem „Institut für Praktische Mathematik" unter Alwin Walther (1898–1967) in Darmstadt entwickelt worden (Mehrtens 1986).

Einige Argumente dieses Artikels sind bereits in Siegmund-Schultze (2009b) entwickelt worden, wo auch die Methodologie der Geschichtsschreibung genauer erörtert wird.

R. Siegmund-Schultze (✉)
University of Agder, Kristiansand, Norway
e-mail: reinhard.siegmund-schultze@uia.no

[1] Wir verwenden das auch zur damaligen Zeit gebräuchliche kritische Wort „Nazis", um uns von dem demagogischen Begriff „Nationalsozialismus" (NS) abzugrenzen.

A. Verdugo Rohrer and J. Zender (eds.), *History of Mathematics and Its Contexts*, Trends in the History of Science, https://doi.org/10.1007/978-3-031-60828-5_14

Warum ist ein Blick auf die Mathematik und die Mathematiker im „Dritten Reich" historisch und psychologisch lehrreich? Verspricht der unterstellte Gegensatz von „rationaler" Mathematik und „irrationaler" Ideologie des Regimes dem Beobachter Spannung und Spannungen? Lässt uns dieser Gegensatz „Widerstand" von Mathematikern erwarten? Können wir damit vielleicht sogar den „Niedergang der deutschen Mathematik" erklären, von einer internationalen Großmacht um 1930 zu einer zweitrangigen Größe im Jahre 1945, zerstört vom „Irrationalismus" der Nazis?

Oder waren die Nazis vielleicht einfach nicht interessiert an dieser Wissenschaft? Anders als Geschichtswissenschaften, Philosophie und Biologie konnte die Mathematik von den Nazis nicht unmittelbar zur Ideologieproduktion[2] verwendet werden. Doch war sie auch nicht ebenso unmittelbar „technologisch anwendbar" wie Physik, Chemie und die Ingenieurwissenschaften. Und der Zweite Weltkrieg war immer noch vor allem ein Krieg der Physiker, nicht der Mathematiker. Ingenieure (Ludwig Prandtl, Wernher von Braun) und Physiker (Carl Ramsauer, Werner Heisenberg) blieben einflussreicher in den Forschungsinstitutionen der Nazis als Mathematiker (Wilhelm Süss, Gustav Doetsch, A. Walther). Die programmierbaren elektronischen Rechenmaschinen waren noch eine Sache der Zukunft. Sie wurden in den 1930er Jahren vereinzelt in England (Alan Turing) und Deutschland (Konrad Zuse) vorausgeahnt. Zuse fand wenig Verständnis in Hitlerdeutschland und baute einen Prototyp in der eigenen Wohnung. Moderne Computer wurden von Amerikanern wie dem ungarisch-deutschen Emigranten John von Neumann gegen Ende des Krieges erstmals konstruiert und angewendet.

Wäre angesichts einer solchen unterstellten Marginalität der Mathematik vielleicht unser Thema uninteressant?

Wie immer liegt die Wahrheit in der Mitte, nämlich in den historischen Details.

Die Mathematik war durchaus nicht durchgehend uninteressant für das Regime. Insbesondere „Angewandte Mathematik" und mathematisch durchwirkte Hybridwissenschaften wie die Aerodynamik (Prandtl) erhielten im NS-Staat besondere Förderung. Einen „Niedergang der deutschen Mathematik" gab es in erster Linie für die Mathematik als Grundlagen- und Universitätswissenschaft. Und dieser Niedergang war vor allem ein Resultat der Massenvertreibung von Wissenschaftlern, darunter von vielen Mathematikern, die die Nazis als „jüdisch" definierten. Der politische und historische Hintergrund der Vertreibungen hat durchaus viel mit Haltungen derjenigen Mathematiker zu tun, die schließlich in Deutschland verblieben. Viele deutsche Mathematiker der Weimarer Republik—wie andere Deutsche der 1920er Jahre auch—hatten jahrelange nationalistische Kampagnen gegen den „Schmachfrieden von Versailles" und gegen die „Kriegsschuldlüge" unterstützt und handelten weder damals noch nach 1933 in einem einfachen Sinne „rational". Dagegen waren die Nazis mit ihrer

[2] Mittelbar aber doch, z.B. als Statistik in der Rassenbiologie.

antisemitischen und sozialdemagogischen Ideologie durchaus „rational" auf die Revision von Versailles und die Vorbereitung von neuen Kriegen orientiert.

Die Nazis konnten dabei auf Unterstützung von Sympathisanten wie des herausragenden Zahlentheoretikers Helmut Hasse (1898–1979) bauen. Hasse bewies 1936 die sogenannte „Riemannsche Vermutung für elliptische Kurven", eine eng mit einer berühmten Vermutung des deutschen Mathematikers Bernhard Riemann aus dem Jahre 1859 zusammenhängende Annahme. Die originale Riemannsche Vermutung ist das achte der 23 mathematischen Probleme, die der deutsche Mathematiker David Hilbert (1862–1943) im Jahre 1900 in Paris aufgestellt hatte. Es ist zugleich eines der wenigen der 23, die noch immer ungelöst sind und für das deshalb das „Clay Institute" in Boston zur Jahrtausendwende eine Million US-Dollar Preisgeld ausgesetzt hat. Der Beweis der ursprünglichen Riemannschen Vermutung würde weitgehende Einsichten über die genaue Verteilung der Primzahlen auf der Zahlengerade ermöglichen. Primzahlen haben bekanntlich in neuerer Zeit in Kryptographie und im Bankwesen erhebliche praktische Bedeutung erlangt. Hasse war mit seinen bahnbrechenden algebraischen und zahlentheoretischen Methoden durchaus ein moderner Mathematiker und war dabei stark von Hilbert und der 1933 aus Deutschland vertriebenen jüdischen Algebraikerin Emmy Noether (1882–1935) beeinflusst. Politisch ordnete sich Hasse dagegen geradezu rechts von den Nazis ein. In einem Interview, das Hasse viel später, um das Jahr 1970 herum, gab, versuchte er anscheinend, an die antisozialistischen Ressentiments des westdeutschen Bürgertums im Nachfeld der Studentenbewegungen zu appellieren, um damit Verständnis für seine zwielichtige Rolle im Dritten Reich zu erlangen:

> My political feelings have never been National-Socialistic but rather 'national' in the sense of the Deutschnationale Partei, which succeeded the Conservative Party of the Second Empire [under Wilhelm II]. I had strong feelings for Germany as it was created by Bismarck in 1871. When this was heavily damaged by the Treaty of Versailles in 1919, I resented that very much. I approved with all my heart and soul of Hitler's endeavors to remove the injustices done to Germany in that treaty. (Segal 1980, 47)

An die Macht gekommen verstanden es die Nazis, die Mehrzahl der Bürger zu Mitschuldigen zu machen, gerade durch deren systematische scheingesetzliche Bevorzugung gegenüber den „Juden". Antisemitismus als Staatsdoktrin in Deutschland nach 1933 war—verglichen mit traditionellem Antisemitismus in anderen Ländern unter Einschluss von akademischem Antisemitismus zum Beispiel in den USA—historisch neu. Staatlicher Antisemitismus schuf ganz andere Bedingungen nicht nur für die Opfer, für die Flüchtlinge und die schließlich Ermordeten, sondern auch für die Täter, die Mittäter, die Angepassten und die wenigen, die Widerstand wagten. Die Nazis definierten, wer als „Jude" zu gelten hatte: weder Konversion zu einer anderen Religion noch außergewöhnliche wissenschaftliche Leistung konnten dieses ideologische Dogma außer Kraft setzen. Wenn man heute die unglaublich anmutende Selbstzerstörung der deutschen Wissenschaft nach 1933 betrachtet, muss man aber immer bedenken, dass Grundlagenforschung auch in anderen Ländern zur damaligen Zeit nicht denselben hohen gesellschaftlichen Stellenwert hatte, den sie heute besitzt.

Die Nazis machten sich das verbreitete anti-intellektuelle Ressentiment gegen die Mathematik (das wir auch heute noch spüren) oftmals zunutze. Ein großes Problem für die Mathematik bestand darin, dass die traditionelle innermathematische Propaganda für die „Schönheit" und „Strenge" dieser Wissenschaft sich wegen ihrer mangelnden Popularität zumindest unter den Bedingungen der Nazi-Diktatur nicht für außerwissenschaftliche Zwecke eignete. Manche Mathematiker glaubten deshalb, ihre reine Grundlagenforschung mit quasimilitaristischem und antisemitischem Gerede über die „Unerbittlichkeit mathematischen Denkens" und gegen die „Spitzfindigkeit jüdischen Denkens" verteidigen und propagieren zu müssen.[3] Der bedeutende Funktionentheoretiker Ludwig Bieberbach (1886–1982), sprach von einer „Entmenschlichung der Mathematik", angeblich verursacht durch die „axiomatische Moderne", die zunehmende Abstraktheit und Verlust des Kontaktes mit den Anwendungen bewirkt habe. Diese Unterstellung erhielt ihre spezielle antisemitische Ausrichtung unter dem Druck der Nazis. Nur wenige gingen aber so weit wie Bieberbach, der behauptete, dass es eine „arteigene Deutsche Mathematik" gäbe und dass diese unvereinbar sei mit „artfremder", insbesondere mit „jüdischer" Mathematik.

Solche verzweifelten und von der Mehrheit der Fachgenossen abgelehnten Theorien gab es auch in anderen Wissenschaften (Physik, Biologie, Chemie). Es gab auch soziale Besonderheiten der Mathematik, verglichen mit den Naturwissenschaften aber noch viel mehr im Vergleich mit Geisteswissenschaften und mit nichtakademischen Berufen. Dies war vor allem der besonders hohe Anteil von führenden Mathematikern, deren Familien einem jüdischen kulturellen oder religiösen Umfeld entstammten. Dies hatte mit intellektuellen Traditionen und der traditionellen Diskriminierung von Juden in öffentlichen Laufbahnen in Militär, Rechtswesen, Verwaltung, Schule zu tun. Es war kein Zufall, dass bis 1933 der relative Anteil jüdischer Mitarbeiter an Universitäten (mit relativ vielen ungesicherten Stellen als Privatdozenten oder Assistenten) viel höher war als im allgemeinen Schuldienst. Wir wollen uns in der folgenden Diskussion auf akademische Mathematiker beschränken, die nach allgemeiner Auffassung den Begriff der „Mathematik",[4] bestimmen. Die politischen Positionen der in dem damaligen mathematischen Hauptberuf des Mathematiklehrers wirkenden Individuen sind durchaus ebenfalls historisch relevant geworden, zum Beispiel in der Militarisierung des Mathematikunterrichts.[5] Der Zusammenhang des Schulunterrichts mit wissenschaftlicher (im Gegensatz zu politischer) Ideologie ist aber schwach, und die internationale Sichtbarkeit der Konflikte im

[3] Mehr in Siegmund-Schultze (2009b).

[4] Die strukturelle Seite der Mathematik: Mehrtens (1986).

[5] Siegmund-Schultze (2008).

Schulwesens des Dritten Reiches ist gering gewesen, unter anderem, weil politisch gemaßregelte Lehrer kaum Möglichkeiten zur Emigration hatten.

So wichtig es für den Historiker ist, bei der Bewertung menschlicher Handlungen und wissenschaftlicher Leistungen die Konsequenzen mitzudenken (von denen wir Auschwitz am Anfang nannten), so essentiell ist es auch, Menschen gerecht in ihrem historischen Umfeld und in ihren Erkenntnismöglichkeiten zu beurteilen. Es ist selbstverständlich, dass Menschen ein Recht auf Überleben und damit auch auf Anpassung an diktatorische Regime haben. Die Frage ist nur, inwieweit Anpassung Blindheit erzeugt und nachträglich von Akteuren oder von Historikern als „notwendig" erklärt, im Extremfall sogar zu „Widerstand" verfälscht wird, wie wir es in der Nachkriegsapologie vieler Deutscher erlebt haben. Wir geben zu, dass der Einzelne Auschwitz nicht voraussehen konnte, aber wir glauben ihm nicht, dass er nichts von der Judenverfolgung der Nazis erlebt hat. Jedoch trifft man sogar in professioneller Geschichtsschreibung, vor allem aber in Selbstdarstellungen der Mathematiker, immer noch auf die Überbewertung menschlicher Sekundärtugenden wie „familiärer Fürsorglichkeit" und „wissenschaftlichen Scharfsinns": gute Mathematiker müssen doch auch gute Menschen sein. Müssen sie?

Wenn wir im Folgenden von „politischen Haltungen von Mathematikern" im Dritten Reich sprechen, so müssen wir versuchen, das „spezifisch Mathematische" darin zu finden. Das ist notorisch schwierig wegen der bereits erwähnten Unpopularität dieser Wissenschaft. Wir werden an einigen individuellen Beispielen das Verhalten von denjenigen Mathematikern, die nach 1933 in Deutschland verblieben, erläutern.[6] Wir werden diese persönlichen Beispiele fünf Gruppen zuordnen, die wir etwas grob und ungefähr als „Täter" (Bieberbach, Teichmüller), „Mittäter" (Blaschke, Doetsch, Hasse, Süss, Lietzmann), „Angepasste" (Behnke, Blaschke, Grunsky, Hasse, E. Hopf, Perron, Reidemeister, Walther), „Widerständige/Diskriminierte" (Grunsky, Moufang, Perron, Reidemeister) und „Opfer" (Landau, Hausdorff) bezeichnen wollen. Wir führen in Klammern Namen von insgesamt 16 fachlich bedeutenden (bis auf eine Ausnahme: Süss) Mathematikern an, die wir den einzelnen Gruppen zuordnen. Es sei aber gleich gesagt, dass die Zuordnung von Personen zu Gruppen von vornherein zu einseitiger Sicht auf jene Individuen führen muss, zumal im Rahmen eines kurzen Essays, wo Biographien nicht im Detail diskutiert werden können, wo im Wesentlichen nur ein oder zwei Äußerungen oder Handlungen der betreffenden Personen erwähnt werden können und wo mit Fußnoten auf ausführlichere Quellen verwiesen werden muss. Die Doppelnennung von Namen in den Klammern weist bereits auf die Aussichtslosigkeit einer klaren Zuordnung zu Gruppen hin. Männer wie Hasse, der 1936 eine. Würdigungsadresse zum 75. Geburtstag seines jüdischen Lehrers Kurt Hensel veröffentlichte, verhielten sich

[6] Die bisher umfassendste Darstellung der Lebens- und Arbeitsbedingungen von Mathematikern im Dritten Reich ist Segal (2003). Die Politik der Deutschen Mathematiker-Vereinigung (DMV) wird ausführlich von Remmert (2004) behandelt.

sehr widersprüchlich. Es sei auch erneut betont, dass alle im Dritten Reich verbliebenen Mathematiker, soweit sie nicht rassistisch oder anderweitig politisch verfolgt wurden, in die Schuld des Regimes verstrickt wurden und allein durch ihr Überleben privilegiert waren. Wir können uns deshalb auch nicht dazu entschließen, Widerständige und aus anderen Gründen diskriminierte Mathematiker (Homosexuelle, Frauen) von vornherein (also ohne zusätzliche Gründe) als „Opfer der Nazis" zu bezeichnen. Wir wissen ja, dass solche Individuen zur gleichen Zeit auch in anderen Ländern benachteiligt wurden, wie das Beispiel des homosexuellen englischen Mathematikers Alan Turing zeigt, oder die Diskriminierung von Frauen, die bis heute anhält. Es muss uns in unserer Erörterung stets um die Spezifik der terroristischen Machtausübung im Dritten Reich gehen, im Vergleich zu welcher auch der deutliche Rassismus gegenüber den Afroamerikanern in den USA zur gleichen Zeit verblasst.

Natürlich gab es Opfer der Nazis unter den in Deutschland verbliebenen Mathematikern. Aber dies waren in erster Linie aus ihren Berufen vertriebene Juden, von denen viele nach Kriegsbeginn untergingen. Das Schicksal der Emigranten, die—obwohl als Überlebende ebenfalls „privilegiert"—allesamt (ob jüdisch oder nicht) in stärkerem Sinne Opfer des NS waren als die in Deutschland verbliebenen Nichtjuden, soll hier nicht diskutiert werden.[7]

Fast alle der im Folgenden zu erwähnenden Mathematiker waren Männer. Während es generelle Kampagnen des Regimes gegen die Berufstätigkeit von Frauen gab,[8] wurden in akademischen Berufen außerdem „Führereigenschaften" verlangt, die im allgemeinen nur Männern zugetraut wurden. Um diese frauenfeindliche Dimension deutlich und als Teil der politischen Bedingungen des Nazi-Hochschulsystems anschaulich zu machen, bringe ich gleich am Anfang das Beispiel einer Mathematikerin. Die Frankfurterin Ruth Moufang (1905–1977) wurde schon in den 1930er Jahren durch wichtige Arbeiten an der Grenze von Geometrie und Algebra bekannt. Als sie 1936 den akademischen Grad der „Habilitation" erreicht hatte wurde ihr der nächste Schritt, die Zuerkennung einer „Dozentur" verweigert. Ein Brief des Reichs- und Preußischen Ministers für Wissenschaft, Erziehung und Volksbildung (Berlin, Unter den Linden 4) vom 9. März 1937 an „Fräulein Dr. phil. nat. habil. Ruth Moufang" nennt die „Gründe":

> Da dem Dozenten im Dritten Reich außer seinen wissenschaftlichen Leistungen wesentlich erzieherische und Führereigenschaften voraussetzende Aufgaben zufallen und die Studentenschaft fast ausschließlich aus Männern besteht, fehlt dem weiblichen Dozenten künftig die Voraussetzung für eine ersprießliche Tätigkeit. Die Reichs-Habilitations-Ordnung hat mit Einführung des Gemeinschaftslagers (Wehrsportlager und Dozenten-Akademie) bereits einen ausschließlich männlichen Hochschullehrernachwuchs im Auge gehabt.

[7] Siegmund-Schultze (2009a).

[8] Unter dem demagogischen Motto „Kampf dem Doppelverdienertum!".

> Bei dieser Sachlage ist es mir leider nicht möglich, Ihnen die Erteilung einer Dozentur in Aussicht zu stellen. Gegen eine forschende Tätigkeit an einer Hochschule oder in einer Forschungsanstalt bestehen jedoch keine Bedenken. (Pieper-Seier 2008, 186)

Für Moufang war damit ihr Lebenstraum, akademischer Lehrer zu werden, vorläufig zerstört. Da sie keine bezahlte Dozentur erlangen konnte, musste sie sich eine Stelle in der Industrie suchen. Erst nach dem Ende des Regimes wurde sie in Frankfurt eine der ersten Mathematikprofessorinnen Deutschlands.

Der oben erwähnte Bieberbach war ein international führender Forscher in Geometrie und Funktionentheorie (Mehrtens 1987). Er löste um 1911 mit geometrischen und gruppentheoretischen Methoden das 18. der 23 Hilbertschen Probleme von 1900, das eng im Zusammenhang mit Anwendungen der Mathematik in der Kristallographie stand. 1916 stellte Bieberbach eine weitreichende Vermutung über die Eigenschaften gewisser Funktionen (Abbildungen) auf, deren Bestätigung wichtig für die Verwendbarkeit des „Abbildungssatzes" von Bernhard Riemann aus dem Jahr 1851 ist. Dieser berühmte Satz hatte theoretisch die Reduzierbarkeit der Untersuchung komplizierter geometrischer Bereiche der Ebene auf den viel einfacheren Kreis gesichert. Die „Bieberbachsche Vermutung" wurde 1984, zwei Jahre nach seinem Tod, von dem französisch-amerikanischen Mathematiker Louis de Branges bewiesen.

Bieberbach war 1933 als Professor in Berlin etabliert, und es ist immer noch nicht vollständig geklärt,[9] warum er sich plötzlich als radikaler Nazi offenbarte und sich an der Propagierung der rassistischen Fiktion einer spezifisch „Deutschen Mathematik" beteiligte. Bieberbach veröffentlichte unter anderem 1934 einen Artikel, in dem er den Boykott von Studenten gegen die Vorlesungen des damals 56jährigen jüdischen Zahlentheoretikers Edmund Landau (1877–1938) in Göttingen 1933 mit den folgenden Worten unterstützte:

> Vor einigen Monaten haben Differenzen mit der Studentenschaft dem Lehrbetrieb des Herrn Landau ein Ende bereitet ... Man hat darin ... ein Musterbeispiel dafür zu sehen, dass Vertreter allzu verschiedener menschlicher Rassen nicht als Lehrer und Schüler zusammenpassen. Im Geistigen prägt sich bekanntlich die Rasse in der Struktur der Persönlichkeit, im Stil der Lebensäußerungen aus. Der Instinkt der Göttinger Studenten fühlte in Landau einen Typus undeutscher Art, die Dinge anzupacken. (Bieberbach 1934, 236)

Bieberbach trug maßgeblich zur Rechtfertigung der Vertreibungen und zur Förderung politisch nazihöriger Wissenschaftler bei. Allerdings wurden rassistische Pseudotheorien wie die „Deutsche Mathematik" vom Regime in späteren Jahren immer weniger unterstützt, als es um die Sicherung der Kompetenz von Wissenschaftlern für die Kriegsvorbereitung ging. Bieberbach

[9] Es gibt Vermutungen, dass die Nazis etwas politisch gegen ihn in der Hand hatten und ihn erpressten.

war der einzige deutsche Mathematiker, der nach dem Krieg nie wieder eine Lehrposition an einer Hochschule erhielt. Sein Opfer Landau starb 1938 in Berlin.

Einer der jungen fanatischen Nazis unter den Studenten, die Bieberbach unterstützten, war der brillante Funktionentheoretiker Oswald Teichmüller (1913–1943). Er war einer der Organisatoren des Boykotts gegen Landau in Göttingen, der schließlich zu dessen Ausscheiden aus der Universität führte. Das Bizarre an Teichmüllers Verbindung zwischen Nazi-Aktivismus und Mathematikbegeisterung kommt wohl nirgends deutlicher zum Vorschein als in dem von Landau erbetenen „Entschuldigungsbrief", den Teichmüller am 3. November 1933, einen Tag nach dem Boykott, an seinen Lehrer Landau schrieb:

> Sie sprachen gestern die Annahme aus, dass es sich um eine antisemitische Demonstration gehandelt habe. Ich stand und stehe auf dem Standpunkt, dass sich eine judenfeindliche Einzelaktion gegen ziemlich jeden anderen eher richten sollte als gegen Sie. Es handelt sich für mich nicht darum, Ihnen als Juden Schwierigkeiten zu machen, sondern lediglich darum, die deutschen Studenten des zweiten Semesters ... davor zu bewahren, gerade in der Differential- und Integralrechnung von einem ihnen ganz fremdrassigen Lehrer unterrichtet zu werden. Ich wage so wenig wie jeder andere Ihre Fähigkeit der rein international-mathematisch-wissenschaftlichen Belehrung von geeigneten Studenten beliebiger Abstammung zu bezweifeln. ... Die Möglichkeit aber, dass Sie den mathematischen Kern ohne eigene nationale Färbung Ihren Hörern vermitteln, besteht so wenig, als es sicher ist, dass ein Gerippe ohne Fleisch nicht läuft, sondern zusammensackt und verwittert. (Schappacher and Scholz 1992, 29/30.)

In der mathematikhistorischen Literatur gibt es gegensätzliche Meinungen, ob man dem damals zwanzigjährigen Teichmüller seine Naivität und Jugend zugutehalten oder vielmehr dem 47jährigen Bieberbach zubilligen sollte, dass er ehrlich um die Zukunft der Mathematik besorgt war. Letzterer lehnte Landaus sprichwörtlichen abstrakten Stil der Präsentation der Mathematik ebenso wie nicht wenige andere Mathematiker innerlich ab. Mir scheint keine dieser „Entschuldigungen" haltbar. Das Naziregime bediente sich in seiner Propaganda rücksichtslos eines diffusen Gemischs konservativer and radikalnationalistischer Vorurteile. Ein solches Gemisch wurde von den „jungkonservativen" Ideologen in der Art von Ernst Jünger (1895–1998), der altersmäßig zwischen Teichmüller und Bieberbach stand, oder von Möller van den Bruck (1876–1925) produziert, der 1923 den Begriff des „Dritten Reiches" schuf. Bekanntlich waren in der verbrecherischen paramilitärischen SS der Nazis junge Akademiker überdurchschnittlich stark vertreten. Der Fanatiker Teichmüller—selbst kein Mitglied der SS—zog schließlich ebenso wie Jonathan Littells brutaler und „wohlmeinender" Max Aue in den Krieg gegen die Sowjetunion, ohne sich um eine Freistellung von der Front zu bemühen, nur dass er anders als Aue nicht überlebte. Einige der wichtigsten Arbeiten von Teichmüller über „quasikonforme Abbildungen" wurden in der von Bieberbach herausgegebenen politisch gefärbten Fachzeitschrift „Deutsche Mathematik" (mit demselben Namen wie die Rassentheorie) veröffentlicht, die im Ausland kaum gelesen wurde. Teichmüllers Arbeiten wurden deshalb international nur langsam wahrgenommen und sein

Engagement für die Nazi-Ideologie hat ihn nicht nur das Leben gekostet sondern auch seinen Nachruhm als Mathematiker beschädigt.

Neben „Tätern" wie Bieberbach und Teichmüller gab es die große Gruppe der „Angepassten", die sich weder aktiv im Sinne der NS-Ideologie äußerten noch allzu deutlich für das Regime handelten. Ein typischer Vertreter dieser Gruppe war der Münsteraner Spezialist für mehrdimensionale Funktionentheorie Heinrich Behnke (1898–1979), der nach dem Kriege Lehrer des späteren Nestors der westdeutschen Mathematiker, Friedrich Hirzebruch (1927–2012) werden sollte. Mathematiker wie Behnke lehnten Bieberbachs Rassentheorien in der Mathematik als inhaltlich völlig haltlos ab. Sie verwiesen in internen Briefen, die wir heute in Archiven finden, darauf, dass einige der führenden nichtjüdischen Mathematiker, wie David Hilbert, abstrakte und unanschauliche „jüdische Mathematik" treiben würden, wenn man an Bieberbachs Klassifikation der „mathematischen Stile" glauben sollte. Dagegen hatten bekannte jüdische Mathematiker, wie Richard Courant (1888–1972) und Richard von Mises (1883–1953), die in die Emigration getrieben worden waren, stets die Rolle von geometrischer Anschauung und von Anwendungen hervorgehoben. In seinen Memoiren von 1978 beschreibt Behnke die Atmosphäre des Terrors, der Denunziationen, der Beschränkung internationaler Kontakte, die unmittelbar nach der Machtergreifung des NS-Regimes im Frühjahr 1933 einsetzte. Persönlich fühlte sich Behnke durch Konsequenzen für ihn und besonders für seinen Sohn bedroht, die seine Ehe mit einer verstorbenen Jüdin haben könnte.

Für die Anfangsjahre der Nazi-Diktatur waren Briefe typisch, die von Entlassungen bedrohte Mathematiker, unter ihnen natürlich viele „Juden" (nach Nazi-Definition), an die Behörden schrieben, um ihren „deutschen Patriotismus" zu bekunden. Während solche Briefe von Juden, wenn sie beispielsweise auf ihren Einsatz für Deutschland im Weltkrieg verwiesen, wegen des unantastbaren Dogmas des Antisemitismus vergeblich bleiben mussten, konnten einige bedrohte „Nichtjuden" zuweilen Gefahr von sich abwenden.

Ein eindrucksvolles Dokument ist ein neunseitiger Verteidigungsbrief, den der Topologe in Königsberg, Kurt Reidemeister (1893–1971), am 13. Mai 1933 an das preußische Nazi-Wissenschaftsministerium schickte. Reidemeister hatte in seiner Wiener Zeit von 1922 bis 1925 Kontakt zu dem linksliberalen Wiener Kreis für empiristische Philosophie gehabt. Seine Schwester Marie (1898–1986), Mitarbeiterin und spätere Ehefrau eines der führenden Philosophen des Kreises, Otto Neurath (1882–1945), floh schon 1934 mit diesem vor dem in Österreich wachsenden Nationalismus und Antisemitismus in die Niederlande. Reidemeister, der von Hilberts axiomatischem Zugang zur Mathematik stark beeinflusst war, ist durch sein Buch von 1932 über mathematische Knotentheorie international bekannt geworden. Dort unterwarf er diese Teildisziplin der Topologie einer modernen, von anschaulichen Argumenten möglichst befreiten, rein gruppentheoretischen Behandlung. Reidemeister hatte 1931 gegen nationalsozialistische Studentenunruhen in Königsberg Position bezogen. Zwei Jahre später wollte ihn das zuständige Ministerium deshalb aus politischen

Gründen entlassen. Reidemeister schrieb in seinem Brief an das Ministerium unter anderem:

> Vor allem Einzelnen erkläre ich, dass ich niemals Mitglied einer politischen Partei gewesen bin und dass ich (meiner Auffassung von politischer Betätigung nach) mich auch niemals politisch betätigt habe.[10]

Reidemeister versuchte dann, seine wirklichen Überzeugungen durch Gebrauch von Nazivokabular wie „soziale Volksgemeinschaft", „großdeutscher Gesichtspunkt" (den er angeblich in Wien vertreten habe), „berufsständische Auffassung" und „Führerprinzip" zu verhüllen. Es gelang Reidemeister aber lediglich—und dies mit Unterstützung von Kollegen, die ein Memorandum an das Ministerium schrieben—seine Entlassung mit politischer „Begründung" zu verhindern und statt dessen an eine andere Universität (Marburg) versetzt zu werden. Für die weiteren 12 Jahre der Nazi-Diktatur musste Reidemeister sich anpassen und war als öffentliche Person weitgehend auf ein totes Gleis geschoben. Er zog sich in eine Beschäftigung mit antiker Wissenschaft und Philosophie zurück.[11]

Ein anderer deutscher Mathematiker mit Ansätzen zu Resistenz gegen die Nazis war der wesentlich jüngere Helmut Grunsky (1904–1986). Die Hervorhebung seines Alters ist wichtig, weil sich jüngere Mathematiker im Allgemeinen stärker an die Nazis anpassen mussten, wenn sie ihre Chancen auf wissenschaftliches Fortkommen wahren wollten. Obwohl Grunsky bemerkenswerte Arbeiten in der komplexen Funktionentheorie verfasst hatte, fand er unter den Bedingungen der Kürzungen an den Hochschulen und der anhaltenden Akademikerarbeitslosigkeit in der Mitte der 1930er Jahre keine Stellung. Er musste sich und seine Familie mit drei Kindern deshalb als Schriftleiter einer mathematischen Referatezeitschrift über Wasser halten, was seiner wissenschaftlichen Entwicklung nicht diente. Das klassische deutsche „Jahrbuch über die Fortschritte der Mathematik" wurde von Bieberbach als Mitglied der Preußischen Akademie der Wissenschaften administriert. Am 11. Januar 1938 erhielt Grunsky einen Brief von seinem Chef Bieberbach, in dem es heißt:

> Besten Dank für Ihre Neujahrswünsche, die ich für Sie und Ihre Frau erwidere. Möchte es Ihnen im neuen Jahr vor allem nun endlich beschieden sein, die Juden aus Ihrem Mitarbeiterstab loszuwerden. Die Durchsicht der bisher erschienenen Hefte des Jahrgangs 1936 – von Heften früherer Jahrgänge will ich im Augenblick nicht reden – zeigt eine

[10] Gabor Szegö Papers, Stanford University Archives, SC 323, box 9, folder 15. Kurt Reidemeister an das preußische Wissenschaftsministerium, 13. Mai 1933, Maschine, 9 Seiten, S. 1. Hervorhebungen von Reidemeister.

[11] Zu den Zusammenhängen zwischen Reidemeisters rationalistischer und teilweise elitärer Weltanschauung und seinen modernistischen Idealen in der mathematischen Forschung siehe (Epple 1999).

> große Zahl an Juden vergebener Referate. Es sind zu viele, als dass ich eine zwingende Notwendigkeit einsehen könnte.
>
> Sie laufen jedenfalls Gefahr, dass Ihr Handeln als mangelnder politischer Instinkt ausgelegt werde. Was soll ich im Falle eines kürzlich von auswärts an mich herangetretenen Angriffs auf die Akademie sagen, der man es zum Vorwurf macht, dass sie (die Akademie) noch immer beim Jahrbuch jüdische Referenten beschäftige? (Siegmund-Schultze 2004, XLVII)

Grunsky, der entlassenen jüdischen Mathematikern Möglichkeiten gab, etwas Geld zu verdienen, wurde fortan von Bieberbach immer weniger gefördert. Dieser verhinderte sogar Grunskys Habilitation in Berlin und seine Beförderung zum Beamten innerhalb der Akademie, zumal Grunsky sich weigerte, politischen NS-Organisationen beizutreten. Grunsky sah sich schließlich gezwungen, seine Stelle beim Jahrbuch zu verlassen. Die für die Habilitation vorgesehene Arbeit hatte Grunsky nicht in Bieberbachs „Deutscher Mathematik" eingereicht, sondern bei der politisch weniger beeinflussten „Mathematischen Zeitschrift", wo sie 1939 erschien. Die dort enthaltenen „Grunskyschen Ungleichungen" halfen de Branges 1984 beim Beweis der „Bieberbachschen Vermutung".

Die von Bieberbachs „Deutscher Mathematik" gepflegten Vorurteile gegen gewisse moderne mathematische Theorien wurden von anderen Mitläufern unter den Mathematikern wie dem einflussreichen Geometer Wilhelm Blaschke (1885–1962) für ihre fachpolitischen Zwecke ausgenutzt. Blaschke spielte beispielsweise kritisch auf die „graue Theorie der reellen Funktionen" an (Blaschke 1941, 44). Einige deutsche Forscher in Wahrscheinlichkeitsrechnung und Statistik hielten sich an alte Begriffe, vermieden die moderne Maßtheorie. Diese Vorurteile beruhten zum Teil auf dem Fehlen von „deutschen Traditionen" in diesen Gebieten, wurden aber verstärkt durch das Abreißen internationaler Verbindungen insbesondere mit den Franzosen und Russen. Der deutsche Mathematiker Eberhard Hopf (1902–1983), der sich zum Erstaunen und zur Enttäuschung seiner amerikanischen Kollegen am MIT zurück nach Deutschland auf eine Professur in Leipzig locken ließ, sagte nach seiner Rückkehr in seinem bahnbrechenden Buch *Ergodentheorie* von 1937, über die in Deutschland vernachlässigte Maßtheorie:

> Statistik ist Maßtheorie... Den Mathematikern, die dem ‚fast alle' oder ‚bis auf eine Nullmenge' keinen Geschmack abgewinnen können, sei entgegnet, dass sich nur das, was in der Natur ‚in der Regel' sich ereignet, mathematisch interpretieren lässt. (Hopf 1937, iii)

Neben Antisemitismus und Antimodernismus waren Auswirkungen der NS-Ideologie auch in der Mathematikdidaktik zu erkennen. Ohnehin war es schwierig den Wert der Mathematik an den Schulen aufrecht zu erhalten, angesichts der starken Betonung körperlicher und rein ideologischer Erziehung. Weitere Probleme ergaben sich aus der Nazi-Doktrin von der Vererbung persönlicher Eigenschaften und Talente, wie es der Göttinger Didaktiker Walther Lietzmann (1880–1959) im Jahre 1935 andeutete:

> Die mathematische Methodik hat ein paar Jahrzehnte lang gegen das Axiom von der mathematischen Begabung gekämpft, ohne die mathematisches Verständnis auf der Schule nicht möglich sei. Müssen wir das nicht … als einen verfehlten Kampf ansehen, weil eben die mathematische Begabung erblich bedingt und also entweder da oder nicht da ist? Was hat unter solchen Umständen überhaupt noch *mathematische Erziehung für Aufgaben?* (Lietzmann 1935, 363)

Lietzmann wurde von den Nazis als vertrauenswürdig und als politisch geschickt genug befunden, die Delegation der deutschen Mathematiker auf dem Internationalen Mathematikerkkongress in Oslo 1936 zu leiten.[12]

Neben Hopf gab es weitere Mathematiker, die die modernen Theorien der mathematischen Forschung trotz offizieller Diskreditierung pflegten.

Es war kennzeichnend für Bieberbachs schwindenden Einfluss, dass im Jahre 1939 Oskar Perron (1880–1975), ein Mathematikprofessor in München, eine ironische Anspielung auf die „Deutsche Mathematik" veröffentlichen konnte. Perron tat dies in der zweiten Auflage (1939) seines Buches „Irrationalzahlen" von 1920, indem er im Vorwort auf Bieberbachs Vorliebe für die Methode anspielte, mit der Georg Cantor und Charles Méray im 19. Jahrhundert die irrationalen Zahlen definiert hatten. Cantor, der deutsche Begründer der Mengenlehre wurde im Dritten Reich als jüdisch angesehen, während Méray als Franzose offensichtlich ebenfalls „undeutsch" war. Im Jahre 1924 jedoch, bevor er seine Sympathien für die Nazis entwickelte, hatte Bieberbach die erste Auflage von Perrons Buch kritisiert, da es die irrationalen Zahlen mit Hilfe des sogenannten „Schnitts" von Richard Dedekind definiert hatte. Jetzt, im Jahre 1939, verwies Perron ironisch auf Bieberbachs Rezension und „verteidigte" seine eigene Vorliebe für den unstrittig „deutschen" Dedekind, der übrigens auch als der wahre mathematische Inspirator der vertriebenen jüdischen Mathematikerin Emmy Noether angesehen werden kann.

> Ich glaube, dass es einem Deutschen, der die Wahl zwischen einem deutschen Erzeugnis und einem an sich ebenso schönen und wertvollen Erzeugnis fremden Ursprungs hat, immer erlaubt ist, der Stimme des Herzens zu folgen und das Deutsche vorzuziehen, weil es deutsch ist. (Perron 1939, vi)

Der Kriegsausbruch im September 1939 schuf natürlich neue Bedingungen für die Mathematik. Ihre mögliche Verwendung im Krieg trat in den Vordergrund und neue mathematische Arbeitsplätze wurden geschaffen.

Zwei Tage nach dem Ausbruch des Krieges, am 3. September 1939, wurde Grunsky als mathematischer Mitarbeiter an der Dechiffrierabteilung des Auswärtigen Amtes eingestellt. Man kann annehmen, dass die Atmosphäre am Auswärtigen Amt, das vom später als Kriegsverbrecher hingerichteten Ribbentrop geleitet wurde, noch stärker politisiert war als an der Preußischen Akademie, Grunskys früherer Arbeitsstelle. Dies und die Geheimhaltungsvorschriften der

[12] Hollings and Siegmund-Schultze (2020).

Dechiffrierabteilung erklären wohl, dass Grunsky am 19. Februar 1940 dann doch einen Antrag auf Mitgliedschaft in der Nationalsozialistischen Deutschen Arbeiterpartei (NSDAP) stellte, dem im April 1940 stattgegeben wurde. Grunsky scheint ein typisches Beispiel dafür zu sein, dass formale Mitgliedschaft in der NSDAP, die nach dem Krieg oft Grund für (meistens nur kurzfristige) Entlassungen war, keine zuverlässigen Aussagen über persönliches Engagement für die Ziele des NS erlaubt. Dies soll nicht verhüllen, dass Grunsky ebenso wie viele andere Mathematiker durch seine Kriegsarbeit mitschuldig an der Verlängerung der Diktatur geworden ist. Wie Grunsky später selbst schrieb, bestand seine außerberufliche „politische Tätigkeit" während des Krieges im Wesentlichen in der Austeilung von Lebensmittelkarten. Es ist wenig hilfreich, wenn man in einer Biographie Grunskys auf der populären mathematikhistorischen Website der St. Andrews University lesen muss, dass er 1946 wegen „verschiedenen Umständen" und „Schwierigkeiten am Ende des Krieges"[13] in Gießen aus dem Universitätsdienst entlassen und mehrere Jahre als einfacher Mathematiklehrer arbeiten musste, bevor er 1950 seine akademische Karriere in Mainz fortsetzen konnte. Solche Formulierungen führen nur zu ungerechtfertigten Vermutungen über besonderes Engagement für die Nazis. In späteren politischen Umbruchsperioden in Deutschland (1989/90) ist formale Parteiangehörigkeit erneut als oberflächliches Argument im akademischen Existenzkampf verwendet worden.

Während des Krieges gelang es Wilhelm Süss (1895–1958), einem ehemaligen Schüler Bieberbachs, der von 1937 bis 1945 Vorsitzender der Deutschen Mathematikervereinigung (DMV) war, zahlreiche Mathematiker für „kriegswichtige" Aufgaben zu reklamieren und so vom Fronteinsatz freistellen zu lassen. Obwohl selbst als Mathematiker wenig produktiv, war Süss sehr geschickt im Umgang mit den NS-Behörden und erreichte beispielsweise noch im vorletzten Kriegsjahr 1944 die Gründung eines „mathematischen Forschungsinstituts" in Oberwolfach im Schwarzwald; dies in gehöriger geographischer Distanz von der anrückenden Roten Armee. Dieses Institut wurde eine Überlebensquelle für die deutsche Mathematik nach dem Kriege und ist ein heute international sehr bekanntes Tagungsinstitut. Um der DMV Einfluss bei den Nazi-Behörden zu sichern, sorgte Süss schon seit 1937 in vorauseilendem Gehorsam für den Ausschluss jüdischer Mitglieder. Im März 1938 schrieb er an seine Kollegen im Vorstand der DMV, darunter Helmut Hasse:

> Dass wir passende Gelegenheiten wahrnehmen, um unsre jüdischen und von den Emigranten die unliebsamen Mitglieder möglichst bald los zu werden, darin sind wir wohl alle einig. Herr Hasse hat als Schatzmeister gewisse Möglichkeiten, die sich benutzen lassen. (Remmert 2004, 223).

[13] "Various circumstances" and "difficulties at the end of the war". https://mathshistory.st-andrews.ac.uk/Biographies/Grunsky/. Letzter Zugriff Januar 2024.

Und im Dezember 1938 fügte er hinzu:

> Wenn wir einer Aufforderung des Ministeriums durch eigene Schritte zuvorkommen wollen, so müssen wir nun rasch handeln. (Remmert 2004, 225)

Während des Krieges halfen Süss und Hasse bei Versuchen des Wissenschaftsministeriums, die DMV zum Zentrum der europäischen Mathematik auszubauen und Mathematiker aus okkupierten Gebieten zur Kollaboration zu bewegen.[14] An seine amerikanischen Kollegen von der American Mathematical Society, die im Februar 1940 einen Aufruf an alle Kollegen im Ausland gerichtet hatten, sich in Kriegszeiten auf ihre Menschlichkeit zu besinnen, schrieb Süss am 30. September 1940 einen insinuierenden Brief, der alle Schuld am Krieg Frankreich und England zuschob:

> Der Vorstand der Deutschen Mathematikervereinigung hat mit großem Interesse von dem Inhalt dieses Aufrufs Kenntnis genommen und empfindet es mit dankbarer Freude, dass die American Mathematical Society sich der unglücklichen und zerstörenden Wirkung bewusst geworden ist, welche die Kriegserklärungen Frankreichs und Englands auf die wissenschaftliche Arbeit der unter dem Kriege leidenden Völker besitzt. Unsere Vereinigung würde sich glücklich schätzen, wenn es gelungen wäre, die notwendige Revision des Versailler Vertrages, der so viel Unglück über die Menschen gebracht hat, auf friedlichem Wege durchzuführen. Ich gebe der Hoffnung Ausdruck, dass trotz des betonten Kriegswillens unseres letzten Kriegsgegners, England, der Krieg ohne allzu starke Beeinträchtigung der wissenschaftlichen Werte und der Gelehrten im Kriegsgebiet bald beendigt werden kann.[15]

Auch im Zweiten Weltkrieg nutzte und schürte also das Regime die alten Ressentiments, die im Ersten Kriege und in seinen politischen Nachbeben entstanden waren.

Süss‘ Politik wurde auch von der breiten Masse der Mitglieder der DMV getragen, und nach dem Krieg fand er viel Unterstützung von seinen Kollegen und wurde nach kurzer Schamfrist wieder in seine Professur in Freiburg eingesetzt.

Wir haben gesehen, dass Grunsky, Perron und Reidemeister, anders als die große Gruppe der „Angepassten", dem Nazi-Regime einigen Widerstand entgegensetzt hatten und dass einige Mathematiker (Grunsky, Moufang) Nachteile erleiden mussten.

Dennoch können wir diese Nachteile nicht mit dem Leid der wirklichen „Opfer" vergleichen, das den in Deutschland verbliebenen jüdischen Mathematiker und später den Mathematikern unter deutscher Okkupation zugefügt wurde. Unter den Letzteren gab es die größten Todesraten, wobei die

---

[14] Siegmund-Schultze (2002).

[15] Archives American Mathematical Society, Providence, RI. Brown University, Marshall Harvey Stone Papers, box 41, folder 9A.

Besatzer in ihrem Vernichtungswillen bei einigen Völkern, besonders den Polen, keine Unterschiede zwischen „Nichtjuden" und Juden machten.

In Deutschland verblieb unter anderem Edmund Landau, über den wir als Opfer eines Vorlesungsboykotts in Göttingen 1933 gesprochen haben. Landau starb im Februar 1938 Berlin, vertrieben aus Göttingen in die Reichshauptstadt. Obwohl wir wenig über Landaus konkrete Todesumstände wissen, haben wir keinen Zweifel, dass der erst 61jährige begeisterte akademische Lehrer partiell an gebrochenem Herzen starb.[16] Der Novemberpogrom 1938 und der Holocaust blieben ihm erspart.

Der berühmte Pionier der modernen Mengenlehre Felix Hausdorff (1868–1942) in Bonn sah dagegen den Holocaust unmittelbar vor sich. Er ging deshalb zusammen mit seiner Frau und seiner Schwägerin in den Tod. Am Abend vor seinem Selbstmord, am 25. Januar 1942, schrieb er einen Brief an seinen Rechtsanwalt, in dem es unter anderem heißt:

> Was in den letzten Monaten gegen die Juden geschehen ist, erweckt begründete Angst, dass man uns einen für uns erträglichen Zustand nicht mehr erleben lassen wird.
>
> Verzeihen Sie, dass wir Ihnen über den Tod hinaus noch Mühe verursachen; ich bin überzeugt, dass Sie tun, was Sie tun können (und was vielleicht nicht sehr viel ist). Verzeihen Sie uns auch unsere Desertion! Wir wünschen Ihnen und allen unseren Freunden, noch bessere Zeiten zu erleben.
>
> Ihr treu ergebener Felix Hausdorff. (Neuenschwander 1996, 263/64)

Hausdorffs Voraussicht war zutreffend, und es bleibt uns hier zum Abschluss unserer Bemerkungen nur noch, auf die Reflexionen von Mathematikern, Historikern und Bürokraten nach dem Krieg, also in „besseren Zeiten", kurz einzugehen, in Zeiten, die die in Deutschland verbliebenen Juden nicht mehr erleben konnten. Seit die alten Täter, Mitläufer und Angepassten und auch viele ihrer Schüler nicht mehr unter uns sind, hat sich viel verändert, auch in der Historiographie des Dritten Reichs. Im Jahre 2021, drei Jahre nach dem 150. Geburtsjahr von Felix Hausdorff, ist die neunbändige mustergültig kommentierte Ausgabe seiner Werke vollendet worden.

Doch die Anfänge der „Vergangenheitsbewältigung" nach dem Kriege waren schwierig. Die Bereitschaft, die Leiden von Emigranten anzuerkennen, war unter den deutschen Mathematikern nicht sehr weit verbreitet, und Selbstmitleid dominierte. Der nach Amerika vertriebene ehemalige Frankfurter Topologe Max Dehn (1878–1952) erhielt beispielsweise wegen des Widerstands von Bürokraten im zuständigen westdeutschen Ministerium keine reguläre Pensionierung. Ruth Moufang, die nach dem Kriege eine neue Chance erhielt, schrieb am 16. September 1948 an Dehn, dass sie nicht glaube, dass Antisemitismus bei der schleppenden Behandlung seiner Ansprüche eine Rolle spiele. Moufang zeigte

[16] Siegmund-Schultze (2011).

sich dann aber generell skeptisch und durchaus vorausschauend hinsichtlich des mangelnden Willens zu wirklicher „Vergangenheitsbewältigung":

> Von den fixen Ideen der Nazizeit ist keineswegs jetzt alles verschwunden. Und die Entnazifizierung war ein Fehlschlag, sie bewirkte weder eine moralische noch eine faktische Säuberung (z.B. sind im Unterrichtsbetrieb 90% der Entlassenen wieder im Amt). Die Deutschen werden es nie fertig bringen, sich vernünftig zu benehmen. ... das wird schlimmer werden, wenn der Einfluss der Besatzung kleiner geworden sein wird. (Siegmund-Schultze 1998, 286)

Noch in seinen Memoiren von 1978 ging Behnke so weit, Emigranten wie seinem ehemaligen Schüler und Koautor eines bekannten mathematischen Lehrbuches, Peter Thullen (1907–1996), vorzuwerfen, sie hätten damals den leichteren Weg gewählt:

> So wurde uns die Würde des Amtes und die Würde des Menschen genommen. Waren nicht die glücklicher, die ihre Stelle niederlegen mussten? Das wurde häufiger gesagt, und doch haben die meisten Kollegen ihr Amt verteidigt. Natürlich macht es Eindruck, wenn man stolz erhobenen Hauptes einfach geht, bevor die braune Bürokratie einen stellt und vielleicht dann heraussetzt. In gewissen Fällen kostet solches freiwillige Verschwinden nicht viel, z. B. wenn man ein junger erfolgreicher Mathematiker ist. Aber mir war diese Ungebundenheit unheimlich. (Behnke 1978, 129).

Thullen hatte aus katholischem Glauben die Nazis verabscheut und war „freiwillig" in die Emigration nach Südamerika gegangen. Dort fand er nie wieder Gelegenheit zu mathematischer Forschung. Thullen hat diese Stellungnahme, in der sogar das Naziwort vom „ungebundenen Intellekt" nachhallt, seinem verehrten Lehrer Behnke sehr übel genommen. Empörende Äußerungen wie diese von Behnke waren typisch für viele andere unsensible und selbstgerechte Bemerkungen von deutschen Mathematikern nach dem Kriege.

Auch Bieberbach zeigte wenig Reue. Unerreicht in seiner Demagogie und zugleich eine schlagende Widerlegung der Annahme, dass rationale Mathematiker auch in politischen Dingen zumindest Grundregeln logischen Denkens befolgen, ist ein Zitat aus einem Brief von Bieberbach von 1946 an seinen Kollegen Konrad Knopp, der Bieberbachs „Deutsche Mathematik" intern kritisiert hatte:

> Ich bestreite aber, dass die von mir entwickelte Auffassung eine Leistung für den NS darstellt. Denn die ironische Ablehnung, die diese Arbeiten im In- und Ausland fanden ist gewiss kein Aktivposten für den NS, dem man sie nur zu bereitwillig in die Schuhe schob.[17]

Etwa um dieselbe Zeit äußerte sich ein Benachteiligter des NS-Regimes, Kurt Reidemeister, über die „Entwöhnung von der Wahrheit" infolge der Herrschaft

[17] Bieberbach an Konrad Knopp, Oberaudorf 13. Juni 1946, Nachlass Ludwig Bieberbach, Deutsches Museum München.

der Nazi-Ideologie, und dachte dabei sicherlich auch an seine eigene verbale Anpassung im Jahre 1933, als seine berufliche Existenz bedroht war:

> Die allgemeine Folge der Weltanschauungspflege war, dass man sich allmählich der Wahrheit entwöhnte. Die Menschen im nationalsozialistischen Staat waren weit davon entfernt, die staatlich sanktionierte Lehre, die auf so vielen Schulungskursen gelehrt, in so vielen Schriften vorgetragen wurde, für bare Münze zu nehmen. Aber sie fanden sich bereit, sie einzukassieren und in das Kleingeld des eigenen Geredes umzuwechseln. (Reidemeister 1947, 35/36)

Vielleicht haben wir zeigen können, dass das Regime mehr forderte und mehr bekam als verbale Anpassung, dass mehr als persönliche Schicksale und individuelles Überleben auf dem Spiel standen, nämlich Fachinteressen und politische Vorurteile gesellschaftlicher Gruppen (in unserem Falle von Mathematikern), aber auch, dass nicht alle dieser Interessen und Vorurteile denen der NS-Verbrecher zuwiderliefen.

Damit wäre der Bogen geschlagen zur problematischen gesellschaftlichen Funktion der Grundlagenwissenschaften und ihrer Vertreter in der modernen Gesellschaft, zum ewigen Problem individueller und korporativer Verantwortung für die Zukunft der Menschheit.

Literatur

Behnke, Heinrich (1978): Semesterberichte. Ein Leben an deutschen Universitäten im Wandel der Zeit, Göttingen: Vandenhoeck & Ruprecht.

Bieberbach, Ludwig (1934): Persönlichkeitsstruktur und mathematisches Schaffen, Unterrichtsblätter für Mathematik und Naturwissenschaften 40, 23–43.

Blaschke, Wilhelm (1941): Mauro Picone: Appunti di Analisi Superiore (Review), in Jahresbericht der Deutschen Mathematiker-Vereinigung 51, II, 44–45.

Epple, Moritz (1999): Die Entstehung der Knotentheorie. Kontexte und Konstruktionen einer modernen mathematischen Theorie, Braunschweig: Vieweg.

Hollings, Christopher D.; Siegmund-Schultze, Reinhard (2020): Meeting under the Integral Sign? The Oslo Congress of Mathematicians on the Eve of the Second World War, Providence: American Mathematical Society.

Hopf, Eberhard (1937): Ergodentheorie, Berlin: Springer.

Lietzmann, Walther (1935): Die geistige Haltung des Mathematikers, Vererbung oder Erziehung?, Zeitschrift für mathematisch-naturwissenschaftlichen Unterricht 66, 361–365.

Mehrtens, Herbert (1986): Angewandte Mathematik und Anwendungen der Mathematik im nationalsozialistischen Deutschland, Geschichte und Gesellschaft 13, 31–47.

Mehrtens, Herbert (1987): Ludwig Bieberbach and 'Deutsche Mathematik'. In: Phillips, E. R. (Hrsg.): Studies in the History of Mathematics, New York 1987, 195–241.

Neuenschwander, Erwin (1996): Felix Hausdorffs letzte Lebensjahre nach Dokumenten aus dem Bessel-Hagen-Nachlass, in Edgar Brieskorn (Hrg.), Felix Hausdorff zum Gedächtnis, Band I: Aspekte seines Werkes, Braunschweig: Vieweg 1996, 25–70.

Perron, Oskar (1939): Irrationalzahlen, Berlin: de Gruyter, 2. Auflage.

Pieper-Seier, Irene (2008): Ruth Moufang: Mathematikerin zwischen Universität und Industrie, in R. Tobies (Hrg.): „Aller Männerkultur zum Trotz" – Frauen in Mathematik, Naturwissenschaften und Technik, Frankfurt / New York: Campus, 177–204.

Reidemeister, Kurt (1947): Über Freiheit und Wahrheit, Berlin: Carl Habel.

Remmert, Volker (2004): Die Deutsche Mathematiker-Vereinigung im „Dritten Reich"; Mitteilungen der Deutschen Mathematiker-Vereinigung 12, pp.159–177, 223–245.

Schappacher, Norbert und Scholz, Erhard (Hrg. 1992): Oswald Teichmüller - Leben und Werk; Jahresbericht DMV 94, –9.

Segal, Sanford L. (1980): Helmut Hasse in 1934, Historia Mathematica 7, 4–6.

Segal, Sanford L. (2003): Mathematicians under the Nazis, Princeton and Oxford: Princeton University Press.

Siegmund-Schultze, Reinhard (1998): Mathematiker auf der Flucht vor Hitler. Quellen und Studien zur Emigration einer Wissenschaft, Braunschweig/Wiesbaden: Vieweg.

Siegmund-Schultze, Reinhard (2002): The Effects of Nazi Rule on the International Participation of German Mathematicians: An Overview and Two Case Studies; in K.H.Parshall and A.Rice, eds., Mathematics Unbound: The Evolution of an International Mathematical Research Community, 1800–1945; Providence and London: American Mathematical Society and London Mathematical Society, 335–357.

Siegmund-Schultze, Reinhard (2004): Helmut Grunsky (1904–1986) in the Third Reich: a mathematician torn between conformity and dissent; in Oliver Roth and Stephan Ruscheweyh (eds.), Helmut Grunsky: Collected Papers; Lemgo, Heldermann, pp.XXXI-L.

Siegmund-Schultze, Reinhard (2008): Nazi Rule and Teaching of Mathematics in the Third Reich, Particularly School Mathematics; in E. Barbin, N. Stehliková, and C. Tzanakis, History and Epistemology in Mathematics Education: Proceedings of the 5th European Summer University, Prague 2007; Plzen: Vydavatelský Servis 2008, pp.863–875.

Siegmund-Schultze, Reinhard (2009a): Mathematicians fleeing from Nazi Germany: Individual Fates and Global Impact, Princeton and Oxford: Princeton University Press. (Extended translation of Siegmund-Schultze (1998)).

Siegmund-Schultze, Reinhard (2009b): The historiography and history of mathematics in the Third Reich; in: Eleanor Robson and Jacqueline Stedall (eds.), The Oxford Handbook of the History of Mathematics, Oxford University Press 2009, pp. 853–879.

Siegmund-Schultze, Reinhard (2011): Landau und Schur: Dokumente einer Freundschaft bis in den Tod in unmenschlicher Zeit; Mitteilungen der Deutschen Mathematiker-Vereinigung 19, 164–173. (Englisch: Landau and Schur – Documents of a Friendship until Death in an Age of Inhumanity; EMS Newsletter no. 84 (2012), 31–36).

Mathematics Textbooks as Sources of Research: A Look at Political and Cultural Aspects

Circe Mary Silva da Silva

Introduction

Every pedagogical action is an imposition, and for it to happen, it is necessary to establish a communication relationship (Bourdieu and Passeron 1992). The main cultural agents for this are, at school, teachers and textbooks. They become an instrument of socialization and acculturation of new generations (Silva 2019). Textbooks as important allies in understanding the mentality of the time—*Zeitgeist*—of a people can be a mirror of society (Coeckelberhs 1978), but, more than being a mirror, they can also transform reality, providing an idealized image of that society that serves to educate the new generations (Choppin 2004).

In textbooks we find an important function that is cultural and ideological, which seeks to reflect the norms, habits, morals and ideals valid in a given period, both in textual and imagery modality. They thus become essential vectors of the language, knowledge and values of the dominant class (Choppin 2004).

Visual content works in approximately the same way as textual content. For Burke (2004) images, as well as texts, are "historical evidence", witnesses of the dominant thought in certain periods of the past; through the analysis of such images in textbooks, it is possible to glimpse an overview of the school culture at that time, even more if one considers that the images can also serve to accentuate ideological aspects, etc. The investigative question of this work is: what can the analysis of mathematics textbooks teach us about content, teaching methods, textual or imagery and ideological documents in a given geographic, cultural, social and historical context?

C. M. S. da Silva (✉)
Universidade Federal de Pelotas, Pelotas, Brazil
e-mail: cmdynnikov@gmail.com

A. Verdugo Rohrer and J. Zender (eds.), *History of Mathematics and Its Contexts*, Trends in the History of Science, https://doi.org/10.1007/978-3-031-60828-5_15

The interpretation of ideology involves, in Thompson's view (1995, p. 38), the socio-historical analysis of power relations, thus being intimately connected with a "critical reflection on power relations and domination". Following Thompson's methodological line, symbolic forms are identified in Mathematics textbooks for the early years of schooling.

By putting discourse analysis into practice as a methodology, we understand that language is, in itself, a discursive event; thus, it is not limited in itself, as an abstract system of symbols, but establishes relationships with history and ideology. According to Orlandi (2003, p. 15), "discourse is [...] the word in motion", an idea that metaphorically can be understood as dealing with a subject from various points of view. When interpreting a text, we are already attesting to the presence of ideology. Discursive practices can be of different natures, among which are, for example, images and texts. In the analyzed textbooks, it was exactly the images and texts that became objects of our interest.

Investigations in Brazilian Textbooks of Germanic Immigrants' Authors

Below, I present some results of research that I have carried out in the last ten years with the main focus on textbooks. I am indebted to Gert Schubring for the first teachings on textbook research, while still a doctoral student at the Institute for Mathematical Didactics at the University of Bielefeld. I was unaware of the theories that underlie research in the history of mathematics education, and Schubring introduced me to this fascinating area that I have been working on for decades. The investigation I carried out on mathematics textbooks for the early years, written by Germanic authors, in Brazil, covered the nineteenth and twentieth centuries.

The choice of authors George Augusto Büchler (1884–1962), a German immigrant, and Karl Sölter (no information) was due to the fact that they brought cultural practices from another educational reality to the Brazilian context. This makes it possible to understand how the processes of transmitting ideas occur, which paths are taken from the appropriation of certain knowledge by an individual, in a certain place, to the transfer of this knowledge to another place, to a new geographic, social and cultural reality.

Arithmética Elementar (1919) is a three-volume work by George Augusto Büchler. In this article, only the first volume was analyzed: the one dedicated to the first school year, precisely because it was the one in which the images were most widely used. In Arithmetic for the first school year [*Rechenübungenfür das erste Schuljahr*], written by Karl Sölter, pictures practically replace text.

The contents of Georg Büchler's book cover: counting; written and spoken numbering up to 1000; arithmetic operations of addition, subtraction, multiplication and division, ordinal numbers; metric system; monetary system; and divisibility. At the same time, it brings a moral, disciplining connotation, which is visible throughout the text, since, in parallel with the teaching of mathematical content, the author inserts questions regarding moral norms and good Germanic

Fig. 1 Warehouse in 1919. *Source* Upon investigating on the matter and to the best of our knowledge, this/these image/s are not under copyright, so we are reproducing the originals, which originally appeared here: (Büchler 1919, p. 24). We reproduce these images in the spirit of science dissemination and cooperating with the community, giving respectful visibility to this material

customs: valuing work; respect and obedience to parents; follow the order; economy; cleaning; follow the divine commandments; and knowing how to listen, among others (Silva 2017). In addition, the author incorporates in the illustrations some aspects of Brazilian culture such as guava sweets (Fig. 1), palm trees, wagons, birds and Brazilian customs. Such precepts at the time were part of what parents and schools wanted young people to comply with as children, students and citizens.

During the government of President Getúlio Vargas (1937–1942), in Brazil, a nationalization campaign was implemented in order to prevent German schools from using German as a school language. This was banned from schools, in some places violently, and all German-language books were recalled. In a transitional stage, anticipating the government's retaliation, in order to reduce this tension between the two cultures, an alternative found by some German authors for these schools was to introduce images of patriotic symbols in their schoolbooks, such as the national flag, the map of Brazil, the currency, the Indian and the tropical fruits, even keeping the German language.

One of the primary functions of images in school books, especially those intended for the early years, is to associate an image with a mathematical concept that one wants to introduce. Thus, the numbers are, in general, presented through figures as objects of the child's daily life or the reality of the school and its surroundings. In this way, the author Sölter (1932), brought images like the ones in Fig. 2, in which he shows the classroom environment. In order to introduce the number 1, he inserts a map of Brazil, a national flag, a key, a child, a teacher, and

Fig. 2 Brazilian symbolic elements. *Source* Upon investigating on the matter and to the best of our knowledge, this/these image/s are not under copyright, so we are reproducing the originals, which originally appeared here: (Sölter, Karl. Rechenübungen für das erste Schuljahr. Ijuí: Serrana Bookshop, 1932). We reproduce these images in the spirit of science dissemination and cooperating with the community, giving respectful visibility to this material

beyond that, other elements of the Brazilian reality. Thus, images can also serve to accentuate ideological aspects that one wishes to instill in young people.

In addition, textbooks can serve to show social tensions. In Brazil, the abolition of black slavery took place in 1888, but the racial problem persisted for many years.

Due to the process of devaluation of the black person, afro-descendants tend to introject the dominant view of the white world, seen as superior. As a result, they tend to devalue the black world or assume that being Afro-descendants is insignificant for their lives (Ferreira 2002, p. 75).

In the 1930s, strong discrimination against blacks and Indians persisted in Brazil, and this is found even in textbooks, camouflaged in an apparently naive little story. The first edition, in 1933, for the first school year by Nast and Tochtrop, entitled *Mein Rechnenbuch* was translated into Portuguese *Meu Livro de Contas* (1961), brings a math story with the theme of subtraction involving 10 black children (Fig. 3).

The story told is as follows:

The Story of 10 Little Black Children

The story told is as follows:
The Story of 10 Little Black Children

Fig. 3 Chap.: History of the 10 black children. *Source* Upon investigating on the matter and to the best of our knowledge, this/these image/s are not under copyright, so we are reproducing the originals, which originally appeared here: (Nast and Tochtrop 1961, p. 21). We reproduce these images in the spirit of science dissemination and cooperating with the community, giving respectful visibility to this material

Once upon a time, there were 10 little black children
They were little brothers. 1, 2, 3, 4, 5, 6, 7, 8, 9, plus 1 equal 10.
It's fine. Now listen!
One day they went for a walk.
Do you know what the oldest did?
He hanged himself from a kale plant. Only nine of them remain. There were then 9 little black children. They were all little brothers. 1, 2, 3, 4, 5, 6, 7, 8 and 9. For twelve, three missing.
One day they took another walk.
Do you know what the skinniest one did?
He ate an ear of corn. He died of a stomachache.

There were 8 black children. They were little brothers. 1, 2, 3, 4, 5, 6, 7 and 8
Do you know what they did?
They went to play ball and the fattest one's heart stopped.
So there were 7 little black children. They were brothers. 1, 2, 3, 4, 5, 6 and 7
Do you know where they went?
They went to the witch's house. She was mean and she put one in the boiler.
There were 6 black children. They were brothers. 1, 2, 3, 4, 5, 6
Do you know what they did?
They went to the lake to swim. One died of shortness of breath.

There were then 5 little black children. They were brothers. 1, 2, 3, 4 and 5

Do you know what they did? They had a lot of beer.

For the smallest one, beer was poison.
But he wasn't convinced, he drank too much and died. 1, 2, 3 and 4

There were then 4 little black children. They were brothers. Do you know what they did?
They played soldier and thief and one fell dead to the ground. There were then 3 little black children.

They were brothers. 1, 2 and 3
Do you know what they did?
They wanted to make porridge.

> The hungriest one went to see: "I think..." and wham, he fell into the pan. There were then 2 little black children. They were little brothers. 1 and 1 are 2 Do you know what they drank?
>
> They drank wine and more wine.
> One died and the other one was left alone.
> The last little black boy didn't want to be alone, poor thing. He married a black girl, she had curly hair.
>
> They had ten little children. See how grown -up they are. (Nast and Tochtrop 1961, p. 21)

Both the textual story and the image are shocking, because every time the black boys go for a walk, one of them dies (the deaths are tragic: one hangs himself, another is attacked by a witch, another dies drunk, another drowns, etc.). In the end, only one black boy is left, and this one, when he meets a girl, marries her and generates 10 more black people. The purpose of the story is to present the meaning of subtraction operation. No mitigating factor can be used to disseminate for more than thirty years, in books in German and Portuguese, a story with such a racial connotation. Socially, we still have problems with racism in the country, but the current textbooks have already overcome them.

Investigations into Mathematics Textbooks of the Soviet Period

Since the school aspires to the socialization of young people, one way of transmitting culture is in school benches through the direct action of teachers and school books. Mainly in dictatorial regimes, it is seen that intellectual and political leaders exert their direct influence, prescribing in a certain way the ideology that must be conveyed through school books. The research carried out on mathematics textbooks from the Soviet era covered the 11 works and 6 authors, in the period from 1916 to 1960.

It is worth noting that in the case of the USSR, the Communist Committee acted directly in the production and dissemination of textbooks, highlighting the role of the teacher and the function of books: the teacher should follow the precepts in teaching children—highlight the fundamentals of science, of conscious obedience education and the creation of the communist vision of teaching and work. An orientation for the writing of books was published, in 1932, in a periodical—*The textbook for the Bolshevik: journal* [*За большевистски˘и учебник: журнал*][1] whose objective was to provide norms for the authors of the books regarding the contents and, naturally, the ideology that should be included in these works (Silva and Dynnikov 2014).

The analysis carried out in textbooks before the Soviet revolution until 1960 allowed us to conclude that before 1917 ideological formation in schools explored values of monarchy, religion and individual work. After the revolution, the members responsible for the ideology implanted by the communist party-imposed ideas on the new generations through printed matter, especially textbooks, praising the supremacy of the new political regime, the one that should solve the country's economic and social problems (Silva and Dynnikov 2014).

Some authors of textbooks for elementary education emerged in the 1930s, such as Mathematics teacher Natalia Sergeeva Popova (1885–1970). Popova dedicated a large part of her life to the production of Mathematics textbooks (which had a large circulation of copies and several reprints): she published a collection for the initial grades (from 1st to 4th grades). The 4th edition of the Arithmetic textbook for the first year had a circulation of 800,000 copies and was approved by the Ministry of Education of the USSR, and it is likely that it was published for the first time in the early 1930s. The contents of the book for the first school year basically involved the concepts of numbering and elementary arithmetic operations (Silva 2019).

Popova's mathematics textbooks from 1933 to 1950 reveal that the author strictly followed the party's prescription to include communist ideology in books. For example, in this investigation, an attempt was made to answer the question: "What traits of communist ideology are present in statements of math activities and in images proposed in books by Natalia Popova?".

An example of this is the author's own open statement, in the preface to the first-year book, in 1937, which says: "The book provides a system of elementary knowledge in mathematics for the first year of studies. For this, the tasks of local material were added aiming at the construction of socialism" (Silva 2019). Therefore, it became necessary to carry out an in-depth study of the official documents of the communist party in which such indicators could be extracted.

To understand the most used words at the time that were loaded with ideological values, some clarification of their meanings is needed.

Oktyabryata—children following the ideology of the Communist Party, carrying stars on their chests with a portrait of Lenin (1st to the 3rd school year). The name originates from the first members born at the time of the October Revolution and emerged in 1923. This it was considered a preparatory stage for entry into the category of the "pioneers".

Pioneer—pioneer is a degree subsequent to oktyabryata on the ideological scale (it was made up of students aged 10–15 years, starting from the 3rd school year, in which students were admitted to best of oktyabryata).

Komsomolet—rank later than that of "Pioneer" and earlier than that of Communist Party member (it was made up of young people aged between 16 and 28 years old).

Komunist—communist, the top of the political hierarchy.

Kolkhozes—community associations formed to exploit the means of production, which were collectively managed, the resulting profits being equally distributed and the "good" maintained as the property of all.

Sovkhozes—state institutions that employed their employees with remuneration regulated.

Politkrujok—political seminars.

Obligacia—loan securities to the government.

Thus, the following indicators were identified: **politicization** (expressed mainly in the occurrence of the terms oktyabryata; pioner, komsomolet, komunist, kolkhozes, sovkhozes, politkrujok, obligacia), in addition to symbolic elements such as the hammer and sickle, nails, planes, stars, red scarves, among others; **work and social action** (cooperative, work in kolkhoz, komsomolets, comparison of work in the USSR and other countries); **industrialization** (tractors, planes, automobiles, machinery in general); **armament and red army** (cartridges, soldiers, shots, anti-aircraft defense); and **atheism** (Union of young atheists).

In order to illustrate, I present some statements taken from Popovas' book.

> In the classroom there are 17 oktyabriata. Of them 8 are boys. How many in the room are oktyabriata girls? (Popova 1937, 32).

> Oktyabryata and pioneers together collected 14 rubles for the construction of an airplane. Pioneers collected 9 rubles. How much money was collected by the oktyabriata? (Popova 1937, 35).

> Soviet airmen departed Moscow for the US on July 12, 1937, at 3:21 am and crossed the North Pole on July 13, at 3:14 am. The entire flight lasted 62 hours and 17 minutes. How long did Soviet airmen fly to the North Pole and when did they land in the US? (Popova 1941, p. 93).

> Under **Soviet power**, a **worker** works from 7 am to 3 pm. He has an hour for lunch. How many hours does the worker work now? (Popova 1937, p. 55). (The bold are mine.)

> Kolkhos (collective farm) sent 19 bags of grain to the reception point. In one wagon, there were 4 bags, and in each remaining wagon, 5 bags could fit. How many carts did the kolkhoz send to the reception point? (Popova 1937, p. 40).

> Three Red Army soldiers had 36 cartridges for shooting practice. First, they used 10 cartridges, the second time 11 and the third time 12. How many cartridges are left over? (Popova 1937, p. 59).

> Calculate how many classmates in your class are members of the **Union of Young Atheists** (ЮВБ). Calculate the boys and girls separately (Popova 1933, p. 23). (The bold are mine.)

The ideology captured in the statements of problems and images reinforces the conviction that textbooks are not neutral, and they operate in a relationship of power and domination. In the research carried out with Popova's books, we also

Fig. 4 Nails and hammer, screws and nuts, tractors. *Source* Upon investigating on the matter and to the best of our knowledge, this/these image/s are not under copyright, so we are reproducing the originals, which originally appeared here: (Popova, Natalia. Part I. For 1st grade. 1937, 3, 4). We reproduce these images in the spirit of science dissemination and cooperating with the community, giving respectful visibility to this material

sought to identify which traits of communist ideology are present in the images proposed in this author's books. In the book for the first year, the visual appeal is intense and the images serve to reinforce and propagate communism, using typical symbolism.

In addition to the representations of these objects (Fig. 4), characters appear—such as workers carrying construction mass, farmers plowing the land—tractors and planes, which reflect industrialization and changes in agricultural work. Not only the texts, but also the illustrations can reinforce the ideological content that the authors intend to convey.

Conclusions

Schubring's thoughts were the starting point for questioning textbooks resulting in an in-depth study of the social, political and economic context of the place where these manuals were produced, which allow the creation of indicators for the study of textbooks.

In the case of research on Brazilian books written by Germanic authors, a broad study of the authors and place of production of such works was necessary. When studying the books of the Soviet period, the same happened, it was necessary to know the assumptions of the political regime, the economic and social reality of each period covered.

The book taken as a document is the product of a particular society that produced it, according to relations of forces that held power, as Le Goff (1990) points out. Both in mathematics books produced in Brazil and in those published in the Soviet Union, we find the prevailing school culture at the time. We verified that the school book always has an intentionality, which can be read in its texts and images, and it is a representative mirror of the imaginary of a collectivity. In the case of the analyzed mathematics books, they clearly present ideologies to be imposed on students in the early years.

We found that in addition to specific mathematical knowledge and teaching methods, the school book always has an intentionality, which can be read between its textual lines and images, and it is a representative mirror of the imagination of a collectivity.

References

Bloch, Marc (2001). Apologia da história ou o ofício de historiador. Rio de Janeiro: Zahar.

Bourdieu, Pierre; Passeron, Jean-Claude (1992). A reprodução: elementos para uma teoria do sistema de ensino. Rio de Janeiro: Francisco Alves.

Büchler, Georg (1919). Arithmética Elementar Livro I. São Paulo, Rio de Janeiro: Weiszflog Irmãos.

Burke, Peter (2004). Testemunha ocular. Bauru: EDUSC.

Coeckelberhs, Hilde (1978). Das Schulbuch als Quelle der Geschichtsforschung. In: Internationales Jahrbuch für Gechichts-und Geographie-Unterricht. Berghahn Books. (18), pp. 7–29.

Choppin, Alain (2004). História dos livros e das edições didáticas: sobre o estado da arte. In: Educação e Pesquisa, (30), n. 3, pp. 549–566.

Ferreira, Ricardo Franklin (2002). O brasileiro, o racismo silencioso e a emancipação do afro-descendente. Psicologia & Sociedade, 14(1), pp. 69–86.

Le Goff, Jacques (1996). História e Memória. Campinas: Editora da Unicamp.

Nast, Wilhelm; Tochtrop, Leonhard (1961). Meu livro de contas. 1o vol. 12a ed. São Leopoldo: Editora Rotermund.

Orlandi, Eni Puccinelli (2003). Análise de Discurso: princípios e procedimentos. Campinas: Pontes.

Popova, Natalia Sergueevna (1933). Учебник арифметики для начально˘и школы. Часть II. Для второго года обучения. Издание второе. Москва, Ленинград: Государственное учебно-педагогическое издательство. https://vk.com/topic-142811591_37159703

Popova, Natalia Sergueevna (1937). Учебник арифметики для начально˘и школы. Часть I. Для 1-го класса. Издание шестое. Москва, Ленинград: Государственное учебно-педагогическое издательство, <hxxxttps://sovie https://sovietime.ru/matematika/uchebnik-arifmetiki-1-j-klass-dlya-nachalnoj-shkoly-1937-god-skachat-sovetskij-uchebnik>.

Popova, Natalia Sergueevna (1941). Попова, Наталья Сергеевна. Сборник арифметических задач и упражнени˘и. Часть III. Для 3-го класса начально˘и школы. Издание девятое. Ленинград: Государственное учебнопедагогическое издательство Наркомпроса РСФСР – Ленинградское отделение. https://vk.com/wall94378522?owners_only=1&q=%D0%BF%D0%BE%D0%BF%D0%BE%D0%B2%D0%B0.

Silva, Circe Mary Silva (2017). Imagens nos livros didáticos de matemática: Georg Augusto Büchler e Karl Solter. In: Acta Scientiarum Education. Maringá, (39), n.1, pp. 55–65.

Silva, Circe Mary Silva (2019). Cantinho Vermelho e outros Símbolos: como se ensinou matemática na URSS. In: Bolema, Rio Claro (SP), (33), n. 65, pp. 1087–1108 https://www.scielo.br/j/bolema/a/Z3zJmq9DG5Cb7BYhGgvgdmJ/?format=pdf&lang=pt.

Silva, Circe Mary Silva; Dynnikov, Vladimir, Ivanovitch (2014). Ideologia em problemas matemáticos nos livros didáticos soviéticos da pré-revolução até 1960. Revista Brasileira de Educação. (19), n. 56, jan.-mar. , pp. 201- 258.

Sölter, Karl (1932). Rechenübungen für das erste Schuljahr. Ijuí: Livraria Serrana.

Thompson, John (1995). Ideologia e cultura moderna: teoria social crítica na era dos meios de comunicação de massa. Petrópolis: Vozes.

"a doctissimo matheseos candidato Weierstrass diligenter et accurate expositam"—The First Publication of Karl Weierstraß, on Analytic Factorials

Peter Ullrich

In the last third of the nineteenth century, Karl Weierstraß (1815–1897) was one of the most respected mathematicians in the world. His lectures at the University of Berlin not only attracted students from Germany, but many also came from abroad, even if they had already received their doctorates at home. From this perspective, the academic career of Weierstraß seemed to have gotten off to a bad start, as he had spent more than a decade as school teacher away from the centers of mathematical research and only became a professor in 1856. It was therefore discussed whether something had gone wrong at the end of his studies at Münster in 1841, in particular why the treatise on elliptic functions which he had written for his teacher's exam and which his academic teacher Christoph Gudermann (1798–1851) held in high esteem was not published in the early 1840s, but only within (Weierstraß 1856b) in an abbreviated form and in his collected works in full length (Weierstraß 1894). Still in the second half of the twentieth century, the chairman of the examination commission for Weierstraß was accused of having canceled the most positive sentences from Gudermann's report on that treatise.

It was not until 1989 that Gert Schubring was able to shed new light on this matter (Schubring 1989): On the basis of his archive research, he discovered numerous surprising facts about the studies and the exam of Weierstraß at Münster. In particular, Weierstraß had indeed achieved excellent results in mathematics. However, his oral examination in the natural sciences showed massive deficiencies, with the result that he only received his teacher's examination by a special decision of the Berlin ministry which was responsible for educational affairs (Schubring

P. Ullrich (✉)
Universität Koblenz, Koblenz, Germany
e-mail: ullrich@uni-koblenz.de

A. Verdugo Rohrer and J. Zender (eds.), *History of Mathematics and Its Contexts*, Trends in the History of Science, https://doi.org/10.1007/978-3-031-60828-5_16

1989, pp. 15–18). In addition, Schubring reasoned that Christian Friedrich Wagner (1782–1863), the chairman of the examination commission, had exactly used those sentences from Gudermann's report on Weierstraß for the official certificate on the examination which promoted the career of Weierstraß in the school system rather than hindered it (Schubring 1989, pp. 14–15). (This confirms a story already reported by Wilhelm Killing (1847–1923) (Killing 1897, p. 10).)

In this way, Schubring's findings and his analysis of the documents provided answers to questions about the beginning of the career of Weierstraß, in a partially unexpected way. But they also raised new questions or exacerbated old ones: Schubring poses this "eigentliche, offene Frage" (= "actual, open question") (Schubring 1989, p. 21) by discussing what could have been done to further the academic career of Weierstraß after his teacher's exam (Schubring 1989, pp. 21–22): The treatise (Weierstraß 1894) could have been published directly or submitted as a doctoral thesis to an academic institution that could already award doctoral degrees as early as in 1841 or, only three years later, to the academic institution at Münster itself which had then received this right. Since nothing of this happened, one can get the impression that Gudermann or Weierstraß was no longer convinced of the latter's abilities, perhaps that the failure in the oral examination on the natural sciences shook the self-confidence of Weierstraß in general, perhaps that Gudermann withdrew his support for Weierstraß after this event which had embarrassed him before the Münster faculty.

Some years after (Schubring 1989), however, new evidence turned up (Ullrich 2007): In November 1844, just two years after the leave of Weierstraß from Münster, the first doctoral thesis in mathematics was accepted at the academic institution there, with Gudermann as supervisor. And in this thesis, Weierstraß was praised with the words cited in the first part of the title of the present contribution. Translated into English from (Féaux 1844, p. 8), the full sentence reads:

> From this it can be seen that our function [which is denoted by the symbol $\Gamma()$] is innermost connected with the analytic factorials, which fact, which was recently carefully and accurately explained by the most learned mathematics candidate Weierstrass (Weierstrass, Ueber die Facultäten. Abhandlung im Programm des Gymnasiums zu Deutsch-Krone, Schuljahr 1842/43), we will, however, pass over in silence at this point.

So the doctoral candidate, Bernhard Féaux (1821–1879), referred to the first publication of Weierstraß (Weierstraß 1843) which had been printed in the program of the "Königliche Progymnasium in Deutsch Crone" in West Prussia in 1843 where he worked as teacher. (For school programs in mathematics in general, see Schubring's publications (Schubring 1986, 2021.)

But the above quote also says that Féaux did not need the results published by Weierstraß in his thesis (Féaux 1844) so that Féaux's reference to Weierstraß is not for mathematical but more or less for formal reasons. Furthermore, Féaux was rather meticulous when giving quotes elsewhere in his thesis. But he made the flaw to call the school a "Gymnasium" and not a "Progymnasium". All this gives the impression that it had not been his own idea to quote Weierstraß, but that someone else had advised him to do so. And the obvious candidate for this was Gudermann so that the above quote indicates that he did not only have a positive evaluation of Weierstraß but also expressed it more or less publicly.

Additionally, from the text of (Weierstraß 1843, p. 90) one learns that Weierstraß had had a meeting in person with Leopold August Crelle (1780–1855) who had encouraged him to publish the considerations of (Weierstraß 1843).

These findings again raise new questions:

How did Weierstraß come to the topic of (Weierstraß 1843), analytic factorials? This surely belongs to analysis but has nothing to do with elliptic, hyperelliptic, or Abelian functions and their foundations which are commonly known as research areas of Weierstraß during the 1840s.

Why did Weierstraß publish (Weierstraß 1843) only within a school program and not in the "Journal für die reine und angewandte Mathematik" which was edited by Crelle and which was the best possibility to publish a mathematical text in Germany at that time? Years later, Crelle commented on the publication of an extended version of (Weierstraß 1843) in the "Journal" with very positive words which confirm his interest in the topic (Weierstraß 1856a, p. 1, footnote *).

In the present contribution, we will try to get near to answers of these questions not by presenting new archival findings but by a study on the history of the theory of analytic factorials. If one combines this with details from Gudermann's career, which he might have thought of as a blueprint for that of Weierstraß, one comes to the hypothesis that perhaps Gudermann, perhaps Weierstraß himself had deliberately chosen this topic of research in order to raise the interest of Crelle in order to get a publication for Weierstraß in the "Journal", if not more. More speculative is the conjecture why this did not take place. But there are some clues at a person that has also been a subject of Gert Schubring's research, cf., e.g., (Schubring 1983).

Weierstraß as a School Teacher and the Late Publication of His Treatise for the Teacher's Exam

Detailed representations on the life of Karl Weierstraß before his call to a professorship at Berlin are given in (Elstrodt 2015) (Mittag–Leffler 1923) with further details in (Schubring 1989) and (Ullrich 2007). Therefore, we will give only a rough sketch here in order to put the following into context:

Weierstraß studied mathematics at the "Akademische Lehranstalt in Münster" with Gudermann from the winter term 1838/39 to the winter term 1839/40 (Schubring 1989, pp. 19–20, 29). From summer 1840 onwards he prepared for and also took his exam as a teacher at a Gymnasium which he passed on the basis of a special decision of the ministry dated from June 18, 1841 (Schubring 1989, p. 28). Then he could take his preparatory year at the Gymnasium Paulinum in Münster. After this, he left Münster and took up a position as teacher at the "Königliche Progymnasium in Deutsch Crone" in West Prussia (today: Wałcz in Poland) in autumn 1842. From there he changed to the "Königliche Katholische

Gymnasium zu Braunsberg" in East Prussia (today: Braniewo in Poland) in 1848. During the summer vacation of 1853 he travelled home to Westphalia, gained access to Gudermann's report on his treatise for the teacher's exam (Weierstraß 1894), completed his manuscript (Weierstraß 1854) on September 11, 1853 and submitted it to the "Journal für die reine und angewandte Mathematik". As a consequence of his mathematical results, he received an honorary doctorate from the University of Königsberg and was promoted to "Oberlehrer" (= "senior teacher") and granted a sabbatical year for research purposes from autumn 1855 onwards by the school authorities. Finally, he became professor at the "Gewerbeinstitut" in Berlin and extraordinary professor at the University of Berlin in 1856.

This means that Weierstraß was a school teacher for gross fourteen years. From this sum one has to subtract both the sabbatical year 1855/1856 and more than one year around 1851 in which he was out of duty because of health problems (Elstrodt 2015, pp. 35–36). But even the remaining net value is more than the nine years that his academic teacher Gudermann or the ten years that his later Berlin colleague Ernst Eduard Kummer (1810–1893) was in active service as school teachers.

Usually, the attitude of Weierstraß toward the status of a Gymnasium teacher is depicted as positive with evidence taken from (Killing 1897, p. 11) and (Lampe 1897, p. 38). But these sources refer to only one statement that Weierstraß made on his 80th birthday when he criticized persons who did not feel well within this status. If one reads more carefully, however, for example, translated from (Killing 1897, p. 11):

> The status of a Gymnasium teacher in itself had nothing oppressive for *Weierstraß*.

one gets the impression that he was rather unhappy with his personal fate. In fact, he complained in a letter to Paul du Bois-Reymond (1831–1889) on June 6, 1875, which was published already in 1923 (Weierstraß 1923, p. 209), here translated from the corrected version (Elstrodt 2015, p. 41):

> When I once [...] decided to pursue mathematics as my life's work [...], I initially had to pay for this decision with fourteen years of banishment in the land of the Velates and Obotrites.

(Velates and Obortrites were Slavic tribes that had lived in the area where Weierstraß worked as a teacher.) As also the following sentences in this letter support, it was not the status of a teacher per se that annoyed and burdened him but the external circumstances, in particular the isolation from the centers of mathematical research. At any rate, one learns from the above quote that in later years Weierstraß remembered his time as a school teacher rather negatively, which may have influenced his reports given on other events of this time.

In fact, it was Weierstraß himself who first mentioned in print the delayed publication of his treatise for the teacher's exam, namely as a comment to the publication (Weierstraß 1894) of this treatise in the first volume of his "Mathematische Werke" in 1894, translated from (Weierstraß 1894–1927, Vol. 1, p. 50):

> The above treatise, written in the summer of 1840, was submitted by me to the academic examination commission in *Münster* in the autumn of the same year in order to obtain the Facultas docendi. My revered teacher Gudermann […] judged it very favorably, and it should be printed. However, this was not done for reasons that I do not want to go into detail here.

Both Killing (Killing 1897, p. 9) and Heinrich Behnke (1898–1979) (Behnke 1965, p. 19) rephrase this text with the plausible modification that it had been Gudermann who had urged for the printing. Also these two professors of mathematics at Münster add that this printing would have taken place if Weierstraß could have received a doctorate for his treatise there: The former full university at Münster had been downgraded to a teaching institution in 1818 whose Philosophical Faculty could grant doctorates only from 1844 onward. Note, however, that Weierstraß does not mention the idea of a doctorate based on (Weierstraß 1894) in the text quoted above, also not in its sequel (Weierstraß 1894–1927, Vol. 1, p. 50).

Gudermann and His Academic Career

Little has been mentioned up to now on Gudermann, the academic teacher of Weierstraß. A conference contribution on his life and work will appear soon (Ullrich 2024) where some of the incorrect informations circulating about him will be corrected. At the present occasion, we will therefore only present the essentials of Gudermann's career:

Just alike Weierstraß, Gudermann was born into a non-wealthy family which made it a problem to finance academic studies. Gudermann studied only four semesters at the University of Göttingen, but not with Carl Friedrich Gauß (1777–1855), as it is sometimes claimed, but with Bernhard Friedrich Thibaut (1775–1832). After his studies, Gudermann took his exam as a teacher at a Gymnasium in Berlin in March 1822 and worked in this position at Cleve (today: Kleve) from 1823 to 1832. During this time, he presented himself as a prototype of the scholarly teacher that was the ideal of the Prussian educational system at that time, cf. (Schubring 1989, p. 15, right column): On the one hand, he was successful as a school teacher in the proper sense. On the other hand, he was a very fruitful scientific writer: In the nine years under consideration, he published 1 book, 1 contribution to a school program, and altogether 11(!) articles in Crelle's "Journal für die reine und angewandte Mathematik", several of them consisting of up to nine parts. The pay for this brilliant performance was a call to a professorship at the academic institution at Münster. In order to get this position, however, he still had to take a doctorate. To this purpose, he applied to the Philosophical Faculty of the University of Berlin which in turn asked Crelle concerning his opinion. Crelle, who held Gudermann in high esteem, answered that, because of his publications, Gudermann should receive a doctorate without performing any other exam, which in fact happened on November 29, 1832.

Under the plausible assumption that Gudermann used his personal experiences as a blueprint for a future academic career of Weierstraß, one gets to the following points, some of which may explain the problems und misunderstandings that Gudermann and Weierstraß obviously had with each other:

Luckily enough, Weierstraß fulfilled the basic requirement that he had a promising talent in teaching mathematics at school as one learns from the official documents on his teacher's exam (Schubring 1989, pp. 26–27).

The next task which Weierstraß would have to fulfill as a scholar both in the eyes of Gudermann and in those of the educational system was to publish "enough" scientific papers. Even if Gudermann had praised (Weierstraß 1894) in his report with the words, translated from (Mittag–Leffler 1923, p. 21):

> The candidate thus joins the ranks of the famous inventors on an equal footing

this might have been only one of several publications of Weierstraß to prove his qualification in Gudermann's eyes. Therefore, one can interpret the quote given above that (Weierstraß 1894) should be printed which, however, did not take place (Weierstraß 1894–1927, Vol. 1, p. 50) in the way that Gudermann made an attempt, say, to get the printing paid, but did not insist on it. (As an aside: The father of Weierstraß was surely not be willing to invest money in the printing after his son had failed both his cameralistic studies at Bonn and the oral exam in the natural sciences: There were younger siblings of Karl who deserved an education, too, including his sisters.)

Furthermore, Gudermann will have meant the word "dereinst" (= "one day") in his report on the treatise (Weierstraß 1894), translated from (Mittag–Leffler 1923, p. 22),

> that favorable circumstances may one day permit him to function as an academic lecturer

rather seriously. In fact, even if Féaux had studied at the academic institutions of Münster, Bonn, and Berlin before taking his doctorate at Münster, Gudermann wrote in the files of the doctoral examination, translated from (Ullrich 2007, p. 156):

> The examiner is of the opinion that as the candidate has matured more, he will later be able to promote science even more effectively than he has already done.

The two not yet matured young men Féaux and Weierstraß will have seen things a little bit different, but Gudermann's point of view implies that he will have seen no urgent need for a doctorate for Weierstraß, in particular since he himself had gotten his own in a way which is nowadays called "cumulative". This does not mean that Gudermann was reluctant concerning this kind of degree: In the years between 1844, when the academic institution at Münster received the right to grant doctorates, and 1851, when Gudermann died, the Philosophical Faculty at Münster granted 14 regular doctorates altogether, of which 4 Gudermann had as reviewer, and this out of a group of 5 possible reviewers. (All of Gudermann's doctoral students became Gymnasium teachers later on.)

Therefore, the question already posed in (Schubring 1989, p. 22) why Gudermann did not encourage Weierstraß to submit his treatise (Weierstraß 1894) as a doctoral dissertation at Münster after 1844 may have its answer in a communication problem: Gudermann thought that Weierstraß should serve some further years as a teacher in order to become more mature, whereas Weierstraß did not dare to ask Gudermann, perhaps because the latter had communicated him other plans. (At least, there seems to have been no formal problem coming from the fact that Weierstraß was not a student at Münster any more: Also Féaux did not matriculate at Münster anew when he came back from his studies at Bonn and Berlin in order to take a doctorate at Münster (Ullrich 2007, p. 155).)

Considering the possibility that Weierstraß would have taken a doctorate elsewhere directly after his teacher's exam (Schubring 1989, p. 22), again the argument of maturing might have spoken against this in the eyes of Gudermann. But there may also have been a concrete idea in his mind to promote Weierstraß at a full university: The University of Bonn was the next to Münster, but Gudermann had some problems with the mathematicians there, cf. (Elstrodt 2015, p. 22, footnote 22). Gudermann himself had studied at Göttingen, but his teacher Thibaut had died already in 1832. Berlin, on the other hand, could have been a good choice: Gudermann had taken his doctorate there with the help of Crelle, and the relation of both mathematicians still was very good. (Crelle had tried Gudermann to get a professorship at Bonn—which had caused the problems mentioned above.) And Crelle had the possibility to get mathematical texts printed in his "Journal", be it (Weierstraß 1894) or another, further text with a topic which interested Crelle. Both alternatives would have promoted the career of Weierstraß.

And, indeed, Weierstraß (also?) chose the second possibility with his considerations on analytic factorials (Weierstraß 1843).

The Mathematics and the History of Analytic Factorials

An "analytic factorial" (sometimes also called: "numerical factorial") is a function F which interpolates the usual factorial

$$n! := 1 \cdot 2 \cdot \ldots \cdot (n-1) \cdot n$$

for n a natural number. More precisely, F has to fulfill the functional equation

$$F(z+1) = (z+1) \cdot F(z)$$

subject to the normalization $F(0) = 1$ and is supposed to be differentiable or even analytic at least on the set of real numbers greater or equal to 0 or even on the set of complex numbers. If one admits negative integer numbers as arguments of

F, the functional equation implies that F will have simple poles in these points so that one has to admit that F has isolated singularities in this case.

Thanks to Leonhard Euler (1707–1783)—and Daniel Bernoulli (1700–1787)—one knows that this interpolation problem can be solved analytically. The most prominent analytic expression for this task is the Gamma integral (or Eulerian integral of second kind)

$$\Gamma(z) = \int_0^\infty t^{z-1} e^{-t} dt,$$

which converges if the argument z has a strictly positive real part and can analytically be continued to the Gamma function on the complete complex plane with the exception of the non-positive integer numbers where it has simple poles. Most important, it fulfills the functional equation

$$\Gamma(z+1) = z \cdot \Gamma(z)$$

and the normalization $\Gamma(1) = 1$ so that the function F with $F(z) := \Gamma(z+1)$ is an analytic factorial in the sense defined above. (Both the notation Γ and the shift of the argument by 1 are due to Adrien-Marie Legendre (1752–1833) in 1809.)

However, there is still one problem left which in historical retrospect has led to complications which will turn out to be important for our discussion: The Gamma function surely is one solution of the above problem, but it is not the only one: If F is an analytic factorial and G is an arbitrary function which is analytic on the whole complex plane, has period 1, and fulfills $G(0) = 1$, then the product function $G \cdot F$ is an analytic factorial, too. (In fact, the characterization of the Gamma function is by no means trivial: Even if an analytic function fulfills both the above functional equation and the normalization, then one needs a substantial additional condition in order to conclude that it is the Gamma function, for example, that its logarithm is convex (Theorem of Harald Bohr (1887–1951) and Johannes Mollerup (1872–1937), 1922) or that it is bounded on the strip of complex numbers with real part between 1 and 2 (Theorem of Helmut Wielandt (1910–2001), 1939).)

This last remark was not known to most of the mathematicians who dealt with the problem of analytic factorials from the end of the eighteenth to the middle of the nineteenth century. Some of them did obviously even not know about the results of Euler and Legendre mentioned above. Slightly more general than defined above they tried to generalize the product

$$a \cdot (a+\Delta) \cdot (a+2\Delta) \cdot \ldots \cdot (a+(n-1)\Delta)$$

with given starting value a and increment Δ of the factors from a natural number n of factors to a rational or even real (or complex) number of factors. For this product and its generalizations to non-natural n Crelle and, following him, Weierstraß in (Weierstraß 1843) used the symbol $(a, +\Delta)^n$ which we will use in the sequel, too, even if other authors used notations like $a^{n|\Delta}$.(In (Weierstraß 1856a), Weierstraß simply used the expression $f(u, x, y) = f(a, \Delta, n)$.)

Of course, this general setting can easily be transformed to the one with $a = 1$ and $\Delta = 1$, but most of the authors seemed to like to treat this seemingly more complicated situation. Formulated more or less explicitly, the blueprint for their research was the case $\Delta = 0$, i.e., the usual powers a^n: There one exploits the functional equation $(a^m)^n = a^{m \cdot n}$ in order to get a reasonable definition $a^{m/n} = \sqrt[n]{a^m}$ for rational exponents m/n, then uses a continuity argument for real exponents and finally tries to get an analytic expression.

Among the first mathematicians to do research on analytic factorials were Alexandre-Théophile Vandermonde (1735–1796) and Louis François Antoine Arbogast (1759–1803). But the most cited author was Christian Kramp: He was born in 1760 in Strasbourg, studied medicine there, and at first worked as a physician. Later on, he became teacher for experimental chemistry and physics at the École centrale du Departement de la Roer in Cologne, and in 1809 he was appointed full professor for mathematics at the University of Strasbourg where he died in 1826. In his book (Kramp 1798) which had astronomical and terrestrial refractions as main topic, he devoted one section to the study of what he firstly named "numerical" and later on "analytic" factorials.

Kramp presupposed that for r rational (at least) an analytic factorial $(a, +\Delta)^r$ (our notation) exists which fulfills the following five functional equations:

$$(a, +\Delta)^{q+r} = (a, +\Delta)^q \cdot (a + q\Delta, +\Delta)^r,$$

$$(ka, +k\Delta)^r = k^r \cdot (a, +\Delta)^r,$$

$$(a, +\Delta)^1 = a,$$

$$(a, +\Delta)^r = (a + (r-1)\Delta, +(-\Delta))^r,$$

$$(a, 0)^r = a^r.$$

The last equation obviously says that for $\Delta = 0$ the analytic factorial specializes to the usual powers, but also the first three ones are inspired by properties of the powers. Despite these rather strong assumptions, however, Kramp failed to give an analytic expression for his analytic factorials.

Carl Friedrich Gauß (1777–1855) will have been the most important mathematician who had taken serious notice of Kramp's considerations: In his article on the hypergeometric series (Gauß 1813), he intensively studied the function

$$\Pi(z) := \lim_{n\to\infty} \frac{1 \cdot 2 \cdot \ldots \cdot n}{(z+1) \cdot (z+2) \cdot \ldots \cdot (z+n)} n^z,$$

which in fact amounts to $\Gamma(z+1)$, used it to express the values of the hypergeometric series for $x = 1$, , and claimed the right to name it (which was directed against Legendre). More important for the situation under consideration, however, is that he showed in (Gauß 1813, Sect. 22) that for n a natural number and $\Delta \neq 0$, one has the following equation:

$$a \cdot (a + \Delta) \cdot (a + 2\Delta) \cdot \ldots \cdot (a + (n-1)\Delta) = (a, +\Delta)^n$$
$$= \Delta^n \cdot \frac{\Pi(\frac{a}{\Delta} + n - 1)}{\Pi(\frac{a}{\Delta} - 1)},$$

where the right-hand side makes sense also for general numbers n instead only natural numbers and gives an analytic expression of an analytic factorial. In the announcement of (Gauß 1813) he commented on this, translated:

> The justification type chosen by the author is generally applicable and admits just as clear a meaning as with real values of z even for imaginary ones and there is absolutely no danger of such paradoxes and contradictions, as formerly Mr. Kramp on his numeric factorials, which can be reduced to the above function but seem less suitable for inclusion in analysis, since the former depend on three quantities, whereas the latter depends on only one, but has to been considered as equally general.

One might assume that the findings of Gauß, combined with his authority, would have put the research on analytic factorials to an end. But this was not the case, because other mathematicians, which were more or less connected to the school of combinatorial analysis, wanted to deduce analytic expressions for analytic factorials directly from the definition.

Crelle, for example, even wrote a complete book called "Versuch einer allgemeinen Theorie der analytischen Facultäten [...]" (= "Attempt at a general theory of the analytic faculties [...]") (Crelle 1823). In it he presupposed only the first three of Kramp's five functional equations for $(a, +\Delta)^n$ and a priori restricted to analytic solutions of them. For these, he deduced explicit expressions from the functional equations.

Martin Ohm (1792–1872) criticized the considerations of Crelle in the second part of his book "Versuch eines vollkommen consequenten Systems der Mathematik" (= "Attempt at a perfectly consistent system of mathematics") in the respect that on the one hand the functions that Crelle had studied did not fulfill the fourth of Kramp's functional equations and that on the other hand a series expansion that Crelle had given for his analytic factorials in analogy to the binomial series does not necessarily converge and even in case of convergence has to be corrected by expressions in the sine function (Ohm 1829, p. 89). Both remarks are correct, even if the first is irrelevant since Crelle had never claimed that his analytic factorials should also fulfill the fourth functional equation. The second one is a result of the fact that the interpolation problem for the usual factorial has no unique solution. However, one should pay attention to the tone in which Ohm expressed his criticism, translated from (Ohm 1829, p. 15):

> So if, initially by Kramp implicitly, then by Crelle expressly, the general factorial is designated as a thing, which has the [...] pronounced qualities, this illogical procedure immediately takes revenge by the fact that at Kramp (Réfractions astronomiques et terrestres,

> 1799) very important contradictions emerge that he himself conceded and that are incomprehensible to him, while at Crelle ("Theorie der analytischen Facultäten", 1824) the fewest formulas are true without this being noted by the author, because the same is not, how Kramp has done, engaged in special investigations, but has only heaped general formulas onto general formulas, which therefore, as such, must for the greatest part be regarded as stillborn [...].

More details on the personality of Ohm are presented in (Schubring 1983), in particular the more than dubious circumstances under which he had achieved his academic position as full professor of mathematics at the University of Berlin.

Additionally, it should already be remarked at this place, that Weierstraß quotes the above text in full length and even more in (Weierstraß 1843, p. 88) including a somewhat risky statement made by Ohm shortly before, translated:

> It is not possible to present a concept of the *broken* factorial (faculty), for which all five numbers of § 340 still apply.

which can be read in the way that Ohm did not know about the results of Gauß concerning analytic factorials in (Gauß 1813) at that time.

Crelle took up the topic of analytic factorials again two years later but rather in passing in a longer article on powers and expansions in his own "Journal" (Crelle 1831, pp. 269–276) without mentioning Ohm's criticism or giving new mathematical arguments for his own results.

The Treatise "Bemerkungen über die analytischen Facultäten" by Weierstraß

The common narrative says that Weierstraß was busy with the research on the fundaments of complex analysis in the time directly after finishing his teacher's exam in order to use it later on as a basis for his results on hyperelliptic and general Abelian functions: He wrote up manuscripts on Laurent expansion, Cauchy estimates for the coefficients of power series and the double series theorem, and the definition of analytic functions by algebraic differential equations which were formally published only in the first volume of his "Mathematische Werke" (Weierstraß 1894–1927) in 1894, but also indirectly used at other occasions beforehand, just alike his treatise (Weierstraß 1894), which was published in parts already in (Weierstraß 1856b, pp. 346–379).

From what has been said before, it is obvious on the one hand that the "Remarks on analytic factorials", as the title of (Weierstraß 1843) translates, do not fit into this picture. On the other hand, such a text could be of highest interest for Crelle if it would resolve the stalemate between him and Ohm concerning the analytic factorials. Whoever had the idea, Gudermann or Weierstraß, such a paper could at least serve as a kind of bait so that Crelle would take notice of Weierstraß and then have a benevolent look at his treatise for the teacher's exam (Weierstraß 1894) with respect to a publication in Crelle's "Journal". Even more, (Weierstraß 1843) could have become a publication of its own in that journal. In the best of all possible situations, if also Ohm could be interested in it, this text could become a doctoral thesis for the Philosophical Faculty of the University of Berlin.

This might read like pure speculation, but according to its contents, (Weierstraß 1843) had the potential to fulfill such hopes, and Weierstraß makes this clear in the very first section of the text: As mentioned above, he quotes Ohm's criticism in length, but also mentions that Kramp's fourth functional equation was not part of the definition of analytic factorials by Crelle and that the article (Gauß 1813) by Gauß shows that one can define analytic factorials that fulfill the first three of Kramp's equations (Weierstraß 1843, p. 89). Then he gives Ohm due credit for his remark concerning the missing sine factors in Crelle's expansion of the analytic factorial. In the penultimate paragraph of (Weierstraß 1843, § 1), Weierstraß comes to a compromise that could have the potential for a true conciliation: Ohm's considerations are completely true if one defines an analytic factorial like Crelle, which is, in fact, the same definition that Gauß had given (Weierstraß 1843, p. 90).

The full truth, however, is a little bit more complicated: The missing sine expressions only were the tip of an iceberg: Crelle—and also Ohm—had not realized that Kramp's first three functional equations do not determine a unique solution. Weierstraß explains this problem in detail in (Weierstraß 1843, § 2). Seen from a more technical point of view, Crelle had simply assumed that $(a, +\Delta)^r$ converges to a^r if Δ tends to 0 even if r is not integer (Weierstraß 1843, § 3). Weierstraß analyzes this problem carefully, reduces it to the situation that one only has to consider the limit of $(1, +\Delta)^r$, however, separately for the cases that Δ is positive or that it is negative (Weierstraß 1843, § 4). After these preparations, he is able to deduce the expansions given by Crelle in a correct way (Weierstraß 1843, § 5–8).

In short: Weierstraß gives a correct mathematical exposition of the problem that had raised the controversy between Crelle and Ohm and resolves the problem. For this, however, he had to point at mistakes that each of these two mathematicians had made. An independent referee should have given due credit to Weierstraß, but the question was how Crelle and Ohm would react if they learnt from his results.

The Meeting of Crelle and Weierstraß

In the preface of (Weierstraß 1843) Weierstraß writes, translated from (Weierstraß 1843, p. 90):

> Invited to publish my comments by the highly respected author [=Crelle] of the above-mentioned treatise, to whom I communicated my comments orally, I believe I may use the present opportunity to do so.

This means that Weierstraß had met Crelle in person and before August 1843 when Weierstraß signed the manuscript (Weierstraß 1843, p. 103). Probably, this meeting took place in autumn 1842 when Weierstraß traveled from Münster to Deutsch

Crone in West Prussia and made a stop at Berlin, if only primarily for administrative purposes. For Weierstraß this meeting offered a chance to get his treatise for the teacher's exam (Weierstraß 1894) published and also to get a feedback on his considerations on analytic factorials. Here the fact that Weierstraß dared to approach Crelle proves anew that the self-confidence of Weierstraß had not been shaken by the formal failure of his studies at Bonn and the failure in his oral exam on the natural sciences at Münster. Furthermore, because of the very good relations between Crelle and Gudermann, Weierstraß will not have visited Crelle if he was afraid that Gudermann had a bad opinion of him. Quite the contrary, one could rather assume that Gudermann had arranged the meeting since a publication of whichever treatise of Weierstraß would have been in accordance with Gudermann's idea for his further advancement.

The more surprising it is that (Weierstraß 1843) was not published in Crelle's "Journal" which was the best possibility to publish a mathematical text in Germany at that time but only in the school program of the Progymnasium in Deutsch Crone. If the director of his school had urged Weierstraß to publish a treatise in the school program, the latter could have taken one of his manuscripts on the fundaments of complex analysis which he had already written during his time at Münster. Additionally, if Weierstraß had wanted to improve his text for his first appearance in Crelle's "Journal", he could have submitted a revised version of (Weierstraß 1843) to that journal one or two years later.

More plausible could be the version that in reality Crelle was not too happy with the correction's of Weierstraß on Crelle's considerations so that the above quote is only a diplomatic formulation. But when Weierstraß indeed published an extended version of (Weierstraß 1843) in Crelle's "Journal" thirteen years later (Weierstraß 1856a), Crelle added an annotation to it whose first paragraph reads as follows, translated from (Weierstraß 1856a, p. 1, footnote *) (note that this is not printed in (Weierstraß 1894–1927)):

> It is with true scientific as well as personal satisfaction that I have received the following treatise and, with the permission of its author, have included it in the present journal, as it shows that the so important theory of analytic factorials, which in earlier times was little considered and to the more detailed and general foundation of which I have also endeavored for thirty-two years to contribute, is more and more claiming the attention of analysts and has now stimulated an even more penetrating investigation by such an excellent and astute mathematician as Mr. *Weierstraß*.

Weierstraß confirms this picture even thirty years later, when he published (Weierstraß 1856a) anew within the collection "Abhandlungen aus der Functionenlehre" (Weierstraß 1886). Even if he admits that the theory of analytic factorials was not that important as some mathematicians had thought before, he refers to Crelle as follows (Weierstraß 1886, p. 190/Vol. 1, p. 158, footnote *):

> The reader may note that the present treatise was written as early as 1854, at Crelle's expressed wish, to whom I had communicated my misgivings about his theory of the faculties, after I had already published part of its contents in 1843 (as a supplement to the annual report of the Progymnasium at Deutsch-Crone).

In fact, when Weierstraß wrote up (Weierstraß 1856a), it must have been of importance for him: It does not only have a size of sixty printed pages in quart format, but it is also the first treatise which he finished after his article (Weierstraß 1854) in which he sketched his solution of the Jacobi inversion problem and which finally brought him to a position at a university. And, seen from the hindsight: Weierstraß published his considerations on analytic factorials more often than his famous treatise (Weierstraß 1894) for the teacher's exam: As mentioned before, on the one hand, (Weierstraß 1894) was only published in parts within (Weierstraß 1856b) and in the "Mathematische Werke" (Weierstraß 1894–1927, Vol. 1) in full. On the other hand, (Weierstraß 1843) was published in the school program and in the "Mathematische Werke", and (Weierstraß 1856a) was even published three times: in Crelle's "Journal", in the "Abhandlungen" (Weierstraß 1886), and in the "Mathematische Werke" (Weierstraß 1894–1927), and both papers were printed each time in full length.

A Clash with Ohm?

If, however, both Crelle and Weierstraß were interested in the considerations of Weierstraß on analytic factorials including its publication, why were they not published in Crelle's "Journal" as early as in 1843 or, after a possible revision, within the next years? Admittedly, the following answer is speculative, but there is enough known about both the mathematical abilities and the personality of Martin Ohm, also thanks to articles like (Schubring 1983), that one gets the idea that the reason for this has to do with Martin Ohm:

Since Ohm was the, so to speak, "contrahent" of Crelle in the debate on analytic factorials, it would have been natural for Crelle—or even Weierstraß himself—to forward the new results on analytic factorials to Ohm: If Ohm had been as interested in them as Crelle, then he would have been the perfect reviewer for a publication in Crelle's "Journal". Perhaps Weierstraß even had the much more far-reaching idea that Ohm, who was full professor at the University of Berlin, could become supervisor of a doctoral thesis on the topic of analytic factorials. (Of course, this idea is highly speculative, but one should note that Weierstraß finished his manuscript of (Weierstraß 1856a) on May 20, 1854, which is about seven weeks after March 31, 1854, when he had received an honorary doctorate from the University of Königsberg, just as if he had kept his considerations in reserve to submit it as a doctoral thesis.)

That Ohm had been informed of the considerations of Weierstraß or at least of the essential points of them, becomes plausible from the fact that a few years later Ohm published two articles (Ohm 1848), (Ohm 1850) on analytic factorials in which he defined them by means of the Gamma function, just as Gauß had done in (Gauß 1813), and joined this with his ideas already published in (Ohm 1829). Ohm quotes Gauß and Kramp but neither Crelle nor Weierstraß, even if (Weierstraß 1843) had been published before.

Now, if Ohm had known about the results of Weierstraß on analytic factorials then the events described above can be explained in the way that Ohm reacted on these results in a diametrical opposed way as Crelle did, namely negative in the sense of feeling of being attacked by that nobody from the province: That Ohm could have been a doctoral supervisor for Weierstraß will have only been a vague idea of Weierstraß (and Gudermann?). But maybe his reaction on the findings of Weierstraß could have been so violent that Crelle even restrained from publishing (Weierstraß 1843) in his "Journal": Ohm was under the patronage of several higher Prussian officials, including a Prince, and Crelle needed the support from official Prussian institutions as the title page of each of the volumes of his "Journal" informed the reader.

It has to be admitted that Ohm is something like the ideal scapegoat for almost everything that went wrong with respect to Berlin mathematics in the first half of the nineteenth century. But there are in fact indications that Weierstraß had problems with him already in the early 1840s:

On the one hand, as already mentioned, Weierstraß quoted Ohm's incorrect criticism of (Crelle 1823) from (Ohm 1829, § 340) verbally and in conspicuous length in (Weierstraß 1843, p. 88). Secondly, the story of Weierstraß visiting Jacob Steiner (1796–1863) at the occasion of a training for teachers for physical exercises at Berlin in July 1844 is well known (Geiser and Maurer 1901, p. 329), in particular Steiner's statement that intelligent students would contact him or Gustav Lejeune Dirichlet (1805–1859) among the professors of mathematics at Berlin, while the others would go to Ohm. But the beginning of the reported conversation is of more interest in the present respect: Steiner asks Weierstraß whether he came to Berlin in order to know Ohm, Weierstraß denies. Why should he not go to Ohm if he obviously intended to visit mathematics professors at Berlin (and had to find out that Dirichlet was on leave in Italy).

Conclusion

The considerations in the last sections are based on clues, but cannot be more than hypotheses. If they were true, they would tell a story of Weierstraß coming to Berlin with the hope for at least one publication and perhaps even a doctorate, but who stung into a wasps' nest and received a harsh rejection by a full professor of mathematics from the University of Berlin instead. In order to let this contribution end in a friendlier way, let us quote the second paragraph of Crelles remark on (Weierstraß 1856a) that shows his way to react on Weierstraß finding an error in his considerations (Weierstraß 1856a, p. 1., footnote *):

Regarding my own work on the subject, I would have something to say to justify the incompleteness that Mr. *Weierstraß* notes; but my workforce, weakened by old age and constant illness, is no longer sufficient for such things. So I have to do without it; but that is also acceptable, since I have never considered myself in my scientific endeavors, never strived for fame and praise, but only, to the best of my ability, to promote the truth, and it makes no difference to me who it is who comes closer to the truth: whether I or someone else, if only a closer approximation to the truth is achieved at all.

References

Behnke, Heinrich. 1965. Karl Weierstraß und seine Schule. In H. Behnke, K. Kopfermann (Eds.): Festschrift zur Gedächtnisfeier für Karl Weierstraß 1815–1965. Wissenschaftliche Abhandlungen der Arbeitsgemeinschaft für Forschung des Landes Nordrhein-Westfalen, 33 (pp. 13–40). Köln, Opladen: Westdeutscher Verlag.

Crelle, August Leopold. 1823. Versuch einer allgemeinen Theorie der analytischen Facultäten nach einer neuen Entwickelungs-Methode: vorbereitet durch einen Versuch einer critischen Untersuchung über die Potenzen, Logarithmen und Exponential-Grössn und begleitet von Bemerkungen und Erörterungen, die Winkel-Functionen betreffend. Berlin: Reimer.

Crelle, August Leopold. 1831. Mémoire sur la théorie des puissances, des fonctions angulaires et des facultés analytiques. Journal für die reine und angewandte Mathematik, 7, 253–305, 314–380.

Elstrodt, Jürgen. 2015. Die prägenden Jahre im Leben von Karl Weierstraß. In W. König & J. Sprekels (Eds.), Karl Weierstraß (1815–1897): Aspekte seines Lebens und Werkes—Aspects of his Life and Work (pp. 11–51). Wiesbaden: Springer Spektrum.

Féaux, Bernhard Josef. 1844. De functione transcendente, quae littera Γ( ) obsignatur, sive de integrali Euleriano secundae speciei. Doctoral thesis Münster. Münster: Coppenrath.

Gauß, Carl Friedrich. 1813. Disquisitiones generales circa seriem infinitam. $1+\frac{\alpha\cdot\beta}{1\cdot\gamma}x+\frac{\alpha(\alpha+1)\cdot\beta(\beta+1)}{1\cdot2\cdot\gamma(\gamma+1)}xx+\ldots$ Commentationes societatis regiae scientiarium Gottingensis recentiores, II, commentationes classis mathematicae, 46 pp.; also in Carl Friedrich Gauß: Werke, Vol. III (pp. 123–162). Göttingen: Königliche Gesellschaft der Wissenschaften 1866.

Geiser, Carl Friedrich, Maurer, Ludwig. 1901: Elwin Bruno Christoffel. Mathematische Annalen, 54, 329–341.

Killing, Wilhelm. 1897. Karl Weierstrass. Rede gehalten beim Antritt des Rektorats an der Kgl. Akademie zu Münster am 15. Oktober 1897. Natur und Offenbarung, 43, 1–21.

Kramp, Christian. 1798. Analyse des réfractions astromiques et terrestres. Leipzig: E. B. Schwikkert and Paris: Amand Koenig.

Lampe, Emil. 1897. Karl Weierstraß. Jahresbericht der Deutschen Mathematiker-Vereinigung, 6, 27–44.

Mittag-Leffler, Gösta. 1923. Die ersten 40 Jahre des Lebens von Weierstraß. Acta mathematica, 39, 1–57.

Ohm, Martin. 1829. Versuch eines vollkommen consequenten Systems der Mathematik, Theil 2: Lehrbuch der niedern Analysis; 2. Algebra und Analysis des Endlichen. Berlin: Riemann und Jonas, 2. edition.

Ohm, Martin. 1848. Über das Verhalten der Gammafunktionen zu den Produkten äquidifferenter Faktoren. Journal für die reine und angewandte Mathematik, 36, 277–295.

Ohm, Martin. 1850. Über die Behandlung der Lehre der reellen Factoriellen und Facultäten nach einer Methode der Einschließung in Grenzen. Journal für die reine und angewandte Mathematik, 39, 23–41.

Schubring, Gert. 1983. Das mathematische Leben in Berlin—Zu einer entstehenden Profession an Hand von Briefen des aus Erlangen stammenden Martin Ohm an seinen Bruder Georg Simon. Erlanger Bausteine zur Fränkischen Heimatforschung, Jahrbuch, 30, 221–249.
Schubring, Gert. 1986. Bibliographie der Schulprogramme in Mathematik und Naturwissenschaften (wissenschaftliche Abhandlungen) 1800–1875. Bad Salzdetfurth: Barbara Franzbecker.
Schubring, Gert. 1989. Warum Karl Weierstraß beinahe in der Lehrerprüfung gescheitert wäre. Der Mathematikunterricht, 35, 1, 13–29.
Schubring, Gert. 2021. Analysen der Profile der Mathematik-Schulprogramme in den verschiedenen deutschen Staaten des 19. Jahrhunderts. In N. Ächtler (Ed.), Schulprogramme Höherer Lehranstalten. Interdisziplinäre Perspektiven auf eine wiederentdeckte bildungs- und kulturwissenschaftliche Quellengattung (pp. 305–327). Hannover: Wehrhahn.
Ullrich, Peter. 2007. Karl Weierstraß und die erste Promotion in Mathematik an der Akademie zu Münster. In G. Wolfschmidt (Ed.), "Es gibt für Könige keinen besonderen Weg zur Geometrie", Festschrift für Karin Reich (pp. 145–162). Algorismus, 60. Augsburg: Dr. Erwin Rauner.
Ullrich, Peter. 2024. Christoph Gudermann (1798–1851)—nicht nur der akademische Lehrer von Weierstraß. Contribution to the joint meeting of the Fachgruppe Geschichte der Mathematik of the DMV and the Arbeitskreis Mathematikgeschichte und Unterricht of the GDM in 2017, to appear in the conference volume.
Weierstraß, Karl. 1843. Bemerkungen über die analytischen Facultäten. Jahresbericht über das Königl. Progymnasium in Dt. Crone vom Herbst 1842 bis zum Herbst 1843. Deutsch Crone: P. Garms, pp. 3–17; here in (Weierstraß 1894–1927, Vol. 1, pp. 87–103).
Weierstraß, Karl. 1854. Zur Theorie der Abelschen Functionen, Journal für die reine und angewandte Mathematik, 47, 289–306; also in (Weierstraß 1894–1927, Vol. 1, pp. 133–152).
Weierstraß, Karl. 1856a. Über die Theorie der analytischen Facultäten, Journal für die reine und angewandte Mathematik, 51, 1–60; also in (Weierstraß 1886, pp. 183–260) and (Weierstraß 1894–1927, Vol. 1, pp. 153–221).
Weierstraß, Karl. 1856b. Theorie der Abel'schen Functionen, Journal für die reine und angewandte Mathematik, 52, 285–379; partially also in (Weierstraß 1894–1927, Vol. 1, 297–355).
Weierstraß, Karl. 1886. Abhandlungen aus der Functionenlehre. Berlin: Julius Springer.
Weierstraß, Karl. 1894. Über die Entwicklung der Modular-Functionen. Manuscript Westernkotten 1840. Published in parts in (Weierstraß 1856b, pp. 346–379) and in full length in (Weierstraß 1894–1927, Vol. 1, pp. 1–49).
Weierstraß, Karl. 1894–1927. Mathematische Werke, 7 vols. Berlin: Mayer & Müller.
Weierstraß, Karl. 1923. Letter to Paul du Bois-Reymond, dated June 6, 1875. Weierstraß estate in the Insitut Mittag-Leffler, Djursholm. Printed in: Briefe von K. Weierstraß an Paul du Bois-Reymond, Acta mathematica, 39, 199–225.

The 1824 Calcutta School-Book Society Arabic and Persian Edition of Euclid's *Elements*: An Unusual Approach to Textbook Production in Colonial India

Gregg De Young

The Anglicist—Orientalist Controversy in Indian Education

The publication of the 1824 Calcutta School-Book Society (CSBS) Arabic and Persian editions of Euclid's *Elements* was consistent with the outlook of the Orientalist branch of the education debates that spanned the first three decades of the nineteenth century.[1] When Warren Hastings was named Governor-General of Bengal in 1772, he was already convinced that British control in India would only be maintained in the long run if it was supported by the Indian population. He tried to set policies that would build such support. Officials of the British East India Company were expected to be sufficiently fluent in native languages to allow them to conduct necessary business. The majority of these officials were sympathetic to Hindu and Muslim religion and culture.

Later governors continued the same policy directions. In 1797, the College of Fort William was established. There Arabic, Persian, and Sanskrit were taught, as well as six native vernaculars. Students were also taught Hindu, Muslim, and Indian law, world geography, mathematics, and the basics of modern sciences. A printing press associated with the school printed textbooks as well as copies of the classics of Indian literature and religion.

G. De Young (✉)
The American University in Cairo, New Cairo, Egypt
e-mail: gdeyoung@aucegypt.edu

[1] There has been a large literature devoted to these debates and their relation to colonial policies in India. The relevant documents have been collected and published by Zastoupil and Moir (1999). For a succinct summary of the controversy and its relation to Orientalist thought more generally, see Macfie (2002), Chapter 3. Evans (2002) showed that even though Orientalism and Anglicism had different approaches, they shared a common goal—preservation and enhancement of British power in India.

A. Verdugo Rohrer and J. Zender (eds.), *History of Mathematics and Its Contexts*, Trends in the History of Science, https://doi.org/10.1007/978-3-031-60828-5_17

But beneath the surface, trouble was brewing. It came to the fore in the Vellore Mutiny (July 10, 1806), when sepoy (native Indian) troops turned on their British commanders. The mutiny lasted only one day, but more than 200 British troops and officers died, along with almost double that number of sepoy rebels. Although the mutiny was short-lived, the debate it sparked over how to govern India continued for over a decade. Both sides of the debate agreed that education was an essential key toward effective government. But the Anglicists insisted that the policies of accommodation previously practiced by the East India Company were ineffective. Efforts to revive ancient Indian learning were a failure. The Orientalists responded with the Doctrine of Engraftment—add Western science and learning to the foundation of ancient Indian traditions.[2] But to Anglicists, this was at best a half-way measure. What was needed, in the colorful phrase of Thomas Babington Macauley (1800–1859), a member of the Supreme Council of India, was the production of a "class of persons, Indian in blood and colour, but English in taste, in opinions, in morals, and in intellect" (Macauley 1835, 9). The quickest path to that goal would be to educate in English and to proselytize the Indian population, converting them to Christianity as quickly as possible.

When William Bentinck, long-time critic of Hindu society, became Governor-General of the East India Company (1827–1835), he acted on his utilitarian personal principles to undermine the efforts of the Engraftment party and secured passage of the English Education Act of 1836. These debates took place primarily in England—debates over funding and debates over the permissibility of missionary proselytizing. By the time of Macauley's famous Minute on Education, the tide had turned against the Orientalist position, strengthened in public opinion by such disasters as the Vellore Mutiny.

The Calcutta School-Book Society

It was within the context of this debate over how to educate the Indian populace that the Calcutta School-Book Society was organized in 1817. This association, which included both Europeans and Indians, was to support "the preparation, publication, and cheap or gratuitous supply of... suitable books of instruction for the use of Native Schools, in the several languages, (English as well as Asiatic,) which are, or may be taught in the provinces..." (CSBS 1819, xi).

The CSBS was the first organization of its kind in India,[3] but within a few years, other similar organizations began to appear—in Madras, in Delhi, in Mysore, and elsewhere. As the East India Company took on more and more governmental responsibilities, one of its main concerns was recruiting Native workers with

[2] İzge (2014, 52) considers this to be a second phase of the Anglicist-Orientalist controversy.

[3] Basak (1959, 36–38) has noted interesting parallels between the organization and goals of the CSBS and the Cheap Book Society of Dublin (organized 1814).

appropriate skills and aptitudes. Consequently, there was a renewed push to educate the Native populace. Schools could be set up, but schools needed textbooks, both in English and in local vernacular languages, in order to function effectively.

Of course, textbooks could be (and were) imported from Britain. But these usually assumed that students had an adequate facility in English. Hence, during the ascendancy of the Orientalist approach to Native education, there was a considerable emphasis on translation of British textbooks into local vernaculars as well as the preparation of textbooks composed either by multilingual English employees of the company or by Natives who were fluent in both English and local languages. Vernacular books could be printed locally, but material supplies, such as paper and ink, still had to be imported from England and were not always available.[4] Demand often outpaced the available supply of cheap and appropriate school textbooks.

It was in this context that the Reverend Thomas T. Thomason (1774–1829), then Chaplain to the British East India Company stationed in Calcutta, offered to the Calcutta School-Book Society a geometry textbook that he had prepared, along with a Persian translation of this textbook prepared by a local Indian scholar. His geometry textbook presented a novel solution to the problem of how to prepare a vernacular geometry textbook. Rather than undertake the arduous task of translating a European textbook, such as Simson's *Geometry*, into Arabic, he took an existing Arabic version of Euclid's *Elements*, the classic *Taḥrīr Kitāb Uṣūl al-Handasa li-Uqlīdis*,[5] completed in 646/1248 by Naṣīr al-Dīn al-Ṭūsī (597–674/1201–1274), and edited it so as to remove the later accretions added by al-Ṭūsī and others, thus leaving, he claimed, only the "pure" Euclidean content.

The Arabic Editor: Thomas T. Thomason

Thomas Truebody Thomason (1774–1829) was born in Plymouth and raised in London by his widowed mother. She used a small inheritance to provide him with a modest education. Proficient in French, at 15 he accompanied Dr. Coke, a Methodist missionary, to the West Indies as his interpreter. On his return, thanks to financial support from the Elland Society,[6] he began his preparatory training for study of theology under the tutelage of the Reverend Mr. Clark of Chesham Bois, in Buckinghamshire ("Thomas Thomason" 1851, 40–41). It was here that

---

[4] Although the CSBS was, according to its own rules, prohibited from publishing religious books, in practice the Society worked closely with the Baptist Missionary Press at Serampur. Some of its most popular titles were printed on the missionary presses (Basak 1959, 46–47).

[5] The treatise has been given a variety of titles. For some of the more common variants, see Rosenfeld and Ihsanoğlu (2003, 212). It was still being copied and studied in traditional Islamic schools (called madrasa) during the nineteenth century.

[6] The Elland Society was founded in 1767 to assist the education of young men who wished to pursue ordination in the Church of England in the Northern Provinces. The early history of the society has been summarized by Stowell (1896).

he developed a sound foundation in both Greek and Hebrew languages (Sargant 1834, 16–17).

In 1792, the Society sent him to Magdalen College, Cambridge. Here, he came under the influence of Charles Simeon, an evangelical minister. It was Simeon who urged him to devote more attention to the study of mathematics. Thus, in his final years as an undergraduate, he expended considerable effort in studying Newton's *Principia* (Sargent 1834, 53–54). A fellow student, Charles Jerram, noted: "I could not help observing the proficiency he appeared to have made both in the classics and mathematics" (Sargent 1834, 35), suggesting that even as an undergraduate, Thomason's mental abilities stood out to his contemporaries. During this time, he also began his study of Arabic (Sargent 1834, 69).

Thomason was ordained as Deacon in 1796 at the beginning of his last year at university and served as Curate in Holy Trinity Church under Rev. Simeon. During his last year at Magdalen, he was offered a position as priest to the mission church in Calcutta. After careful consideration, he decided not to accept the offer. Instead, he completed his formal education and received his BA degree. He was ordained as Presbyter in the Church of England. He then matriculated at Queen's College to pursue his MA. He was elected Fellow in 1797 and was appointed a tutor in 1798. In 1799 he was awarded his MA degree and served the parish at Shelford, near Cambridge. At the same time, he married.

By 1805, Thomason was again thinking of "going to the Heathen." He was advised against proceeding with this plan, however, because his wife did not support his ambitions. But when in 1808 he was nominated by Rev. Simeon as chaplain to the East India Company, he accepted the position, since his wife had become reconciled to the prospect of living abroad. In preparation, Thomason immediately began the study of Persian with a "moonshee" (native-speaking language instructor) resident in Cambridge (Sargent 1834, 130–131). On June 10, 1808, Thomason, with his wife and two oldest children, boarded an East India ship, the *Travers*, for the five-month voyage to India. During the uneventful voyage, Thomason immersed himself in the study of Persian (Thomason, letter to his mother, dated 13 November 1808, quoted in Sargent 1834, 142–149). But on November 7, as the *Travers* entered the Bay of Bengal to sail toward Calcutta, it struck some hidden rocks off Cape Negrais and quickly sank. Although few lives were lost, the crew and passengers finally reached Calcutta with little more than the clothes on their backs.

The Thomason family was welcomed to Calcutta by members of their new parish and everyone tried to help them quickly settle into their new home. Despite his efforts to learn the vernacular languages, Thomason's attitude toward the native inhabitants and their culture was surprisingly hostile. Writing to an acquaintance in England after his arrival in India, he described the many difficulties opposing the progress of the Gospel (Sargent 1834, 60–62). He noted that they arose.

> . . . partly from the nature of the climate, and partly from the nature of society, and chiefly from the close intercourse with the wretched natives, who are more degraded than you can

> imagine. . . . The natives swarm around us, and corrupt the minds of children from the earliest years. . . . The better sort of people send their children home at five or six; until then, they are under the charge of native servants, and their minds are poisoned as far as they can be at that tender age. . . . The children cannot even speak their mother tongue; they have to learn English on their way home. . . . It is a fixed rule with us (Thomason and his wife), that the children are always in the presence of one amongst us. We never suffer them, not for a moment, to have any intercourse with the servants alone.

As chaplain, Thomason's primary responsibility was to his parish. His official duties included leading two services each Sunday, along with a midweek service on Thursday evening and another service on Saturday evening for youth and their parents (Sargent 1834, 164–165). In addition, of course, he officiated at weddings and conducted funerals. Within six months of his arrival, attendance at weekly services necessitated enlarging the worship hall. In addition to his pastoral duties, he was involved in revising both the Arabic translation of the Scriptures and the Hindustani translation of the New Testament, as well as serving as Examiner in Arabic at the College of Fort William ("Thomas Thomason" 1851, 41).

At the second annual meeting of the Calcutta School-Book Society (1819), The Rev. T. Thomason was listed as among nearly three dozen new subscribers to the Society with a donation of 100 pounds (CSBS 1819, 9).[7]

In 1825, Thomason's wife fell seriously ill and he decided to sail for England in hopes of recovering her health. She died during the homeward passage and was buried at sea. After his return to England, he became rector of the parish at Cheltenham. In 1828, he decided to return to India, believing that he might be of more use there than in England. Prior to setting sail, he married again and his new wife accompanied him. Arriving at the end of 1828, he was almost immediately struck down with illness. His physicians recommended that he go to Mauritius, hoping that healthier living conditions would allow him to recover. He and his wife landed there on his fifty-fifth birthday, but his condition rapidly deteriorated and he died on Mauritius two weeks later, on June 21, 1829 (Sargent 1834, 125–137; "Thomas Thomason" 1851, 41–42).

The Persian Editor/Translator: Hydur Alee

Not much biographical information is available about Hydur Alee.[8] The first mention of his name that I have found occurs in the minutes of the second annual meeting of the society (CSBS 1819, 8),where it is noted that he had prepared a

[7] This is a substantial contribution, equivalent to about 11,000 pounds today.

[8] To avoid ambiguity and confusion, I shall use only one version of each name rather than reproduce the varied transliterations of the contemporary sources.

review in Persian of three Arabic manuscripts by Tafazzul Hussein Khan (1727–1800) on astronomy, geometry, and algebra.[9] He seems, thus, to be a native Persian speaker who was well-versed in mathematical sciences.

Al-Ṭūsī's *Taḥrīr Kitāb Uqlīdis*

Al-Ṭūsī's treatise,completed in 646/1258, was not a translation, but rather an edition.[10] It is based, he tells us in the introduction, on the revision of the Arabic translation version attributed to Thābit ibn Qurrah (220–288/836–901) who is reported to have edited the more recent translation prepared by Isḥāq ibn Ḥunayn (215–298/830–910) using additional Greek manuscripts. Al-Ṭūsī also had access to what medieval scholars called the version of al-Ḥajjāj (commonly assumed to have been lost), the earlier translator of Euclid's *Elements* into Arabic according to a report penned by the fourth/tenth-century biobibliographer al-Nadīm (2009, III, 207–210).[11]

Al-Ṭūsī, in his *Taḥrīr*, tried to express the content of the *Elements* using a more distinctly Arabic style, abandoning efforts by early translators to retain as much as possible of the Greek syntax. In addition, he added numerous notes on mathematical points, many introduced by the Arabic phrase "*aqūlu*" (I say), suggesting that their content originated with al-Ṭūsī. In fact, many of these notes are drawn from earlier authors who had discussed or commented on the *Elements*. Although al-Ṭūsī does not directly mention his predecessors, De Young (2008–2009) has shown that one of his major sources was the fourth/tenth-century Arabic commentary, *Fī Ḥall Shukūk Kitāb Uqlīdis,* written by Ibn al-Haytham.

---

[9] For biographical information on Tafazzul Hussein (who is sometimes given the descriptive title Nawāb, an Arabic term meaning an agent or representative, because he often acted as go-between, representing the East India Company in negotiations with local Indian rulers), see "An account" (1804) and also the collection of excerpts and anecdotes in Khan (1908), as well as Abid (2006). Additional background sources are noted in Schaffer (2009, note 7).

These manuscripts were in the possession of Tafazzul Hussein's son, Tajummul Hussein Khan of Lucknow. He had apparently loaned them to the CSBS for copying (CSBS 1819, 16). It is not clear whether Hydur Alee was involved in this copying process. The manuscripts represented Arabic translations of: (1) Newton's *Principia Mathematica*, (2) Edmund Halley's Latin edition of Apollonius, *De sectione rationis* (1706), and (3) Thomas Simpson, *Treatise of Algebra* (1745). (For additional information on his mathematical translations, see "An account" (1804, 7).) Hydur Alee's Persian review has been translated into English by Masumian (2021).

[10] Al-Ṭūsī's *Taḥrīr* is often confused with another treatise with the same title published by the Medicean Press in 1594 and wrongly attributed to al-Ṭūsī on its title page (Cassinet 1993; De Young 2003, 2012a). This attribution was apparently added in the printing process because the only surviving manuscript copy (Florence, Laurenziana 50) lacks the ascription to al-Ṭūsī.

[11] Important points of this traditional historical narrative have been called into question by Brentjes (2018). She has argued that, despite the claims to the contrary made in several medieval manuscripts, the Isḥāq-Thābit version is an edited version of the previously presumed missing translation of al-Ḥajjāj; and if Is Ṭāq created a complete and independent translation, it now appears to be missing from the currently known documents.

The *Taḥrīr* became the classic textbook introducing students to higher mathematical studies and it continued to play a central role in mathematical pedagogy for centuries. The study of Euclidean geometry, through the window of al-Ṭūsī's *Taḥrīr*, was frequently accompanied by a collection of shorter treatises, either copied into the same codex or copied independently, on specific topics or specific sections of the text. Although membership within these ancillary texts can vary both geographically and temporally, some seem to have enjoyed a remarkable longevity (Brentjes 2019; Abdeljaouad and De Young 2020, 70–77). Al-Ṭūsī's *Taḥrīr* was translated into Persian by Quṭb al-Dīn Maḥmūd b. Mas'ūd al-Shīrāzī (634–710/1236–1311) and served as the foundation for Euclidean studies in Persian (Brentjes 1998; De Young 2007).

Quṭb Al-Dīn Al-Shīrāzī—The Persian Translator of the *Taḥrīr*

Quṭb al-Dīn al-Shīrāzī succeeded his father as chief ophthalmologist at the Muzzafar Hospital in Shīrāz while still a teenager and held that position for more than a decade. But he thirsted for more knowledge and moved to Marāgha where he was among the earliest students to study with al-Ṭūsī at the observatory and school (founded 659/1259). There, he read medicine and philosophy under al-Ṭūsī and assisted with the astronomical observation program of the observatory for nearly a decade.[12] Some historians consider him the best of the exceptional group of students that worked and studied at the observatory during al-Ṭūsī's lifetime. Quṭb al-Dīn accompanied his teacher on a trip to Khorasan, but stayed behind to study with Najm al-Dīn Kātibī al-Qazvīnī, a religious scholar of high reputation.

After some time, he journeyed westward, spending some time in Baghdad, and eventually moved farther westward to Konya in Anatolia, where he studied with Ṣadr al-Dīn al-Qūnawī, another religious scholar. At this point, al-Shīrāzī's autobiographical narrative breaks off, so we must rely on secondary biographical accounts. Unfortunately, they do not agree on all chronological points, so I shall follow the narrative proposed by Niazi (2013).

Shortly after his arrival in Konya, he was named as judge in Sivas and Malaṭya. Since most of the administrative work of this position would have been done by his assistants, he found ample time for research and writing. His treatise, *Al-Tuḥfa. al-Shāhiyya fī al-Hay'ah*, was completed in 1285 and dedicated to Tāj al-Dīn Mu'taz bin Ṭāhir,[13] who has been identified as the vizier of Amīr Shāh Muḥammad ibn

[12] Al-Shīrāzī wrote only one short autobiographical sketch, focused mainly on his medical education and covering only the first part of his life. The best discussions of the secondary sources for biographical information and summarizing his works are Walbridge (1992) and Niazi (2011; 2013).

[13] Niazi identifies the dedicatee as Mu'īn al-Dīn Sulaymān Parvana, a Persian official appointed as chamberlain to the Seljuk Sultanate of Rūm in Konya, then a vassal state to the Mongol Ilkhanid rulers in Tabriz. This identification seems to rest on two facts: (1) al-Dhahabī's report that it was

Ṣadr al-Sa'īd (Rosenfeld and Ihsanoğlu 2003, 234), perhaps as an attempt to gain court patronage.[14]

Al-Shīrāzī reports that in 681/1282 he was requested to undertake a diplomatic mission from Takūdār Aḥmad to the Mamlūk court. His mission seems to have been both a diplomatic and personal success. He reports that during his visit to the court, he was able to obtain several manuscripts including discussions of Ibn Sīnā's famous medical treatise, *al-Qanūn*, that finally enabled him to understand the obscurities in that treatise. Shortly after this successful visit to Egypt, al-Shīrāzī claimed to have experienced a series of "calamities" that significantly hampered his intellectual efforts for more than a decade until the accession of Ghāzān (694/1295). By this time, it appears probable that al-Shīrāzī had taken up residence in the Ilkhanid capital, Tabriz. During this period, he produced several treatises on illuminationist philosophy as well as the philosophical compendium, *Durrat al-Tāj*.

Al-Shīrāzī's Persian Translation and *Durrat Al-Tāj*

Near the end of his life, Al-Shīrāzī composed a lengthy philosophical compendium called *Durrat al-Tāj li-Ghurrat al-Dubbāj*. He appears to have consciously constructed his compendium on the model of the earlier philosophical compendium, *Kitāb al-Shifā'*, of Ibn Sīnā, except that it was compiled using Persian rather than Arabic texts (De Young 2012b). The two treatises have the same general arrangement—each is divided into four sections:

Logic provides basic tools for philosophical enquiry—syllogisms and argumentation.

Physics considers whatever is composed of matter; the four elements and how they connect together to create the empirical characteristics of sensory objects as well as their becoming and degeneration.

Mathematical sciences consider physical things as abstract or non-tangible.

Metaphysics considers non-material entities, whatever consists of being and its properties, such as substance and accident; potentiality and actuality.

In both treatises, the mathematical section was divided into four parts: arithmetic, geometry, mathematical cosmology (*hay'a*), and music (ratios). In both

"the Parvana" who named al-Shīrāzī as judge in Sivas and (2) the dedication of *Al-Tuḥfa al-Shāhiyya fī al-Hay'ah* to Mu'īn al-Dīn (Niazi 2013, 30). This latter fact raises a problem because the name of the dedicatee reported by Rosenfeld and Ihsanoğlu (2003, 233) is the same as the dedicatee of al-Shīrāzī's Persian translation of al-Ṭūsī's *Taḥrīr Kitāb Uqlīdis*. But in the two manuscripts of al-Shīrāzī's translation currently available to me (Tehran, Majlis Shūrā Library, Sinā 226, folio 4b and Columbia University Library, Plympton or. 282, folio 4a), the name of the dedicatee is stated as Mu'tazz ibn Ṭāhir. It is not known how Niazi would explain the differences in the name of the dedicatee.

[14] For a concise description of forms of patronage and the significance of patronage for the development of mathematical sciences in Islamicate societies, see Brentjes (2008).

treatises, the geometrical section was a single text. In the case of *al-Shifā'* Ibn Sīnā used an epitome of the *Elements* containing many features of an Arabic tradition that medieval authors associated with the translation of al-Ḥajjāj (De Young 2023). In the case of *Durrat al-Tāj*, al-Shīrāzī used a lightly edited version of his own Persian translation of al-Ṭūsī's *Taḥrīr*. As we have seen, al-Ṭūsī based his *Taḥrīr* on what has traditionally been known as the Isḥāq-Thābit version of the Arabic *Elements*, but where the two traditions diverge, he sometimes added editorial notes to describe these differences. Many of these notes had been removed by al-Shīrāzī during the translation process.

Al-Shīrāzī's compendium carries a dedication to the Amīr Dubbāj ibn Fīlshāh, a relatively minor warlord controlling the region of Gīlān (in what is now northeast Iran). It is tempting to read this dedication as a ploy for patronage, but we are specifically told that Amir Dubbāj commissioned the work from al-Shīrāzī.

Thomason's Justifications for His Edition

In his introductory letter to W. B. Bayley, the President of the CSBS, Thomason outlined three reasons why he thought that his edition of Euclid would make an important contribution to the goals of the society.[15]

The first justification was that an introduction to Euclidean geometry was entirely consistent with the stated goals of the CSBS—to make available to students the elementary works of science.[16] Moreover, a knowledge of the first six books of Euclid would suffice for further study in natural philosophy and astronomy.

The second justification was that too much time in Native education was currently spent in "subtleties of logic." Thomason was not opposed to students having the ability to analyze arguments logically. But, he insisted that this should not be the end or goal of education. Logic was a tool to be used in acquiring scientific knowledge, not an end in itself. So he hoped that his geometry textbook would help to provide the first steps in moving from logic to science.

His third justification was implicit in his conclusion: "we must not only supply the means, by the distribution of books, but also point out the way...." One technique that could help to "point the way" would be to refer to a recognized scholar such as al-Ṭūsī, "whose compositions in Geometry, Astronomy, and Natural Philosophy have ranked him amongst the most eminent authors in the annals of Arabic literature." Thomason thought that the most effective way to point that students forward toward the acquisition of science was through an introduction

[15] Thomason's letter was quoted in Appendix IV of the Second Report of the Calcutta School-Book Society (CSBS 1819, 37–39).

[16] Gupta (2014, 60) notes that in its first year of operation, the CSBS produced seven titles, all in Bengali, and all devoted to arithmetic and science.

or preface that would exhort teachers and schoolmasters to continually encourage the students to move forward from "the porch into the temple of science."[17] He believed that reminding students how al-Ṭūsī had built on his study of logic and geometry to make many important contributions to science of his day would inspire students to emulate his example.

Thomason's Editing Procedures

Writing to the secretary of the Calcutta School-Book Society, Thomason stated (CSBS 1819, 37) that his goal was to produce "a revised copy of [al-Ṭūsī's] work … stripped of its appendages, that is, of all that is not Euclid's... as far as the end of the sixth book; and in doing this to guard the accuracy of the text and exhibit the work in as clear and perspicuous a form as possible for the greater assistance of Native Students...." This procedure seems a brilliant solution to the task of providing essential mathematical instruction to "native students" in clear and precise Arabic that has stood the test of ages.[18]

But how do we distinguish what is Euclid's from what is only Euclidean—additional propositions or demonstrations in Euclidean style built on but not actually part of Euclid's original text? When dealing with a text like the *Taḥrīr Uqlīdis*, one could perhaps rely on al-Ṭūsī's own linguistic conventions, since most additions are introduced by "I say..." (*aqūlu*) or by phrases such as "By another approach" (*wa-bi-wajh ākhar*). Or the editor might think back to his own education in England and its geometrical content (often in the form of Simson's *Geometry*, whether in Latin or in English (1756), or the somewhat more recent *Geometry* published by John Playfair (1795).[19] I shall argue that in preparation of the CSBS Arabic edition, Thomason probably employed both techniques.

At the end of his cover letter, Thomason notes that "In the process of revision it occurred to me, that with a view to recommend and facilitate yet further the study of Geometry, it might be expedient to clothe this elementary treatise in a Persian

[17] See, for example, the introduction to Cornelius van Dyck's Arabic translation of Playfair's *Geometry* (De Young, 2014, 507–508). Ultimately, Thomason's edition was published without an introduction.

[18] Thomason's approach to producing a vernacular version of a geometry textbook is in distinct contrast to the approach to textbook preparation taken by the American missionary, Cornelius van Dyk, who published a geometry textbook in the Levant (1856) prepared by translating Playfair's *Geometry* into Arabic. (DeYoung 2014).

[19] Simson's *Geometry* first appeared in Latin but was then brought out in English (1756) and went through many editions over the course of the next century. For purposes of this paper, I will refer to the 1781 edition. Playfair's *Geometry* (1795) was published only in English but also went through multiple editions, both official and pirated. For purposes of this paper, I will refer to the 1806 American printing.

dress" (CSBS 1819, 38).[20] This Persian translation was prepared by "Hydur Alee, a member of our community" (CSBS 1819, 38).

Since a Persian translation of al-Ṭūsī's *Taḥrīr* had already been created by al-Shīrāzī, it may seem surprising that Thomason did not simply follow the same procedure he had used in editing al-Ṭūsī's Arabic *Taḥrīr*. Instead, he arranged for Hydur Alee to prepare a new Persian translation based on his Arabic edition. I think his reason was precisely because Hydur Alee had not been educated in England and so had not studied geometry using the traditional British approach exemplified in Simson and Playfair or similar British textbooks. He had most probably read Euclid using al-Ṭūsī's Arabic *Taḥrīr* or its Persian translation in *Durrat al-Taj* just as had many others in the Islamicate world. For Thomason, the key to recognizing what was the "true Euclid" apparently lay in his own personal experience— which for Thomason almost certainly included reading either Simson's or Playfair's *Geometry*.

That Thomason must have read Simson's *Geometry* or its restatement by John Playfair, is indicated by the fact that he apparently referred to these—at least in some cases—in order to judge what was genuine Euclid. (See, for example, point 10 in the following section.) It appears that Thomason was not only removing al-Ṭūsī's mathematical and editorial comments from the *Taḥrīr*, he was also silently editing the text itself. This editing was usually done, it seems, in reliance on the textbooks of Simson and/or Playfair. It was probably for this reason that Hydur Alee was requested to translate Thomason's Arabic edition of al-Ṭūsī's *Taḥrīr* rather than to edit either al-Shīrāzī's existing Persian translation of the *Taḥrīr* or the geometrical section of *Durrat al-Tāj* using the same principles that Thomason had announced for producing his edition.

Features of Thomason's Edition of the *Elements*: Additions to Al-Ṭūsī's Text

De Young (2012c, 6–10) has related the significance of some of the physical characteristics of the printed textbook to its colonial context. In this section, I focus primarily on internal content of Thomason's treatise, noting (a) changes that Thomason introduced into al-Ṭūsī's Arabic text, apart from omission of al-Ṭūsī's alternative demonstrations, and (b) where Thomason has apparently abandoned temporarily his stated editing principle to remove all that was not Euclid. I have focused solely on the Arabic text in this section because the Persian translation prepared by Hydur Alee is either a direct translation or close paraphrase of Thomason's Arabic edition.

[20] Thomason's original plan was to present the two versions in parallel columns. They were ultimately printed in two independent volumes.

The Arabic edition generally follows the Arabic text of al-Ṭūsī's *Taḥrīr*.[21] I begin by considering changes that Thomason introduced into al-Ṭūsī's text:

Def. I, 3: Al-Ṭūsī's formulation is similar to that found in the original Greek: "A straight line is a line that lies evenly with the points on itself" (Heath 1956, I, 153). In Thomason's Arabic edition, this definition has become "A straight line is the shortest distance between two points." This alternative version of the definition has been known since antiquity and is usually attributed to Archimedes. It is found in the Arabic discussions of al-Nayrīzī (Arnzen 2002, 6) and Ibn al-Haytham (Sude 1974, I, 33 and II, 12). But neither of these sources is likely to have been known to Thomason. This Archimedean definition does not occur in Simson (1781, 9) or in Playfair (1806, 1), two of the most prominent English geometry textbooks of the time. On the other hand, the Archimedean definition appears in Legendre's French textbook (1794, 1). That Thomason had access to Legendre's textbook is certainly not impossible.

Def. I. 21: Al-Ṭūsī stated: "Also, among them are the right-angled and the obtuse-angled if there occurs in it (the triangle) a right or an obtuse angle; but the acute-angled if they (a right or obtuse angle) do not occur [in them]." This was replaced in Thomason's Arabic edition (1824, 4) by the statement: "Also the [triangle] is right-angled if a right angle occurs in it; and [it is] obtuse-angled if an obtuse angle occurs in it; and [it is] acute angled if all the angles are acute." These statements are very similar to the definitions given by Simson (1781, 12) and Playfair (1806, 4–5). Cf. the statement in the Greek: "Of trilateral figures, a right-angled triangle is that which has a right angle, an obtuse-angled triangle that which has an obtuse angle, and an acute-angled triangle is that which has its three angles acute" (Heath 1956, I, 154).

Def. I, 22: Thomason replaced al-Ṭūsī's definition: "the rectangle (or oblong figure—*al-mustaṭīl*) is that whose angles are right and whose sides are not equal to one another" with the statement: "the rectangle (or oblong figure—*al-mustaṭīl*) is that whose angles are right and whose opposite pairs of sides are equal to one another" (Thomason 1824, 4). This statement of the definition is closer to that of Legendre (1794, 3) than to that of Simson (1781, 12) or Playfair (1806, 5), each of whom stated this definition in a form identical to that of al-Ṭūsī.

Proposition II, 14: Al-Ṭūsī appears to have forgotten (or perhaps he assumed) the construction of the line from the center of the semicircle (H) to point T on the circumference. Thomason (1824, 70) inserted this instruction into his Arabic

[21] I have throughout compared the Arabic text of the Calcutta edition to the text of the *Taḥrīr* preserved in British Library, add. 23,387, whose colophon reports that the manuscript was copied in 656/1258—that is, ten years after the treatise was completed by al-Ṭūsī. The copyist, unfortunately, is prone to homeoteleuton, in which the eye of the copyist skips to an identical or similar word later on the page being copied, thus creating the appearance of a small lacuna. In cases of doubt, I have consulted additional copies of the text, including the facsimile of Istanbul, Feyzullah 1359 (al-Ṭūsī 2012) and the Tehran 1298/1880 lithograph edition, as well as the manuscript Munich, Bayerische Staatsbibliotheek, *cod. arab.* 2697.

edition. He could be following either Simson or Playfair in making this correction. Or it is possible that either Thomason or Hyder Allee noticed the missing construction and inserted it. (If they had checked *Durrat al-Tāj*, they would have found that the construction is either omitted or assumed there as well.)

Proposition III, 9: Two versions of the demonstration were given in the Greek manuscripts. Heath (1956, II, 22) noted that Heiberg (and August before him), as editors of the Greek text, accepted the first version and placed the alternative demonstration in an appendix. Al-Ṭūsī included both the accepted version and the alternative demonstration, which he attributed to Thābit. Simson (1781, 76) gave only the alternative demonstration in his textbook, as did Playfair (1806, 79). Thomason (1824, 80–81) also opted for the alternative demonstration. Thomason's choice of the alternative form suggests that he was looking either to Simson or Playfair to define what is due to Euclid.

Proposition III, 10: There were two versions of the demonstration given in the Greek manuscripts. Heath (1956, II, 23) noted that both August and Heiberg placed the second or alternative demonstration in an appendix. Al-Ṭūsī included both the accepted Greek version and the alternative demonstration, which he ascribed to "some manuscripts" of Thābit. Simson (1781, 76–77) and Playfair (1806, 80) gave only the alternative Greek version. Thomason (1824, 81–82) followed the choice of Simson and Playfair, but used the diagram found in al-Ṭūsī, not the diagram used by Simson and Playfair.

Proposition III, 28 (Greek III, 29): Thomason (1824, 96) moved al-Ṭūsī's statement "the diagram (*al-shakl*) is just like what preceded" from the end of the proposition to immediately follow the enunciation. This alteration of the text was consistent with changing views about diagrams and their role in the demonstration of mathematical propositions. In traditional Greek and Arabic texts the diagram typically was placed near the end of the proposition and sometimes even followed the text of the proposition (De Young 2005, 165–166). This implied a shift in how the diagram was perceived to function, from metonym or recapitulation (in ancient texts) to aid for visualization of the geometric relations in more modern texts.[22]

Proposition III, 32 (Greek III, 33): This proposition was III, 33 in Simson and Playfair. There are three possible scenarios with regard to the given angle: it may be acute, or right, or obtuse. This was the order in which they were demonstrated in the Greek (Heath 1956, II, 67–71). Al-Ṭūsī demonstrated them in the opposite order, explaining that the perpendicular connecting the point of tangency with the center of the circle lay between the sides of the obtuse angle. If the given angle was right, the perpendicular was superimposed on one side of the angle (as the diameter of the circle). And in the case of an acute angle, the perpendicular fell outside the two sides of the angle. Thomason (1824, 100–101) adopted the right, obtuse, and acute order of demonstration following Simson (1781, 95–96) and Playfair (1806, 100–101). The diagrams were also labeled following the pattern used by Simson or Playfair, rather than that used by al-Ṭūsī.

[22] On the geometrical diagram as metonym, see Netz (1998, 37–38).

Proposition III, 34 (Greek III, 35): This was proposition III, 34 in al-Ṭūsī, but it was III, 35 in Simson and Playfair. In the Greek, two possible scenarios were demonstrated: (a) when the two chords intersect one another at the center and (b) when they do not intersect at the center (Heath 1956, II, 71–72). Simson and Playfair added two additional scenarios: (c) when a chord passes through the center and bisects the chord that does not pass through the center; and (d) when a chord passes through the center and does not bisect the chord that does not pass through the center. Thomason (1824, 102–104) included the demonstrations for all four scenarios as found in Simson (1781, 97–98) and/or Playfair (1806, 102–103) to edit the demonstration of al-Ṭūsī. In this case, he did not simply translate the demonstrations of Simson or Playfair. He used the phrasing and diagrams of al-Ṭūsī as much as possible, but edited them so as to produce the demonstrations for the four scenarios.[23]

Book IV definitions: The first two definitions were those of Euclid. Thomason (1824, 108) has added three additional definitions: (a) if each of the sides of a rectilinear figure touches a circle, it is said to circumscribe it and [the circle] is said to be inscribed in it; (b) if the circumference of a circle passes through all the angles of a rectilinear figure, it is said to be inscribed within the circle; and (c) if the endpoints of a straight line within a circle touch the circumference of the circle, it is said to be within [the circle]. These interpolated definitions correspond to definitions 4, 6, and 7 of Simson's *Geometry* (Simson 1756, 102–103).[24] These definitions are not found in al-Ṭūsī's *Taḥrīr*.

Proposition IV, 10: Thomason (1824, 115–116) has edited out the alternative demonstration that is included by al-Ṭūsī. The demonstration of Thomason is consistent with the demonstration found in Simson (1781, 110–111) or Playfair (1806, 117–118).

Proposition V, 10: Thomason (1824, 133) added, following this proposition: "I say: This only occurs in magnitudes that are of the same species (or rank)." This statement was not present in al-Ṭūsī but does occur in *Durrat al-Tāj*. As an insertion into the text, this statement is set off from the remainder of the proposition by a lozenge-shape and is typeset centered on the line rather than right-justified as is the proposition text. This explanatory statement was not present in either Simson or Playfair.

Proposition VI, 8: Thomason omitted the next three propositions found in al-Ṭūsī's *Taḥrīr*. These propositions explain: (VI, 9) We want to find a line intermediate according to the ratio between two given lines; (VI, 10) we want to find a third line for two given lines that are in a ratio; (VI, 11) we want to find a fourth line for three given lines in a ratio. The reasons for this omission are unclear to me because these propositions were not omitted from *Durrat al-Tāj* nor were they

[23] Heath (1956, II, 73) has pointed out that the demonstrations for these four scenarios were earlier discussed in Campanus of Novarre's Latin edition (Busard 2005, I, 137–138).

[24] The same definitions appear in Playfair's version of the *Elements* (1806, 107–108).

omitted by either Simson or Playfair. But Legendre has placed these propositions among the "Problémes annexes au Livre III."

Proposition VI, 24: Thomason (1824, 173–174) has edited the enunciation, giving it a different formulation from al-Ṭūsī: "If there be constructed on half the line a parallelogram, it is the greatest of all the parallelograms applied to that line and deficient from its completeness by a surface similar to the surface constructed on half that line and whose position is as its position." The origin for this editing has not been determined, but may be related to the alternative demonstration given by al-Anṭākī.[25]

Features of Thomason's Edition of the *Elements*: Retained Comments from Al-Ṭūsī

Although Thomason claimed that the purpose of his editing was to remove from al-Ṭūsī's treatise "all that is not Euclid," he was curiously inconsistent in pursuit of this goal. Sometimes he has retained notes from al-Ṭūsī that we would have expected him to have removed during his editing process. In this section, I review these retained features of al-Ṭūsī's treatise.

Proposition I, 15: Thomason (1824, 12) included al-Ṭūsī's corollary: "This proposition is established for all angles surrounding a point, wherever the point may be and however many angles there may be [surrounding it]." This note may have been retained because it corresponded conceptually to Corollary 2 in Simson (1781, 25) and Playfair (1806, 19).

Proposition I, 16: Thomason (1824, 24) retained the concluding porism statement of al-Ṭūsī: "It may be shown from that that it is not possible to extend from a point to a [straight] line two lines bounding with it (the straight line) two equal angles on the same side." This corollary does not appear in either Simson or Playfair.

Proposition I, 22: Thomason (1824, 29) has included the explanatory note of al-Ṭūsī:

> I say: he only set out the condition of any two lines being longer than the third on account of the necessity of the sides of the triangle being thus. That is none other than the necessity for the intersection of the two circles.

[25] Al-Anṭākī's commentary is known today in only one incomplete manuscript (Oxford University, Bodleian Library, Huntington 70, folio 24a). Although Thomason is unlikely to have had access to this surviving manuscript, it is possible that he may have encountered a still unknown copy in some Indian library (many of which are still not adequately cataloged). More plausible, though, is the hypothesis that he encountered the demonstration (or at least the alternate form of the enunciation) copied into the marginalia of a manuscript of al-Ṭūsī's *Taḥrīr*. Thomason's version of the enunciation, for example, is quoted in a marginal note in Munich, Bayerische Staatsbibliothek, *cod. arab.* 2697, folio 70b.

For if the sum of A, B be not longer than G, let TH be equal to HD or longer than it. Since circle KTL, which is bounded by circle KDL either is tangent or is not tangent to it interiorly. And if the sum of B, G be not longer than A, circle KDL similarly is bounded by circle KTL; and if the sum of A, G be not longer than B, then ZH is equal to the sum of ZD, HT or longer than it. And therefore there does not exists between the two circles a surrounding nor do they cut one another, but rather they are either tangent or not tangent to one another exteriorly.

Proposition I, 24: Euclid's text demonstrated only the most difficult of the three possible scenarios. Proclus, in his commentary on book I of the *Elements*, demonstrated all three scenarios (Morrow 1970, 263–268). Al-Ṭūsī also discussed all three possible scenarios, introducing the remaining two scenarios by "*aqūlu.*" Based on his stated editing principle, we would have expected Thomason (1824, 33–34) to have omitted these two.[26]

Proposition I, 35: In this proposition, Euclid has again demonstrated only the most difficult of the three possible scenarios. It was Proclus who demonstrated the other two scenarios (Morrow 1970, 310–316). Al-Ṭūsī placed a summary of these two scenarios in a note following the proposition. Thomason (1824, 41) included this note in his edition, although neither Simson nor Playfair discussed the additional scenarios.

Proposition I, 39: Thomason (1824, 44) included a note found in al-Ṭūsī's *Taḥrīr*: "But if E falls outside of BD, the demonstration is just as has occurred." The demonstration of this alternative scenario was given by Proclus (Morrow 1970, 322). Neither Simson nor Playfair included this scenario.

Proposition I, 41: Thomason (1824, 45–46) included a note from al-Ṭūsī: "Likewise, if the two of them are on bases that are equal to one another. And the author of the book (*Ṣāḥib al-Kitāb*) will use it in proposition three of book XII." This note does not appear in the earliest dated manuscript of al-Ṭūsī's *Taḥrīr*, although it is found in some later copies. I suspect that a marginal gloss was imported into the text by a copyist. Neither Simson nor Playfair included this note.

Proposition I, 42: Thomason (1824, 47) included a brief note from al-Ṭūsī: "I say: There are different scenarios occurring because EZ may either be superimposed on AE or may fall on one of its two sides." Neither Simson nor Playfair included this note.

Proposition I, 47: Thomason (1824, 52) included a brief note from al-Ṭūsī[27]: "I say that this proposition is nicknamed 'the bride'." Thomason did not set off this editorial note from the main text of the proposition, unlike his procedure with some other retained notes. This statement was not widely found in Arabic discussions of Euclid and was not taken over into the Latin transmission (Kunitzsch 1993, 208).

Proposition II, 4: Following the porism at the end of the proposition, Thomason (1824, 57) has added the statement: "And the meaning of the falling is that some

[26] Heath (1956, I, 297) points out that Simson was able to sidestep the issue of these two added scenarios by inserting the statement: "Of the two sides DE, DF let DE be the side which is not greater than the other" (Simson 1781, 31).

[27] Kunitzsch (1993, 208) suggests that this name may be of Greek origin.

of the diameters of these parallel-sided [figures] are diameters of squares. And that the squares occurring in the squares is on account of the superimposition of the two sides on the two sides on which their diameters fall." This statement may have originated as a marginal gloss that a copyist thought should be inserted into the text. It is not at all in the style of Euclid, nor was it included in the corollary by either Simson (1781, 54) or Playfair (1806, 53), so we would have expected Thomason to have excluded it from his edition.

Proposition II, 6: Euclid stated, at the end of the construction instructions, "we complete the figure." In his edition, Thomason (1824, 58–59) added "by the fact that we connect the diagonal and we extend BH to O and LH to K." This may have been another marginal gloss that a copyist added to the manuscript that Thomason was using.

Following the demonstration, al-Ṭūsī inserted an alternate demonstration, followed by a mathematical note. Thomason (1824, 59) has removed the alternate demonstration from his edition, but has retained the mathematical note:

> It is possible to express this proposition and the one preceding it by a single statement, namely that one says line AB being bisected at G and there being marked off from it [segment] BD from what follows B in one of the two directions, however it may fall, then the product of AD into DB is either subtracted from square GB or added to it to produce square GD and the demonstration continues from here.

Proposition II, 10: Thomason (1824, 65) omitted al-Ṭūsī's alternative demonstration, as was his custom, but retained al-Ṭūsī's mathematical note following it.

> It is possible to state this proposition and that preceding it in a single expression, namely that one says that line AB is bisected at G and there is taken from it BD from what follows B in one of the two directions, then the squares AD, BD are equal to half the squares AG, GD—and the demonstrations proceed from there.

Proposition II, 13: Following the demonstration, Thomason (1824, 69) has retained al-Ṭūsī's mathematical note:

> I say [that] for this proposition there are different scenarios because if angle G is right, the perpendicular is superimposed on side AG; and that which falls between the angle and the square is the base itself. And if [the angle] is obtuse, the perpendicular falls outside [the triangle] in the direction of G; and the falling is larger than the base. And if [the angle] is acute, the perpendicular falls within the triangle; and the falling is [only] some of the base, as is drawn in the book.

Thomason also retained a second mathematical note from al-Ṭūsī:

> It is possible to state this proposition and that which precedes it be a single expression, namely that one says [that] for any triangle, if the difference between the square on the chord of its angle that is not right and the two squares on its sides is double the product of the base

> into that which occurs between the angle and the foot of the perpendicular from the line of the base. Thereafter the proof is commensurable with respect to its argumentation.

Proposition III, 1: Thomason (1824, 72) retained the porism following al-Ṭūsī's demonstration: "From this it may be shown that the two chords do not intersect at right angles and one of them bisects the other unless one of the two of them passes through the center of the circle; and by another expression: There cannot be extended a perpendicular from the bisection of a chord unless it passes through the center."

Thomason (1824, 73) also retained a mathematical note inserted by al-Ṭūsī: "But if the center specified on AB is other than point H, such as point Z, the contradiction is from another point of view, namely the bisecting of the line in two places, namely H, Z."

Proposition III, 8: Thomason (1824, 79–80) retained a mathematical note from al-Ṭūsī about combining this proposition with the previous proposition.

> I say that it is possible to state this proposition and that which preceded by a single expression, namely that one says [that] any point that is not the center of a circle, from which there extend lines to its circumference, the longest of the lines is that which passes through the center after its extension from the point and before its ending at the circumference; and the shortest of them is that which does not pass through it and is [extended] rectilinearly; and those that are nearer to the longer are longer and those that are closer to the shortest are shorter and there are not among them two that are equal to one another except the two on either side of one another [in a straight line]. The proof follows.

Proposition III, 9[28]: Following the demonstration, Thomason (1824, 81) has included a mathematical note introduced by "*aqūlu*" in the style of al-Ṭūsī:

> I say [that] this proposition has different scenarios. For ETB either falls between EZ, EH or on one of the two of them or outside the two of them. These are the three scenarios. As for the first, it is what occurs in the book [of Euclid]; as for the second and the third, it is necessary for the two of them to be the equal of the lines extended from one of the two sides of the longest – and it is impossible also since the two are not equal to one another unless the two are on the two sides of it; and if ETB is superimposed on EA in the first scenario, on account of the necessity of it being the longer of the remaining, together with its being equal to the two of them, and it is necessary in the second scenario also.

This note is not found in any manuscripts of al-Ṭūsī's *Taḥrīr* that I have consulted. It is, however, present in *Durrat al-Tāj* (Istanbul, Damat Ibrahim 16, folio 66b). Perhaps it originated as a marginal gloss that was incorporated into the text by a later copyist. Neither Simson nor Playfair discussed these scenarios.

Proposition III, 19 (Greek III, 20): Thomason (1824, 89) has retained the mathematical note of al-Ṭūsī: "I say that for this proposition there are different scenarios, because AD falls either between sides AB, AG as in the text or it is superimposed

[28] See also the note on this proposition in the previous section.

on one of the two of them or it is outside of them, like this. And each scenario is clear from what has been shown." Both Simson (1781, 85) and Playfair (1806, 89) demonstrated the first and third cases mentioned by al-Ṭusı.

Proposition III, 20 (Greek III, 21): Thomason (1824, 90) has retained the mathematical note from al-Ṭūsī:

> I say: This is when the segment is greater than a semicircle. As for when it is not thus, the principle is not demonstrated by this approach since there does not exist here a central angle on arc GD. The approach here is to demonstrate that angles EGA, EDA that occur in segment EGDA, which is greater than a semicircle are equal to one another and the two complementary angles at H are equal to one another. Thus there remain in triangles AHD, EHG angles DAH, GEH equal to one another.

This mathematical note corresponds to the second part of the demonstration given by both Simson (1781, 86) and Playfair (1806, 89–90). Thomason may have retained al-Ṭūsī's note reason for this reason.

Proposition III, 24 (Greek III, 25): Thomason (1824, 93) has retained al-Ṭūsī's mathematical note. "I say that for this proposition there are different scenarios, because AE either falls outside of the segment, or is superimposed on AD and united to ED, or falls inside the segment. The first is what is intended in the book and the remaining two are thus and the two of them are evident."

In this proposition also, both Simson (1781, 88–89) and Playfair (1806, 92–93) discuss all three cases mentioned by al-Ṭūsī.

Proposition III, 30 (Greek III, 31): Thomason (1824, 98) omitted al-Ṭūsī's alternate demonstration but included his mathematical note:

> I say: Conversely, if angle D from triangle ABD is right and we draw on AB a semicircle, it passes through point D. But if not, we do not extend AD to the circumference and we connect between it (point D) and B, then the exterior and interior [angles] of the triangle produced are both right. This is a contradiction. And this converse is used frequently.[29]

This converse note does not appear to correspond to anything found in either Simson or Playfair. Why Thomason chose to include it in his edition is unknown.

Proposition III, 35 (Greek III, 36): Thomason omitted al-Ṭūsī's editorial note but included (1824, 106) his statement of the corollary: "From this it may be shown that any pair of line extended from a point and tangent to one and the same circle on its two sides are equal to one another." But he omitted al-Ṭūsī's demonstration of the corollary. This corollary has probably been preserved because it is similar to the corollary stated by Simson (1781, 100) and Playfair (1806, 105).

Proposition IV, 5: Thomason (1824, 112) retained al-Ṭūsī's mathematical note on differences of position in this demonstration: "For this proposition there are different scenarios: For if the meeting of the two perpendiculars is at Z, it is either

---

[29] Thomason has omitted the concluding clause: "And in this proposition he used the lemma (*muqaddimah*) demonstrated in the first proposition of book V".

outside the triangle, as drawn in the text, and that is when the angle BAG is obtuse, or inside it, and that is when it is acute, or on side BG, when it is right, thus." It is probable that Thomason retained this note because its content is similar to the corollary that Simson (1781, 107) and Playfair (1806, 112–113) attached to this proposition.

Proposition IV, 10: Thomason (1824, 116) retained al-Ṭūsī's note on terminology: "This triangle is known as the triangle of the pentagon." Neither Simson nor Playfair used it.

Proposition IV, 15: Thomason (1824, 120–121) retained al-Ṭūsī's porism: "It may be shown that the side of the hexagon is equal to half the diameter of the circle; and it is possible to construct about a circle a hexagon and [to construct] a circle in a hexagon or about it, just as occurred in the case of the pentagon." This note was probably retained because its content is similar to the corollary that Simson (1781, 117) and Playfair (1806, 125) have attached to this proposition.

Definition V, 2 (Greek definition V, 3): Thomason (1824, 122) placed first the standard Greek definition, following al-Ṭūsī. He then gave an alternative formulation of the definition that al-Ṭūsī had attributed to Thābit. This alternative formulation is closer to the formulation found in Simson (1781, 119) and Playfair (1806, 130). Perhaps it was for this reason that Thomason decided to keep both formulations of the definition.

Proposition V, 10: Following the proposition, Thomason (1824, 133) inserted a brief note: "I say [that] this only occurs between magnitudes of the same kind (or species)." The note is set off from the remainder of the proposition text by a lozenge-shape that typically separates propositions and is centered on the page rather than being right-justified like the remainder of the proposition. This note was not present in the earliest dated manuscript of the *Taḥrīr* (British Library, add. 23,387, folio 84a), but it is found in some later copies. For example, the statement is present in Istanbul, Feyzullah 1359 (al-Ṭūsī 2012, folio 58a). The statement was also included in *Durrat al-Tāj* (Istanbul, Damad Ibrahim 815, folio 71b). It may have been a marginal gloss that a copyist inserted into the text itself. This condition does not appear in either Simson's or Playfair's work.

Proposition V, 14: Following the demonstration, Thomason (1824, 137–138) placed al-Ṭūsī's mathematical note, set off from the remainder of the proposition by a lozenge-shaped marker: "It is well known that this principle also is specific to magnitudes of the same kind (or species), for if the first two be not of the same kind then there is no relationship between the two of them with respect to largeness and smallness and equality together with the existence of the relation in which they occur." This comment also is not found in in the earliest dated manuscript of the *Taḥrīr* (British Library, add. 23,387, folio 82b), but it is found in some later copies. For example, the statement is added in the margin of Istanbul, Feyzullah 1359 (al-Ṭūsī 2012, folio 59a). The statement was also included in *Durrat al-Tāj* (Istanbul, Damad Ibrahim 815, folio 71b). It may have been a marginal gloss that a copyist inserted into the text itself. In this proposition also, the condition is not stated explicitly by Simson or Playfair.

Proposition V, 16: Thomason (1824, 139) included al-Ṭūsī's mathematical note (introduced by "*aqulu*") but set it off from the proposition by a lozenge-shape that usually signals the end of a proposition. "He (Euclid) sets a condition in it that the four [magnitudes] be of the same kind (or species). For if they are set in a ratio but they are of two different species (or ranks)—for example the ratio of a line to a line is as the ratio of a surface to a surface—alternation cannot occur here." Thomason may have retained this note because both Simson (1781, 139) and Playfair (1806, 146) had placed this condition in the enunciation, although it is not present in the Greek text (Heath 1956, II, 164).

Definition VI, 5: Al-Ṭūsī ascribed this definition to Thābit. He gave two versions of the definition, stating first that "the ratio compounded from ratios is that resulting from the increasing of some magnitudes of these ratios by others." The second version, we are told, appeared in some manuscripts: "A ratio is divided into ratios when some of these ratios are partitioned they produce others." Thomason (1824, 149) included the first, but not the second, version in his list of definitions. Thomason added at the end: "I mean, from the multiplying (*ḍarb*) of some of them into others." This explanatory phrase is not in al-Ṭūsī's *Taḥrīr* This phrase probably came from *Durrat al-Tāj* (Istanbul, Fazil Ahmed Paşa 867, folio 92b). Neither Simson nor Playfair included it among their definitions.

Proposition VI, 1: Thomason (1824, 151) included a proof that "if triangles or parallelograms are to one another as their bases, their altitudes are equal." Its inclusion may be due to the corollary of Simson (1781, 157) and Playfair (1806, 157): "Triangles, and also parallelograms, that have equal altitudes, are to one another as their bases."

Features of the Persian Translation

The decision to produce a Persian translation of the Arabic was, in hindsight, very useful. The Persian translation often employs more contemporary terminology than that found in *Durrat al-Tāj*. And the Persian translation regularly proved more popular with readers than did the Arabic. (There is some indication that the Persian translation was reprinted at least once.) But by the 1860s, sales of both editions were in sharp decline. In the last half of the nineteenth century, it was now English textbooks, either produced in England for the Indian market (Aggarwal 2007, 11–14) or produced locally and cheaply for local schools.

Table 1 illustrates the differences between al-Shīrāzī's Persian translation from al-Ṭūsī's Arabic and Hyder Alee's translation from the Arabic edition of Thomason. These differences may suggest another reason why Hyder Alee was asked to translate the Arabic edition rather than edit the version found in *Durrat al-Tāj*—al-Shīrāzī did not just translate the Arabic text into Persian. He also edited the text just as Ibn Sīnā had done before, condensing the text by combining the enunciation with the specification section of the Euclidean proof.

This example also reveals that Hyder Alee has used native Persian words and constructions rather than rely on the Arabic loan words that al-Shīrāzī often seems

Table 1 Comparison between the translations of al-Shīrāzī and Hyder Alee

Proposition V, 13 (-Greek V, 12) enunciation	
Hydar Alee's translation from Arabic edition	Al-Shīrāzī's translation from al-Ṭūsī's Arabic
Whenever (*vaqtīki*) there are some (*chand*) magnitudes that are mutually proportional, then (*pas*) one (*yek*) antecedent in relation to (*be-suvī*) its consequent is as all (*hama*) the antecedents to all (*hama*) the consequents	If (*chūn*) there are proportional magnitudes, such as (*chūn*) A, B, G, D, E, Z that are in the same ratio (*yek nisbet*), the ratio of an antecedent (*muqaddim-i vāhid*) to (*bā*) a consequent, such as A to B, is the same as (*chunān bā shad ki*) the ratio of all (*jamīʿ*) the antecedents to all (*jamīʿ*) the consequents—I mean, for example (*chūn*), the ratio of A, G, E to B, D, Z

to prefer. The increased use of Persian makes the resulting treatise seem less like a translation and hence makes it easier for Persian-speaking readers to comprehend the content of the treatise.

The Fate of Thomason's Arabic Edition and Its Persian Translation

Thomason's Arabic edition had an initial print run of 2000 copies, of which 118 were apparently still in stock in 1844; the Persian translation had an initial print run of 2060 copies, of which 886 were still in the book depository (CSBS 1845, 37). In 1852, the CSBS directors decided to print a second edition of the Persian translation, with a print run of 1000 copies (CSBS 1851, 13–14). It appears unlikely that the print order was completed because in 1852 the inventory showed that there were only 196 copies of both the Arabic and the Persian versions in stock (CSBS 1852, 24–25). The proceedings make no further mention of this planned reprinting. Table 2 shows the annual distribution of each between 1845 and 1860, drawn from the inventory records published in the Proceedings volumes.

Concluding Thoughts

Thomason took an unusual approach to the task of creating a geometry textbook in Arabic for use in Indian schools.[30] Rather than translate a modern European textbook, he edited a manuscript of an existing Arabic textbook that was still

[30] Although unusual, Thomason's approach was not unique. The Rev. Robert May had produced an elementary arithmetic text in Bengali, titled *Gonito*, using verse (or rhymed prose), a traditional Bengali style. The approach turned out to be very popular—the first print run of 500 copies sold out almost immediately, prompting the Secretary to lament that they had not published 5000 copies (Gupta 2014, 59; Basak 1959, 50). Rhymed prose has a long history in traditional mathematics instruction, although frequently ignored by historians (De Young 2016).

Table 2 Number of copies supplied annually between 1845 and 1859

Year	Number of Arabic copies supplied	Number of Persian copies supplied
1845	1	41
1846	8	1
1847	39	3
1848	21	0
1849	2	0
1850	6	13
1851	13	43
1852	6	51
1853	0	7
1854	6	30
1855	0	62
1856	0	6
1857	0	3
1858	0	31
1859	7	10

regularly used in mathematics instruction. (Mathematics instruction was typically in Arabic, or at least used sources written in Arabic, even though the prevailing vernacular of the court and everyday life was Persian.) Thus, almost everyone in the intended audience would have recognized the name of al-Ṭūsī, even though they might not be able to explain much about what he had actually said or done in mathematics. Use of a traditional textbook would also simplify the process of producing the modern vernacular Arabic textbook because the prose would already be in an easily recognized and accessible style.

As we have seen, Thomason's stated editing principles are simple—remove from al-Ṭūsī "all that is not Euclid." But in practice, his procedures were somewhat more complex because he sometimes added material to al-Ṭūsī's text and more often retained notes from al-Ṭūsī that are not Euclid's and that are not represented in the textbooks of Simson and Playfair. As is so often the case, Thomason does not explain why he made these exceptions to his stated procedures.

Hyder Alee also does not simply translate Thomason's Arabic edition into Persian, but sometimes paraphrases the Arabic. He is especially likely to replace Arabic pronouns with their antecedent nouns. The result is usually a cleaner and clearer statement. He may have been influenced in part by the style used by al-Shīrāzī while translating al-Ṭūsī's *Taḥrīr*.

Thomason's Euclid edition represented a specific response to a perceived educational need. After completing his edition, Thomason reportedly intended to follow up on this first mathematical publication by preparing "in Arabic a compendium of Plane and Spherical Trigonometry" (CSBS 1819, 16). This would have seemed a logical next step for someone educated using Simson or Playfair since both add an introduction to plane and spherical trigonometry to their geometry textbooks. Unfortunately, this planned work was never completed because of the illness of his wife, his return to England, and his own untimely death.

Thomason's Arabic edition was able to appeal to both Anglicist and Orientalist sentiments of that time. On the one hand, it was an edition of a treatise in an oriental language that already had a long and illustrious history in Eastern culture. And on the other hand, the underlying goal for Thomason was not the preservation of ancient cultures or a return to a time when Eastern culture was in the ascendant, but rather the education of contemporary Indians so as to raise them above their current level of moral and intellectual degradation, as he described it in his private correspondence. This improvement of Indian character through the study of mathematics and natural sciences would, of course, also be of great benefit to the colonial exploitation of India by providing the East India Company with employees equipped with the intellectual tools needed for administrative work.

Thomason's edition of Euclid's *Elements* seems to have had less success than one might initially suppose, given that Persian was the official language of the Mughal court and that Arabic was the language of Islamic religious revelation and elaboration on it. Arabic was also the traditional language of mathematics instruction throughout the Islamicate world. The importance of al-Ṭūsī's *Taḥrīr* in mathematics instruction is underlined by the role played by Muḥammad Barakat's commentary on book I of the *Elements* in the Dars-i Niẓāmī curricular reforms in the traditional Islamic schools (madrasa) in India later in the nineteenth century (De Young 2012c, 10–13; 2012d). But the CSBS records do not suggest that there was any significant interaction between Calcutta and the Farangi Mahall reformist group in Lucknow.

References

Abdeljaouad, Mahdi and De Young, Gregg (2020). Al-Shīrāzī's "proofs" of Euclid's postulates in Arabic and Persian. *SCIAMVS* 21: 1–82.

Aggarwal, Abhilasha (2007). Mathematical books for and in India in the nineteenth century. *Bulletin of the British History of Mathematics Society* 22: 11–21.

Arnzen, Rüdiger (2002). *Abū l-'Abbās an-Nayrīzīs Exzerpte aud (Ps.-?) Simplicius' Kommentar zu den Definitionen, Postulaten und Axiomen in Euclids Elementa I*. Köln, R. Arnzen.

"An Account of the life and character of Tufuzzel Hussein Khan," (1804). *Asiatic Annual Register or View of the History of Hindustan of Asia for the Year 1803*. "Characters", pp. 1–8. London, Cadell and Davies.

Abid, Arif (2006). Nawab Tafazzul Hussain Khan, 1727–1800: A Poisoned chalice. Online at: https://3quarksdaily.com/3quarksdaily/2006/03/nawab_tafazzul_.html. Accessed 23 August 2023.

Basak, N. L. (1959). Origin and role of the Calcutta School-Book Society in promoting the cause of education in India, especially vernacular education in Bengal (1817–1835). *Bengal Past and Present* 77: 30-69.

Brentjes, Sonja (1998). On the Persian transmission of Euclid's *Elements*. In: Vesel, Ževa, *et al.* (eds), *La science dans le monde iranien à l'époque islamique*, pp. 73–94. Louvain, Peeters.

Brentjes, Sonja (2001). Observations on Hermann of Carinthia's version of the *Elements* and its relation to the Arabic transmission. *Science in Context* 14: 39–84.

Brentjes, Sonja (2006). An exciting new Arabic version of Euclid's *Elements*: MS Mumbai, Mullā Fīrūz R.I.6. *Revue d'histoire des mathématiques* 12: 169–197.

Brentjes, Sonja (2008). Patronage of the mathematical sciences in Islamicate societies. In: E. Robson and J. Stedall (eds), *Oxford Handbook of the History of Mathematics*, pp. 301–337. Oxford, Oxford University Press.

Brentjes, Sonja (2018). Who translated Euclid's *Elements* into Arabic? In: J. Hämeen-Attila and I. Lindstadt (eds), *Translation and Transmission: Collection of Articles*, pp. 21–54. Münster, Ugarot-Verlag.

Brentjes, Sonja (2019). MS Munich, Bayerische Staatsbibliothek, *codex arab.* 2697 and its properties. *Micrologus* 27: 443–466.

Busard, Hubert L. L., ed (2005). *Campanus of Novara and Euclid's Elements.* Wiesbaden, Franz Steiner.

Cassinet, Raymonde (1993). L'aventure de l'édition des Éléments d'Euclide en arabe par le Societé Typographique Médecis vers 1594. *Revue française d'histoire du livre* 78–79: 5–51.

CSBS (1819), *Calcutta School-Book Society's Proceedings—Second Year*. Calcutta, Calcutta School-Book Society. The *Proceedings* (but not the appendices) are reprinted in *The Friend of India*, II (1819), pp. 443–449.

CSBS (1845). *Thirteenth proceedings of the Calcutta School-Book Society*. Calcutta, Calcutta School-Book Society.

CSBS (1852). *Fifteenth proceedings of the Calcutta School-Book Society*. Calcutta, Calcutta School-Book Society.

CSBS (1853). *Sixteenth proceedings of the Calcutta School-Book Society*. Calcutta, Calcutta School-Book Society.

De Young, Gregg (2003). The *Taḥrīr* of Euclid's *Elements* by Naṣīr al-Dīn al-Ṭūsī: Redressing the balance. *Farhang: Quarterly Journal of Humanities and Cultural Studies* 15–16: 117–143.

De Young, Gregg (2005). Diagrams in the Arabic Euclidean tradition: A preliminary assessment. *Historia Mathematica* 32: 129–179.

De Young, Gregg (2007). Quṭb al-Dīn al-Shīrāzī and his Persian translation of Naṣīr al-Dīn al-Ṭūsī's *Taḥrīr Uṣūl Uqlīdis*. *Farhang: Quarterly Journal of Humanities and Cultural Studies* 10 (61–62): 17–75.

De Young, Gregg (2008–2009). The *Taḥrīr Kitāb Uṣūl Uqlīdis* of Naṣīr al-Dīn al-Ṭūsī: Its Sources. *Zeitschrift für Geschichte der arabisch-islamischen Wissenschaften* 18: 1–71.

De Young, Gregg (2012a). Further adventures of the Rome 1594 redaction of Euclid's *Elements*. *Archive for History of Exact Sciences* 66: 265–294.

De Young, Gregg (2012b). Euclidean geometry in two medieval Islamic philosophical compendia. *CSHPM/SCHPM Proceedings* 25: 63–71.

De Young, Gregg (2012c). Nineteenth century traditional Arabic geometry textbooks. *International Journal for the History of Mathematics Education* 7(ii): 1–34.

De Young, Gregg (2012d). Muḥammad Barakat's commentary on *Elements* I: An unintended mathematics textbook. *Gaṇita Bhāratī* 34: 161–191.

De Young, Gregg (2014). Translating Playfair's *Geometry* into Arabic: Mathematics and missions. In: Sidoli, Nathan and Van Brummelen, Glenn (eds), *From Alexandria, Through Baghdad: Surveys and Studies in the Ancient Greek and Medieval Islamic Mathematical Sciences in Honor of J. L. Berggren*, pp. 503–527. Berlin, Springer Verlag.

De Young, Gregg (2016). An Arabic introduction to Euclidean geometry in didactic verse. *SCIAMVS* 27: 161–215.

De Young, Gregg (2023). MS Paris, BULAC Ara 606 and the medieval Ḥajjāj tradition. To appear in *Historia Mathematica*.

Evans, Stephen (2002). Macaulay's Minute revisited: Colonial language policy in nineteenth century India. *Journal of Multilingual and Multicultural Development* 23: 260–281.

Gupta, Abhijit (2014). The Calcutta School-Book Society and the production of knowledge. *English Studies in Africa* 57: 55–65. DOI: https://doi.org/10.1080/00138398.2014.916908

Halley, Edmund (1706). *Apollonii Pergaei de sectione rationis*. London, E Theatro Sheldoniano.

Heath, Thomas L. (1956). *The Thirteen Books of Euclid's Elements*. New York, Dover.

İzge, Mahmut (2014). *British Colonial Education Policy: The Orientalist-Evangelist Controversy in India (1780–1835)*. Istanbul, Sakarya University MA thesis.

Khan, Muhammed Ali, ed. (1908). *Life of Tuffuzzool Hussain Khan.* Secundarabad, Cheekety Veerunnah.

Kunitzsch, Paul (1993). "The Peacock's Tale:" On the names of some theorems of Euclid's *Elements.* In: Folkerts, Menzo and Hogendijk, Jan (eds), *Vestigia Mathematica: Studies in Medieval and Early Modern Mathematics in Honour of H. L. L. Busard*, pp. 205–214. Atlanta, Rodopi.

Legendre, Adrien-Marie (1794). *Éléments de géométrie.* Paris, Firmin Didot.

Macauley, Thomas Babington (1835). *Minute on Education.* [s. n.]

Macfie, Alexander (2002). *Orientalism.* London, Routledge.

Masumanian, Adib , tr (2021). Account of three mathematical works of Tufuzzool Hoosyn Khan [Tafazzul Hussein Khan], deceased, drawn up by Mowluwee Hydur 'Ulee [Maulvi Haidar 'Ali]. Calcutta, Calcutta School-Book Society, 1819. Available online: https://adibmasumian.files.wordpress.com/2021/09/maulvi_haidar_ali_trans_09072021.pdf. Accessed 23 August 2023.

Al-Nadīm (2009). *Kitāb al-Fihrist.* A. F. Sayyid (ed). London, Furqān Islamic Heritage Foundation.

Netz, Reviel (1998). Greek mathematical diagrams: Their use and their meaning. *For the learning of mathematics* 18(3): 33–39.

Niazi, Kaveh (2011). *A Comparative Study of Quṭb al-Dīn al-Shīrāzī's Texts and Models on the Configuration of the Heavens.* Columbia University PhD thesis.

Niazi, Kaveh (2013). Quṭb al-Dīn al-Shīrāzī as depicted in early historical sources. *Tarikh-e Elm* 11: 23–39.

Playfair, John (1795/1806). *Elements of Geometry, Containing the First Six Books of Euclid.* Philadelphia, F. Nichols.

Rosenfeld, Boris and Ihsanoğlu, Ekmeleddin (2003). *Mathematicians, Astronomers, and Other Scholars of Islamic Civilization and Their Works (7th—19th c.).* Istanbul, IRCICA.

Sargent, John (1834). *The Life of the Reverend T. T. Thomason, MA.* London, Seeley and Burnside.

Schaffer, Simon (2009). The Asiatic enlightenments of British astronomy. In: Schaffer, Simon (ed), *The Brokered World: Go-betweens and Global Intelligence, 1770–1820*, pp. 43–104. Sagamore Beach, Science History.

Simpson, Thomas (1745). *Treatise of Algebra.* London, Collingwood.

Simson, Robert (1756/1781). *The Elements of Euclid, viz. The First Six Books together with the Eleventh and Twelfth.* Edinburgh, J. Balfour.

Stowell, T. Alfred (1896). The Elland Clerical Society. *The Churchman* 10(8): 417–426.

Sude, Barbara (1974). Ibn al-Haytham's Commentary on the Premises of Euclid's Elements (*Sharḥ Muṣādarāt Kitāb Uqlīdis fī al-Uṣūl*). Books I–VI. Princeton University PhD Thesis.

"The Reverend Thomas Thomason" (1851). In: *The Bengal Obituary*, pp. 40–42. London, Thacker.

Thomason, Thomas, ed (1824). *Sitta Maqālāt min Kitāb Taḥrīr al-Uqlīdis.* Calcutta, CSBS.

Al-Ṭūsī, Naṣīr al-Dīn. (2012). *Eukleides'in Elemanlar Kitabının Tahrîri, Tahrîru Usûli'l-Hendese ve'l-Hisâb.* Istanbul, Türkiye Yazma Eserler Kurumu Başkanlığı. Facsimile edition of MS Feyzullah 1359 of al-Ṭūsī's *Taḥrīr.*

Walbridge, John (1992). *The Science of Mystic Lights: Quṭb al-Dīn al-Shīrāzī and the Illuminationist Tradition in Islamic Philosophy.* Cambridge, Harvard Center for Middle Eastern Studies.

Zastoupil, Lynn and Moir, Martin (1999). *The Great Indian Education Debate: Documents relating to the Orientalist-Anglicist Controversy, 1781–1843.* Richmond, Curzon.

Curriculum Vitae

22. 2. 1944	Born in Arnswalde (Germany/prov. Pomerania)
Autumn 1949 to March 1963	Attending schools in Oberbozen (Italy), Rome (Italy) and Wiesbaden (Germany)
March 1963	*Abitur* at the Dilthey-Schule Wiesbaden (*humanistisches Gymnasium*)
Summer 1963 to Summer 1969	Studies in mathematics and physics at Mainz University and at Bonn University: *Diplom* in mathematics
October 1969 to January 1972	Tutor at the Mathematical Institute of Bonn University
February 1972 to June 1973	Member of a curriculum project in mathematics for the upper grades of the *Gesamtschule* (comprehensive school) Gelsenkirchen, developing conceptions for teaching analysis
Since 1973, until 2009	Member of the *Institut für Didaktik der Mathematik* of the Bielefeld University, an interdisciplinary research institute for mathematics education founded the same year-first as research assistant and since 1977 as department head
1977	Ph. D. thesis on the genetic principle (under aspects of philosophy and of history of science); Dr. math. at the Mathematics Department of the Bielefeld University Begin of research in history of mathematics and in history of mathematics education, and of lecturing on these subjects
1980	Organizer of the Symposium *Comparative Study of the Development of Mathematical Education as a Professional Discipline in Different Countries* within the 4th International Congress on Mathematical Education, Berkeley
1981 to 1983	Guest at the *Interdisziplinäres Zentrum für Hochschuldidaktik* of Bielefeld University (center for research in higher education) to direct the research and development project on the emergence of the profession of mathematics teachers in Prussia. My book "Die Entstehung des Mathematiklehrerberufs" is a result of this project

(continued)

A. Verdugo Rohrer and J. Zender (eds.), *History of Mathematics and Its Contexts*, Trends in the History of Science, https://doi.org/10.1007/978-3-031-60828-5

(continued)

1984–1990	Directing the *chronique historique* of the journal *Recherches en Didactique des Mathématiques*
1989	Organizer of the Symposium: *The Prussian institutional reforms in science and education—their challenge to national policies reassessed*, as part of the XVIIIth International Congress of History of Science (Hamburg/Munich) The volume "'Einsamkeit und Freiheit' neu besichtigt", edited by me, contains the Proceedings of this Conference
1992	Three months as guest professor at the Pedagogical University in Maputo/Moçambique, giving courses in history of mathematics for students and in mathematics education for lecturers
1993	Member of the Scientific Committee for the First European Summer University "Histoire et Epistémologie dans l'Education Mathématique", 19–23. 7. 1993, Montpellier (France)
1994	Member of the Scientific Committee for the Conference "Réformer l'Enseignement Scientifique: Histoire et Problèmes Actuels", 17–19.1.1994, Paris
May 1994	Organizer of the International Conference "150 Jahre 'Lineale Ausdehnungslehre'. Hermann G. Graßmann—Werk und Wirkung", 23–28. 5. 1994, Rügen (Germany) The Proceedings of this Conference, edited by me, have been published by Kluwer Acad. Publ
July/August 1995	One month as guest professor at the Mathematics Department of the Pontifícia Universidade Católica de Rio de Janeiro
April 1996	Guest lectures at the Seminar for History of Exact Sciences, University of Zaragoza (Spain)
July 1996	Member of the Scientific Committee for the *Deuxième Université d'Été Européenne Histoire et Epistémologie des Mathématiques* and the HPM ICME-8 Satellite Meeting *História e Educação Matemática*, Braga (Portugal)
April 1997	Participation at the 2° Seminário Nacional de História da Matemática, Aguas de São Pedro/SP
July 1997	Lecturer at the Third European Summer School for Graduates in Mathematics History (ERASMUS Intensive Programme), at Enfield, England
March 1999	Participation at the 3° Seminário Nacional de História da Matemática, Vitória/ES
June 1999	History Courses at the ERASMUS Intensive Programme "New Directions in Didactics of Mathematics", Nicosia (Cyprus)
1999	Member of the Scientific Committee for the *Troisième Université d'Été Européenne Histoire et Epistémologie des Mathématiques* in Louvain/Leuven (Belgium), July 1999
November/December 2000	Guest professor at the Universidade Federal de Espírito Santo, Vitória (ES), Brazil (one month)

(continued)

(continued)

2001 to 2003	German coordinator of the DAAD/CAPES exchange program PROBRAL for scientists and doctoral students: "Relation between History of Mathematics and its Teaching—Modernisation of science and its transmission" Regular one month guest stays at the Brazilian partner universities
2002	Member of the *Comité Scientifique* for the *7e Colloque Maghrébin d'Histoire des Mathématiques Arabes*, Marrakesch, June 2002
2002	*Habilitation* and *Privatdozent* (livre docência) for History of Mathematics
2004	Member of the Program Committee for *History and Pedagogy of Mathematics*, ICME 10 Satellite Meeting Monday 12–Saturday 17 July 2004. Uppsala, Sweden
July 2004	Prêmio Jabutí in the category of exact sciences for the book *Análise Histórica de Livros de Matemática. Notas de Aula*, published in 2003
2004	Member of the *Comité Scientifique* for the 8e *Colloque Maghrébin d'Histoire des Mathématiques Arabes*, Tunis, December 2004
2004	Co-Chair of Topic Study Group 28: The history of teaching and learning mathematics, ICME 10—Copenhague, July 2004
2006	Member of the *Comité Científico* of HTEM III: História e Tecnologia no Ensino da Matemática, São Paulo/SP, 25–27. 5. 2006
2006	German coordinator of the DAAD/CAPES exchange program PROBRAL for scientists and doctoral students: History and Epistemology of Mathematics for innovative teaching of Mathematics—new approaches
2007	Co-organizer of the Second International Meeting: Semiotic Approaches to Mathematics, the History of Mathematics and Mathematics Education, July 2007, Melle/Germany
2007	Member of Scientific Committee for the 5th European Summer University History and Epistemology of Mathematics, July 2007, Prague
2008	Member of the *Comité Científico* of HTEM IV: História e Tecnologia no Ensino da Matemática, Rio de Janeiro, May 2008
2008	Member of the Program Committee for *History and Pedagogy of Mathematics*, ICME 11 Satellite Meeting July 2008. Mexico City, Mexico
2009	Co-organizer of the international workshop: On-going research into the history of mathematics education, 20–23 June 2009, Reykjavik (Islândia)

(continued)

(continued)

2009	Co-organizer of the Third International Meeting: Semiotic Approaches to Mathematics, the History of Mathematics and Mathematics Education (Semmhisted), Aristotle University of Thessaloniki, July 16–17, 2009
Since 2000	Member of the Advisory Board of HPM, The International Study Group on the Relations between the History and Pedagogy of Mathematics
March 2009	Retirement from Bielefeld University
Since November 2009	Visiting Professor at the Programa de Pós-Graduação em Ensino de Matemática, Universidade Federal do Rio de Janeiro, Rio de Janeiro
July 2010	Member of Scientific Committee for the 6th European Summer University History and Epistemology of Mathematics (ESU 6), July 2010, Vienna—Austria
June 2011	Member of the *Comité Científico* of HTEM V: História e Tecnologia no Ensino da Matemática, Recife, 25–31 June 2011
October 2011	Co-organizer of the Second International Conference on the History of Mathematics Education, 2–5 October 2011, Lisbon (Portugal)
July 2012	Member of the Program Committee for *History and Pedagogy of Mathematics*, ICME 12 Satellite Meeting July 2012. Daejeon, South Korea
July 2013	Member of the Comité Científico of HTEM VI: História e Tecnologia no Ensino da Matemática, Sao Carlos (SP), July 2013
October 2013	Member of the Comité Scientifique for the 11e Colloque Maghrébin d'Histoire des Mathématiques Arabes, Alger, October 2013
September 2013	Co-organizer of the Third International Conference on the History of Mathematics Education, September 2013, Uppsala (Sweden)
July 2014	Member of the Scientific Committee for the 7th European Summer University History and Epistemology of Mathematics (ESU 7), July 2014, Copenhagen–Denmark
September 2015	Co-organizer of the 4th International Conference on the History of Mathematics Education, September 2015, Turin (Italy)
May 2016	Member of the Comité Scientifique for the 12e Colloque Maghrébin d'Histoire des Mathématiques Arabes, Marrakech (Morocco), May 2016
July 2016	Co-Organizer of the Thematic Afternoon "The Legacy of Felix Klein", within the 13th International Congress on Mathematics Education (ICME 13), July 2016, Hamburg (Germany)

(continued)

(continued)

May 2017	Chair of the International Programme Committee for the II International Conference on Mathematics Textbook Research and Development (ICMT-2), 7–11 May 2017, Rio de Janeiro (Brazil)
September 2017	Co-organizer of the 5th International Conference on the History of Mathematics Education, September 2017, Utrecht (Netherlands)
September 2019	Co-organizer of the 6th International Conference on the History of Mathematics Education, September 2019, Luminy (France)
September 2022	Co-organizer of the 7th International Conference on the History of Mathematics Education, September 2022, Mainz (Germany)

Publications of Gert Schubring

Books and Monographs

Mathematiker über die Mathematik. Hrsg. Michael Otte unter Mitwirkung von H.N. Jahnke, Th. Mies und G. Schubring (Berlin/Heidelberg/New York: Springer 1974).

Dokumentation der Mathematik-Lehrpläne in der Bundesrepublik Deutschland (Stand: September 1977). Unter Mitarbeit von H.G. Braun und Ute Thiede. Schriftenreihe des IDM, 12/1977 (Universität Bielefeld 1977).

Das genetische Prinzip in der Mathematik-Didaktik (Stuttgart: Klett 1978).

International Bibliography of Journals in Mathematical Education. In cooperation with Jutta Richter. Schriftenreihe des IDM, 23/1980 (Universität Bielefeld 1980).

Die Entstehung des Mathematiklehrerberufs im 19. Jahrhundert. Studien und Materialien zum Prozeß der Professionalisierung in Preußen (1810–1870) (Weinheim/Basel: Beltz 1983).

Bibliographie der Schulprogramme in Mathematik und Naturwissenschaften (wissenschaftliche Abhandlungen) 1800–1875 (Bad Salzdetfurth: Franzbecker 1986).

Editor, *Quellen zur Geschichte des Mathematikunterrichts, der Zeitschrift Der Mathematikunterricht*, 34 (1988), *34:1.*

CONDORCET—Moyens d'apprendre á compter sûrement et avec facilité. Appareil critique—études, notes, commentaires, bibliographie. Charles Coutel, Nicole Picard, Gert Schubring, eds. (Paris: ACL-éditions 1988).

Editor, *'Einsamkeit und Freiheit' neu besichtigt. Universitätsreformen und Disziplinenbildung in Preußen als Modell für Wissenschaftspolitik im Europa des 19. Jahrhunderts.* Proceedings of the Symposium of the XVIIIth International Congress of History of Science at Hamburg-Munich, 1–9 August 1989 (Stuttgart: Franz Steiner Verlag 1991).

Die Entstehung des Mathematiklehrerberufs im 19. Jahrhundert. Studien und Materialien zum Prozeß der Professionalisierung in Preußen (1810–1870). Zweite, korrigierte und ergänzte Auflage (Weinheim: Deutscher Studien Verlag 1991),

A. Verdugo Rohrer and J. Zender (eds.), *History of Mathematics and Its Contexts*, Trends in the History of Science, https://doi.org/10.1007/978-3-031-60828-5

Die Mathematiker, Astronomen und Physiker an der Universität Jena [1558–1914]. Kommentierte Edition des Manuskripts von Fritz Chemnitius (1930). Reihe Algorismus, Heft 7 (München 1992).
Editor, *Hermann Günther Graßmann (1809–1877): Visionary Mathematician, Scientist and Neohumanist Scholar. Papers from a Sesquicentennial Conference* (Boston Studies in the Philosophy of Science, volume 187). (Dordrecht/Boston/London: Kluwer, 1996).
Analysis of Historical Textbooks in Mathematics. Lecture Notes (Rio de Janeiro: Pontifícia Universidade Católica do Rio de Janeiro, Departamento de Matemática, 1997). Second, revised edition 1999.
Análise Histórica de Livros de Matemática. Notas de Aula (Campinas: Editora Autores Associados, 2003).
One Hundred Years of L'Enseignement Mathématique. Moments of Mathematics Education in the Twentieth Century. Proceedings of the EM-ICMI Symposium Geneva, 20–22 October 2000, Daniel Coray, Fulvia Furinghetti, Hélène Gispert, Bernard R. Hodgson, Gert Schubring, eds. (Geneva: L'Enseignement Mathématique, 2003). Monographie No 39 de L'Enseignement Mathématique.
A algebrização do conceito do limite no século XVIII. (Rio de Janeiro: Pontifícia Universidade Católica do Rio de Janeiro, Departamento de Matemática, Pré-publicação Mat. 10/2003).
Le Retour du Refoulé. Der Wiederaufstieg der synthetischen Methode an der École Polytechnique. Reihe Algorismus, No. 46 (Augsburg: Erwin Rauner, 2004).
Conflicts between Generalization, Rigor and Intuition. Number Concepts Underlying the Development of Analysis in 17th–19th Century France and Germany. Sources and Studies in the History of Mathematics and Physical Sciences (New York: Springer, 2005).
Editor, *Paedagogica Historica*, Special Issue: History of Teaching and Learning Mathematics, , 2006, *XLII*: IV&V.
Semiotics in Mathematics Education. Epistemology, History, Classroom, and Culture, Luis Radford, Gert Schubring & Falk Seeger, eds. (Sense Publ.: Rotterdam, 2008).
"Dig where you stand". Proceedings of the conference "On-going research in the History of Mathematics Education", Bjarnadóttir, K., Furinghetti, F., & Schubring, G. (eds.) (Reykjavik: University of Iceland—School of Education, 2009).
Die Debatten um einen Mathematiklehrplan in Westfalen 1834. Eine regionale Sozialgeschichte der Einführung von Mathematik als Hauptfach (Münster: WTM Verlag, 2010).
Signifying and meaning-making in mathematics teaching and learning: Semiotic perspectives, Luis Radford, Gert Schubring & Falk Seeger (guest-editors), Educational Studies in Mathematics, 2011, vol. 77: 2–3.
Turning points in the history of mathematics teaching—studies on national policies, Gert Schubring, Fulvia Furinghetti & Man-Keung Siu (guest-editors), ZDM—The international journal on Mathematics Education, 2012, vol. 44: 4.

Os números negativos—Exemplos de obstáculos epistemológicos? (Rio de Janeiro: LIMC editora, 2012).

"Dig where you stand." 2. Proceedings of the Second International Conference on the History of Mathematics Education, Lisbon, October 2–6, 2011, eds. Kristín Bjarnadottir, Fulvia Furinghetti, José Matos, Gert Schubring (Caparica: UIED, 2012).

O papel da visualização na história dos números complexos (Natal: Sociedade Brasileira de História da Matemática, 2013).

Handbook on the History of Mathematics Education, Alexander Karp & Gert Schubring, eds. (New York: Springer, 2014).

"Dig where you stand." 3. Proceedings of the Third International Conference on the History of Mathematics Education, Uppsala, September 25–28, 2013, eds. Kristín Bjarnadottir, Fulvia Furinghetti, Johan Prytz, Gert Schubring (Uppsala: Uppsala University—Department of Education, 2015).

Felix Klein, *Elementary Mathematics from a Higher Standpoint. Volume I: Arithmetic, Algebra, Analysis.* Translated by Gert Schubring (Berlin & Heidelberg: Springer, 2016).

Felix Klein, *Elementary Mathematics from a Higher Standpoint. Volume II: Geometry.* Translated by Gert Schubring (Berlin & Heidelberg: Springer, 2016).

Felix Klein, *Elementary Mathematics from a Higher Standpoint. Volume III: Precision Mathematics and Approximation Mathematics.* Translated by Marta Menghini in collaboration with Anna Baccaglini-Frank; Mathematical advisor for the English translation: Gert Schubring (Berlin & Heidelberg: Springer, 2016).

O Curso de Análise de Cauchy: uma edição traduzida e comentada, Schubring, Gert & Tatiana Roque, eds. Coleçao História da Matemática, vol. 1 (Rio de Janeiro: Editora da SBM, 2016).

"DIG WHERE YOU STAND" 4. Proceedings of the Fourth International Conference on the History of Mathematics Education, September 23–26, 2015, at University of Turin, Italy, Kristín Bjarnadóttir, Fulvia Furinghetti, Marta Menghini, Johan Prytz & Gert Schubring, eds. (Roma: Edizioni Nuova Cultura, 2017).

Os números negativos—Exemplos de obstáculos epistemológicos? Segunda edição (São Paulo: editor Física, 2018).

Die Geschichte des IDM Bielefeld als Lehrstück. Ein Forschungsinstitut in einer Universität (Aachen: Shaker Verlag, 2018).

Thematic Issue: Recent advances in mathematics textbook research and development, Gert Schubring & Lianghuo Fan, guest editors, ZDM—Mathematics Education. Issue 5-2018.

Proceedings of the Second International Conference on Mathematics Textbook Research and Development (ICMT-2) Gert Schubring, Lianghuo Fan, & Victor Giraldo, eds. (Rio de Janeiro: Instituto de Matemática, Federal University of Rio de Janeiro, 2018).

The Legacy of Felix Klein, Hans-Georg Weigand, William McCallum, Marta Menghini, Michael Neubrand & Gert Schubring, eds. (New York: Springer, 2019).
Editor, *Interfaces between Mathematical Practicesand Mathematical Education.* International Studies in the History of Mathematics and its Teaching. Cham: Springer, 2019.
"Dig where you stand" 5. Proceedings of the Fifth International Conference on the History of Mathematics Education, September 19–22, 2017, at Utrecht University, the Netherlands, Kristín Bjarnadóttir, Fulvia Furinghetti, Jenneke Krüger, Johan Prytz, Gert Schubring & Harm Jan Smid, eds. (Utrecht: Utrecht University, 2019).
Folkerts, Menso & Gert Schubring, *Adolph Tellkampf (1798–1869). Elementarmathematik und ihre Grenzen* (Augsburg: Erwin Rauner, 2020).
"Dig where you stand" 6. Proceedings of the Sixth International Conference on the History of Mathematics Education, September 16–20, 2019, at the CIRM (Luminy) France, eds. Évelyne Barbin, Kristín Bjarnadóttir, Fulvia Furinghetti, Alexander Karp, Guillaume Moussard, Johan Prytz, Gert Schubring & Harm Jan Smid (Münster: WTM Verlag, 2020).
Geschichte der Mathematik in ihren Kontexten—neue Zugänge (Basel: Birkhäuser, 2021).
Analysing Historical Mathematics Textbooks. International Studies in the History of Mathematics and its Teaching (Cham: Springer, 2022).
O Curso de Análise de Cauchy: uma edição traduzida e comentada. Coleçao História da Matemática, vol. 1., Schubring, Gert & Tatiana Roque, eds. Segunda edição; revisão de Gert Schubring (Rio de Janeiro: Editora da SBM, 2022).

Chapters in Books

"On Education as a mediating Element between Development and Application: The Plans for the Berlin Polytechnical Institute (1817–1850)", *Epistemological and Social Problems of the Sciences in the Early Nineteenth Century*, ed. H.N. Jahnke/M. Otte (Dordrecht: Reidel 1981), 269–284.
"The Conception of Pure Mathematics as an Instrument in the Professionalization of Mathematics", *Social History of Nineteenth Century Mathematics*, ed. H. Mehrtens, H. Bos, I. Schneider. (Basel: Birkhäuser 1981), 111–134.
"Die Mathematik an der Ecole Normale des Jahres III—Wissenschaft und Methode", *Wissen und Bewußtsein. Studien zur Wissenschaftsdidaktik der Disziplinen*, Hrsg. F. Schmithals. (Hamburg: Arbeitsgemeinschaft für Hochschuldidaktik, 1982), 103–133.
"Pläne für ein Polytechnisches Institut in Berlin", *Philosophie und Wissenschaft in Preußen*, Hrsg. F. Rapp, H.W. Schütt. (Technische Universität Berlin 1982), 201–224.

"Gegensätze in den Mathematikauffassungen von Amateuren und Professionellen: eine Fallstudie zu Kreisquadrierern im 19. Jahrhundert", *Mathematikdidaktik -Bildungsgeschichte—Wissenschaftsgeschichte*, Hrsg. H. G. Steiner, H. Winter. Schriftenreihe: Untersuchungen zum Mathematikunterricht, Band 12. (Köln: Aulis 1985), 65–69.

"Zur Geschichte der Entstehung des Mathematiklehrerberufs—Systematische Hauptlinien und methodische Probleme inhaltsspezifischer Bildungsgeschichte", *Mathematikdidaktik -Bildungsgeschichte—Wissenschaftsgeschichte*, Hrsg. H. G. Steiner, H. Winter. Schriftenreihe: Untersuchungen zum Mathematikunterricht, Band 12. (Köln: Aulis 1985), 95–100.

"Mathematisch-naturwissenschaftliche Fächer", *Handbuch der deutschen Bildungsgeschichte, Band III, 1800–1870*, Hrsg. K.-E. Jeismann, P. Lundgreen (München: Beck 1987), 204–220.

Introduction: "Un savant des lumières—Un livre élémentaire pour la république", *CONDORCET—Moyens d'apprendre à compter sûrement et avec facilité. Appareil critique—études, notes, commentaires, bibliographie.* (Paris: ACL-éditions 1988). 157–175.

"Pure and Applied Mathematics in Divergent Institutional Settings in Germany: the Role and Impact of Felix Klein", *The History of Modern Mathematics. Volume II: Institutions and Applications* eds. David Rowe, John McCleary (Boston: Academic Press 1989), 171–220.

"The Rise and Decline of the Bonn Natural Sciences Seminar (Conflicts between Teacher Education and Disciplinary Differentiation)", *Science in Germany. The Intersection of Institutional and Intellectual Issues.* ed. Kathryn M. Olesko, *Osiris* (Second series), 1989, *5*, 56–93.

"Theoretical Categories for Investigations in the Social History of Mathematics Education and Some Characteristic Patterns", *Mathematics, Education and Society*, C. Keitel, P. Damerow, A. Bishop, P. Gerdes (eds.), Science and Technology Education Document Series No. 35 (Paris: UNESCO 1989), 6–8.

Spanish translation von A. Orellana und L. Rico: "Categorias teoricas para la investigacion en la historia social de la ensenanza de la matematica y algunos modelos caracteristicos", *epsilon*, revista de la sociedad andaluza de educacion matematica "Thales", no. 19, 1991, 100–104.

"Factors determining theoretical developments of mathematics education as a discipline—Comparative historical studies of its institutional and social contexts", *Proceedings of the 2nd TME-Conference: Foundations and Methodology of the Discipline Mathematics Education (Didactics of Mathematics)*, July 1985 Bielefeld, eds. H.-G. Steiner, A. Vermandel. Antwerpen 1988, 161–173.

"Differenzierung und Institutionalisierung von Wissen—Die Wirkung von Lehrplänen am Beispiel der Entstehung der Schulmathematik". *Zugänge zur Geschichte staatlicher Lehrplanarbeit*, Stefan Hopmann (Hrsg.). IPN Kiel 1988, 143–167.

"Die Mathematik—ein Hauptfach in der Auseinandersetzung zwischen Gymnasien und Realschulen in den deutschen Staaten des 19. Jahrhunderts",

Bildung, Staat und Gesellschaft im 19. Jahrhundert. Mobilisierung und Disziplinierung. Hrsg. K.-E. Jeismann. (Stuttgart: F. Steiner 1989), 276–289.

"La réforme du savoir savant: la contribution de Condorcet au premier concours des 'livres élémentaires'", *Condorcet, Mathématicien, économiste, philosophe, homme politique*, éds. Pierre Crépel, Christian Gilain (Paris: Minerve 1989), 44–51.

"Zur strukturellen Entwicklung der Mathematik an den deutschen Hochschulen 1800–1945", *Mathematische Institute in Deutschland 1800–1945*, Hrsg. Winfried Scharlau (Braunschweig: Vieweg 1990), 264–278.

"Das mathematisch-Unendliche bei J. F. Fries", *Konzepte des mathematisch Unendlichen im 19. Jahrhundert*, Hrsg. G. König (Göttingen: Vandenhoeck u. Ruprecht 1990), 152–164.

"Probleme vergleichender Analyse historischer Lehrbücher", *Mathematikdidaktik, Bildungsgeschichte, Bildungsgeschichte II* Hrsg. Hans-Georg Steiner (Köln: Aulis 1990). 123–136.

"Wilhelm Diesterweg. Ein Elementarmathematiker in der entstehenden Forschungsuniversität", *Adolph Diesterweg. Wissen im Aufbruch*, Hrsg. Arbeitsgruppe Diesterweg. Katalog zur Ausstellung zum 200. Geburtstag (Weinheim: Deutscher Studien Verlag 1990), 75–83.

"Spezialschulmodell versus Universitätsmodell: Die Institutionalisierung von Forschung", *'Einsamkeit und Freiheit' neu besichtigt*, Hrsg. G. Schubring (Stuttgart: Franz Steiner Verlag 1991), 276–326.

"Zur Modernisierung des Studiums der Mathematik in Berlin, 1820–1840", *AMPHORA. Festschrift für Hans Wußing zu seinem 65. Geburtstag*. Hrsg. S.S. Demidov et al. (Basel: Birkhäuser 1992), 649–675.

"Wissenschaftstransfer: Kultureller Austausch oder wissenschaftlicher Imperialismus?", *Wissenschafts- und Technologietransfer zwischen Industrieller und Wissenschaftlich-Technischer Revolution,* Hrsg. K.-P. Meinicke, K. Krug (Stuttgart, GNT-Verlag 1992), 24–35.

"The German mathematical community", *Möbius and his band. Mathematics and astronomy in Nineteenth-century Germany*, eds. John Fauvel, Raymond Flood and Robin Wilson (Oxford, Oxford University Press 1993), 21–33. German (not authorised) translation in: *Möbius und sein Band. Der Aufstieg von Mathematik und Astronomie im Deutschland des 19. Jahrhunderts* (Basel etc.: Birkhäuser, 1994), 31–46. Japanese translation (: 1995), 24–41.

"Report on the Symposium: The Prussian Institutional Reforms in Science and Education. Their challenge to National Policies Reassessed", *XVIIIth International Congress of History of Science, 1st–9th August, 1989. Final Report*, eds. F. Krafft, C.J. Scriba (Stuttgart: F. Steiner 1993), 109–111.

"Germany to 1933", *Companion Encyclopedia of the History and Philosophy of the Mathematical Sciences*, Volume 2, ed. Ivor Grattan-Guinness (London and New York: Routledge 1994), 1442–1456.

"Euklid versus Legendre in Italien", *Mathematik erfahren und lehren.* Festschrift für Hans-Joachim Vollrath, Hrsg. Günter Pickert, Ingo Weidig (Stuttgart: Klett 1994), 188–194.

"Fachlehrerausbildung im 19. und 20. Jahrhundert: Das Beispiel Rechen- und Mathematikunterricht", *Revision der Lehrerbildung. Neue Überlegungen anläßlich des Kongresses zu 200 Jahren Lehrerbildung in Oldenburg*, Hrsg. Hilke Günther-Arndt, Hans-Dietrich Raapke (Oldenburg: Carl von Ossietzky Universität Oldenburg, 1995), 106–115.

(gemeinsam mit Kurt-R. Biermann) "Einige Nachträge zur Biographie von Karl Weierstraß", *History of Mathematics: State of the Art.* Flores quadrivii—*Studies in Honor of Christoph J. Scriba*, eds. Joseph W. Dauben et al. (San Diego: Academic, 1996), 65–91.

"Changing cultural and epistemological views on mathematics and different institutional contexts in 19th century Europe", *L'Europe mathématique—Mythes, histoires, identités. Mathematical Europe—Myths, History, Identity*, eds. Catherine Goldstein, Jeremy Gray, Jim Ritter (Paris: Éditions de la Maison des Sciences de l'Homme, 1996), 361–388.

"La réforme de l'enseignement des mathématiques en Allemagne dans les années 1900–1914 et son rôle dynamique dans le mouvement international de réforme", *Les Sciences au Lycée. Un siècle de réformes des mathématiques et de la physique en France et à l'étranger*, éds. Bruno Belhoste, Hélène Gispert, Nicole Hulin (Paris: Vuibert, 1996), 235–248.

"Introduction—Reflections on the Complex History of Grassmann's Reception", *Hermann Günther Graßmann (1809–1877): Visionary Mathematician, Scientist and Neohumanist Scholar. Papers from a Sesquicentennial Conference*, ed. Gert Schubring (Dordrecht/Boston/London: Kluwer, 1996), ix-xxix.

"Remarks on the Fate of Grassmann's *Nachlaß*", *Hermann Günther Graßmann (1809–1877): Visionary Mathematician, Scientist and Neohumanist Scholar. Papers from a Sesquicentennial Conference*, ed. Gert Schubring (Dordrecht/Boston/London: Kluwer, 1996), 19–26.

"The Cooperation between Hermann and Robert Grassmann on the Foundations of Mathematics", *Hermann Günther Graßmann (1809–1877): Visionary Mathematician, Scientist and Neohumanist Scholar. Papers from a Sesquicentennial Conference*, ed. Gert Schubring (Dordrecht/Boston/London: Kluwer, 1996), 59–70.

"The Impact of the Napoleonic Structural Reforms of the Educational System in Europe", a cura di Luigi Blanco e Luigi Pepe, *Stato e pubblica istruzione. Giovanni Scopoli e il suo viaggio in Germania (1812). Annali dell'Istituto storico italo-germanico in Trento 1995, XXI* (Bologna: il Mulino, 1996), 435–443.

"Zur Bedeutung von Johann Georg Tralles als erstem Mathematikprofessor der Universität Berlin", *Natur, Mathematik und Geschichte. Beiträge zur Alexander-von-Humboldt-Forschung und zur Mathematikhistoriographie*, Hrsg. H. Beck, R. Siegmund-Schultze, Chr. Suckow, M. Folkerts. *Acta Historica Leopoldina*, 1997, *27*: 325–338.

"L'interaction entre les débats sur le statut des nombres négatifs et imaginaires et l'émergence de la notion de segment orienté", *Le nombre, une hydre à* n *visages: entre nombres complexes et vecteurs*, Hrsg. Dominique Flament (Paris: Éditions de la Maison des Sciences de l'Homme, 1997), 1–14.

"Johann Georg Tralles: Der erste Ordinarius für Mathematik an der Universität Berlin—Eine Edition seiner Antrittsvorlesung 1810", *Mathematik in Berlin. Geschichte und Dokumentation.* Zweiter Halbband, Hrsg. Heinrich Begehr (Aachen: Shaker Verlag 1998), 297–343.

"Philosophie der Mathematik bei Fries", *Jakob Friedrich Fries—Philosoph, Naturwissenschaftler und Mathematiker*, Hrsg. Wolfram Hogrebe, Kay Hermann (Frankfurt am Main: Peter Lang, 1999), 175–193.

"Theoretical Categories for Investigations in the Social History of Mathematics Education and some Characteristic Patterns", *A Multidimensional Approach to Learning in Mathematics and Sciences*, ed. Athanasios Gagatsis (Nicosia, 1999), 217–224. [revised version of the IDM occasional paper 109, 1988]

"Institutional history of mathematics", *The History of Mathematics from Antiquity to the Present: A Selective Annotated Bibliography*, ed. Joseph W. Dauben, revised CD-ROM edition by Albert C. Lewis, in cooperation with the ICHM (American Mathematical Society, 2000).

"Recent research on institutional history of science and its application to islamic civilization", *Science in Islamic Civilisation*, eds. Ekmeleddin Ihsanoglu, Feza Günergun (Istanbul: Research Centre for Islamic History, Art and Culture, 2000), 19–36.

"Wessel et Argand (1798–1806): des mathématiciens marginaux surmontant l'obstacle central de l'acceptation des nombres complexes", *La Enseñanza de las Ciencias: Una Perspectiva Historica*, eds. Elena Ausejo, Carmen Beltrán, Cuadernos de Historia de la Ciencia. 11-II (Zaragoza, Universidad de Zaragoza, 2000), 585–603.

"History of Mathematics for Trainee Teachers". Chapter 4 of *History in Mathematics Education. The ICMI study. What engine of wit: History in Mathematics Education*, John Fauvel, Jan van Maanen (eds.) (Dordrecht: Kluwer, 2000), 91–142.

"Recent research on institutional history of science and its application to Islamic civilization", ed. Athanasios Gagatsis, *Learning in Mathematics and Science and Educational Technology.* Vol. II (Nicosia: Intercollege Press, 2001), 57–76.

"Mathematik und Naturwissenschaften zwischen Spezialschul-Struktur und Forschungs-Imperativ", *Humboldt International. Der Export des deutschen Universitätsmodells im 19. und 20. Jahrhundert,* Hrsg. Rainer C. Schwinges (Veröffentlichungen der Gesellschaft für Universitäts- und Wissenschaftsgeschichte 3), Basel 2001, S. 403–443.

"Argand and the early work on graphical representation: New sources and interpretations", *Around Caspar Wessel and the Geometric Representation of Complex Numbers.* Proceedings of the Wessel Symposium at The Royal Danish Academy of Sciences and Letters, Copenhagen, August 11–15 1998: Invited Papers. Matematisk-fysiske Meddelelser 46:2, Jesper Lützen (ed.) (C. A. Reitzel: Copenhagen, 2001), 125–146.

"Debatten um den westfälischen Mathematiklehrplan 1834. Eine Fallstudie zum Verhältnis von Mathematik und Kultur", *Beiträge zur Didaktik der Mathematik für die Primarstufe. Festschrift für Siegbert Schmidt*, Hrsg. Werner Weiser, Bernd Wollring (Hamburg: Kovač, 2001), 202–216.

"Aspetti istituzionali della matematica", *Storia della scienza*, ed. Sandro Petruccioli, Vol. VI: *L'Etá dei Lumi* (Roma: Istituto dell'Enciclopedia Italiana, 2002), 366–380.

"O primeiro movimento internacional de reforma curricular em matemática e o papel da Alemanha", *Euclides Roxo e a modernização do ensino da Matemática no Brasil,* org. Wagner R. Valente (São Paulo: Sociedade Brasileira de Educação Matemática, 2003), 11–45.

New edition: idem (Brasilia: Editora Universidade de Brasilia, 2004), 11–43.

"Il calcolo vettoriale di Grassmann", *Storia della scienza*, ed. Sandro Petruccioli, Vol. VII *L'Ottocento* (Roma: Istituto dell'Enciclopedia Italiana, 2003), 147–149.

"Neues über Legendre in Italien", *Mathematik im Fluss der Zeit.* Algorismus, Heft 44, Hrsg. W. Hein, P. Ullrich (Augsburg: ERV Rauner, 2004), 256–274.

"Das Bonner naturwissenschaftliche Seminar (1825–1887)—Eine Fallstudie zur Disziplinendifferenzierung", *Christian Gottfried Nees von Esenbeck. Politik und Naturwissenschaften in der ersten Hälfte des 19. Jahrhunderts*, Hrsg. D. von Engelhardt, A. Kleinert, J. Bohley. *Acta Historica Leopoldina* (Halle/S.), 2004, *43*: 133–148.

(with Paolo Freguglia) "Il calcolo geometrico", *Storia della scienza*, ed. Sandro Petruccioli, Vol. VIII 1870–1950 (Roma: Istituto dell'Enciclopedia Italiana, 2004), 88–92.

"Der Briefwechsel Quetelet—Gauß: Magnetismus und Sternschnuppen. Eine Edition", *Mathematics Celestial and Terrestrial. Festschrift für Menso Folkerts*, Hrsg.J. W. Dauben, St. Kirschner, A. Kühne, P. Kunitzsch, R. P. Lorch. Acta Historica Leopoldina Nr. 54 (Halle/S.: Deutsche Akademie der Naturforscher, 2008), 789–807.

"The origins and early incarnations of ICMI", *The First Century of the International Commission on Mathematical Instruction (1908–2008). Reflecting and Shaping the World of Mathematics Education*, eds. Marta Menghini, Fulvia Furinghetti, Livia Giacardi e Ferdinando Arzarello (Roma: Enciclopedia Italiana, 2008), 113–130.

"Mathematisches Publizieren zwischen Schule und Hochschule", *Publikationsstrategien einer Disziplin—Mathematik in Kaiserreich und Weimarer Republik*, Hrsg. Volker R. Remmert & Ute Schneider (Wiesbaden, Harrassowitz, 2008), 52–72.

"Processes of Algebraization in the History of Mathematics: The Impact of Signs", *Semiotics in Mathematics Education. Epistemology, History, Classroom, and Culture*, eds. L.Radford, G. Schubring & F. Seeger (Sense Publ.: Rotterdam, 2008), 139–155.

"Mathematics in Naples. An extraordinary case of institutional development", *La Historia de la Ciencia y de la Técnica: Un Arma Cargada de Futuro. Ensayos en Homenaje a Mariano Hormigón*, eds. M.A. Velamazan, F. Vea, J. Cobos & C.

Martin (Cádiz: Diputación Provincial de Cádiz, FPC, Servicio de Publicaciones, 2008), 143–153.
"Preface", *Matematikundervisningen i Sverige för 100 år sedan. Redovisningar fårn folkskole- till universitetsnivå*, ed. Bengt Johannson. Årsböcker i svensk undervisningshistoria, 211 (Uppsala, 2009), 11–13.
"A origem da geometria de Legendre e o seu impacto internacional", em: Adrien-Marie Legendre, Elementos de Geometria. Reedição da primeira tradução brasileira de 1809, Luis Carlos Guimarães (org.) (Rio de Janeiro: Editora LIMC, 2009), 353–384.
"The way from the combinatorial school to the reception of the Weierstrassian analysis". *Dalla pecia all'e-book. Libri per l'Università: stampa, editoria, circolazione e lettura.* Atti del Convegno internazionale di studi, Bologna, 21–25 ottobre 2008, a cura di Gian Paolo Brizzi, Maria Gioia Tavoni. Bologna: CLUEB, 2009, pp. 431–442.
"Mathematik, Statistik und Naturwissenschaften in den Studien Prinz Alberts", *Die Studien des Prinzen Albert an der Universität Bonn (1837–1838*). Prinz-Albert-Forschungen, Band 5. Hrsg. Franz Bosbach (Berlin/New York: de Gruyter, 2010), 149–164.
"André Revuz, l'Allemagne Fédérale et l'« Institut für Didaktik der Mathematik » (IDM)", *Hommage à André Revuz—L'engagement universitaire, L'héritage didactique* (Paris: LDAR, Université Paris-Diderot, 2010), 146–149.
"Methodological remarks about the use of history of mathematics in teaching mathematics—which history for teachers?", *Mathematics and its didactics—fourty years of commitment. In occasion of the 65 years of Bruno d'Amore*, ed. Silvia Baragli (Bologna: Pitagora Editrice, 2011), 213–215.
"Lettres de mathématiciens français à Weierstraß—documents de sa réception en France", *L'aventure de l'analyse, de Fermat à Borel. Mélanges en l'honneur de Christian Gilain*, éd. Suzanne Féry (Nancy: Presses Universitaires de Nancy, 2012), 567–594.
"A Matemática Elementar de um Ponto de Vista Superior: Felix Klein e a sua Atualidade", *O saber do professor de Matemática: ultra-passando a dicotomia entre didática e conteúdo*, Victor Augusto Giraldo e Tatiana Marins Roque (org.s). (Rio de Janeiro: Editora Ciência Moderna, 2014), 39–56.
"On Historiography of Teaching and Learning Mathematics". In: *Handbook on the History of Mathematics Education* Alexander Karp & Gert Schubring (eds.) (New York: Springer, 2014), 3–8.
"Mathematics Education in Catholic and Protestant Europe". In: *Handbook on the History of Mathematics Education*, Alexander Karp & Gert Schubring (eds.) (New York: Springer, 2014), 130–143.
"Mathematics Education in Germany (Premodern Times)". In: *Handbook on the History of Mathematics Education* Alexander Karp & Gert Schubring (eds.) (New York: Springer, 2014), 241–255.
Gert Schubring & Alexander Karp, "History of Mathematics Teaching and Learning". In: *Encyclopedia of Mathematics Education,* Stephen Lerman (ed.) (New York: Springer, 2014), 260–267.

"From Pebbles to Digital Signs. The Joint Origin of Signs for Numbers and for Script—Their Intercultural Standardization and Their Renewed Conjunction in the Digital Era". In *Applied Virtuality Book Series Symbolizing Existence—Metalithikum III*, Bühlmann, Vera & Ludger Hovestadt (eds.) (Basel: Birkäuser, 2016), 114–126.
"Preface to the 2016 Edition". In: Felix Klein, *Elementary Mathematics from a Higher Standpoint. Volume I: Arithmetic, Algebra, Analysis.* Translated by Gert Schubring (Berlin & Heidelberg: Springer, 2016).
"Conflits sur les pratiques des mathématiques entre professionnels et amateurs: les cas des "quarreurs du cercle" et des "Fermatistes". In: *Circulation: mathématiques, histoire, enseignement*, Jérôme Auvinet, Guillaime Moussard & Xavier Saint Raymond (eds.) (Limoges, PresJornal Internacional de Estudos em Educaçao Matemáses universitaires de Limoges, 2018), 199–219.
"The impact of teaching mathematics upon the development of mathematical practices". In: *Interfaces between Mathematical Practices and Mathematical Education,* Gert Schubring (ed.) (Cham: Springer, 2019), 123–141.
"Klein's conception of ''Elementary Mathematics from a Higher Standpoint''". In: *The Legacy of Felix Klein,* Hans-Georg Weigand, William McCallum, Marta Menghini, Michael Neubrand & Gert Schubring (eds.) (New York: Springer, 2019), 167–180.
"The German IMUK subcommission". In*: National Subcommissions of ICMI and their Role in the Reform of Mathematics Education,* Alexander Karp (ed.*)* (Cham: Springer, 2019), 65–91.
Gert Schubring, Vinicius Mendes Pereira & Thiago Oliveira, "The Dissemination of Descriptive Geometry in Latin America". In: *Descriptive Geometry, The Spread of a Polytechnic Art*, É. Barbin, Marta Menghini & Klaus Volkert (eds.) (Cham: Springer, 2019), 377–400.
"The Myth of the Polytechnic School", in: Évelyne Barbin, Marta Menghini & Klaus Volkert (eds.), *Descriptive Geometry, The Spread of a Polytechnic Art* (Cham: Springer, 2019), 403–421.
"Patterns for studying history of mathematics education: A case study of Germany". In: *Researching the history of mathematics education,* Fulvia Furinghetti & Alexander Karp (eds.) (New York, Springer, 2018), 241–259.
"Computation Devices in 19th Century Mathematics Instruction". In: *Computations and Computing Devices in Mathematics Education Before the Advent of Electronic,* Alexej Volkov & Viktor Freiman (eds.), *Calculators* (New York: Springer, 2019), 365–384.
"Analysen der Profile der Mathematik-Schulprogramme in den verschiedenen deutschen Staaten des 19. Jahrhunderts". In: *Schulprogramme Höherer Lehranstalten. Interdisziplinäre Perspektiven auf eine wiederentdeckte bildungs- und kulturwissenschaftliche Quellengattung*, Norman Ächtler (Hg.) (Hannover: Wehrhahn, 2021), 305–327.
"The History of IMUK: The First Phase as IMUK and CIEM" (2022). In: *The International Commission on Mathematical Instruction, 1908–2008: People,*

Events and Challenges in Mathematics Education, Fulvia Furinghetti & Livia Giacardi (eds.) (Cham, Springer Nature), 3–42.

Biographical Essays: “Heinrich Behnke”, 382–386; “Bent Christiansen”, 417–420; “Henri Fehr”, 446–449; “Hans Freudenthal“, 450–456; “Felix Klein”,17 497–509; “Walther Lietzmann”, 522–527; “David Eugene Smith“, 601–605; “Hans-Georg Steiner”, 611–617; “Emanuel Beke” (together with Sándorné Kántor Tünde Varga), 650–653; “Wilhelm Wirtinger”, 705–707. In: *The International Commission on Mathematical Instruction, 1908–2008: People, Events and Challenges in Mathematics Education*, Fulvia Furinghetti & Livia Giacardi (eds.) (Cham, Springer Nature).

Articles in Journals and Periodicals

Kapazitaetsberechnungen, ZRS, Clearingstellen. Fachbereichs-Materialien Mathematik-Naturwissenschaften, Band 2. (Bonn: Verband Deutscher Studentenschaften, 1972).

Ottse, Michael, Hans Niels Jahnke & Gert Schubring, “Zu einigen Hauptaspekten der Mathematikdidaktik”. *Schriftenreihe des IDM*, Band 1, Bielefeld, 1974, 4–84.

Otte, Michael, Thomas Mies, Veronika Reiß, Heinz Steinbring, Gert Schubring, Dankwart Vogel. “Tendenzen und Probleme der Mathematiklehrerbildung”. Schriftenreihe des IDM, Band 6, Bielefeld, 1975, 8–142.

“Das historisch-genetische Prinzip in der Mathematik-Didaktik—Zur Bedeutung der Geschichte der Mathematik”, *Zentralblatt für Didaktik der Mathematik*, 1977, *9*, 209–213.

Hültenschmidt, Erika & Gert Schubring. “Bemerkungen zu reduktionistischen Wissenschaftsauffasssungen”, *SoPo–Sozialisitsche Politik*, 1978, I: *45*, 150–158; II: *46*, 141–151. Errata Teil I: *46*, 175.

“Die Konzeption der reinen Mathematik und ihre Bedeutung für eine Anwendungsorientierung der Mathematik”, *6th International Congress of Logic, Methodology and Philosophy of Science.* Abstracts Section 13 (Hannover 1979), 177–181.

“Bedingungen der Professionalisierung von Wissenschaft. Eine vergleichende Übersicht zu Frankreich und Preussen”, *lendemains* Nr. 19, 1980, 125–133.

“Mathematics and Teacher Training: Plans for a Polytechnic in Berlin”, *Historical Studies in the Physical Sciences*, 1981, *12/1*, 161–194.

“Gegenständliche und soziale Momente des Wissens als Kategorien für Untersuchungen zur Geschichte der Mathematik-Didaktik”, *Journal für Mathematik-Didaktik*, 1981, *2*, 1–34.

“Ansätze zur Begründung theoretischer Terme in der Mathematik—Die Theorie des Unendlichen bei Johann Schultz (1739–1805)”, *Historia Mathematica*, 1982, *9*, 441–484.

"L'histoire de l'enseignement des mathématiques", *Contributions à la Seconde Ecole d'Eté des Didactiques des Mathématiques (1982)*, IREM d'Orléans (éd.). IREM, Université d'Orléans 1983, 167–170.

Seminar-Institut-Fakultät: Die Entwicklung der Ausbildungsformen und ihrer Institutionen in der Mathematik (1810–1945). Methodologische Probleme der Sozialgeschichte der Mathematik (Vortrag auf dem XVI. Intern. Kongreß für Wissenschaftsgeschichte, Bukarest 1981). Reihe: Diskussionsbeiträge zur Ausbildungsforschung und Studienreform des IZHD, Universität Bielefeld, Nr. 1/ 1983. (56 S.)

"Das mathematische Leben in Berlin. Zu einer entstehenden Profession anhand von Briefen des aus Erlangen stammenden Martin Ohm an seinen Bruder Georg Simon", *Erlanger Bausteine zur Fränkischen Heimatforschung*, Jahrbuch 30 (1983), 221–249.

"Comparative Study of the Development of Mathematics Education as a Professional Discipline in Different Countries. General Trend Report", *Proceedings of the Fourth International Congress of Mathematics Education*, ed. M. Zweng (Birkhäuser: Boston 1983), 482–484.

"Comparative Study of the Development of Mathematics Education as a Professional Discipline: Western Germany", *Proceedings of the Fourth International Congress of Mathematics Education*, ed. M. Zweng (Birkhäuser: Boston 1983), 487.

"Introduction à la chronique historique sur l'enseignement des mathématiques", *Recherches en Didactiques des Mathématiques*, 1983, *4*, 325–331.

Greek translation: "Eisagogí sto Istorikó Chronikó páno sti Didaskalia ton Mathematikón (Istorikó Chronikó I)", *Stoicheía Istorías tis Mathematikís Ekpaídeusis* Tmima Mathematikón—Aristoteles Panepistímion Thessaloníki, Hrsg. Athanasios Gagatsis (Thessaloniki 1993), 10–15.

"Die Promotion von P.G. Lejeune Dirichlet. Biographische Mitteilungen zum Werdegang Dirichlets", NTM *(Schriftenreihe für Geschichte der Naturwissenschaften, Technik und Medizin)*, 1984, *21/1*, 45–65.

"Martin Ohm und Friedrich August Pfeiffer". Eine Ergänzung zu "Das Mathematische Leben in Berlin", *Erlanger Bausteine zur Fränkischen Heimatforschung*, Jahrbuch 31 (1984), 203–205.

"Le Développement des Mathématiques et des Sciences en France et en Prusse (1800–1830)", *Sciences et Techniques en Perspective*. Ed. Université de Nantes, Institut de Mathématiques. 1984, *vol. V*, 42–59.

"Chronique Historique", *Recherches en Didactique des Mathématiques*, 1984, *5*, 235–237. Greek translation: "Istorikó Chronikó II", *Stoicheía Istorías tis Mathematikís Ekpaídeusis* Tmima Mathematikón—Aristoteles Panepistímion Thessaloníki, Hrsg. Athanasios Gagatsis (Thessaloniki 1993), 16–18.

"Essais sur l'histoire de l'enseignement des mathématiques, particulièrement en France et en Prusse", *Recherches en Didactique des Mathématiques*, 1984, *5*, 343–385. Greek translation: "Dokímio páno stin istoría tis didaskalías ton Mathematikón (Kyríos sti Gallía kai stin Prosía)", *Stoicheía Istorías tis Mathematikís*

Ekpaídeusis Tmima Mathematikón—Aristoteles Panepistímion Thessaloníki, Hrsg. Athanasios Gagatsis (Thessaloniki 1993), 75–130.
"Le retour du refoulé—Les débats sur "La Méthode" didactique à la fin du 18ème et au debut du 19ème siècle: Condillac, Lacroix et les successeurs", *IIème Ecole d'Eté de Didactique des Mathématiques. 2–13 juillet 1984, Orléans. Recueil des Textes et Comptes-Rendus* (IMAG, Saint-Martin d'Hères 1984), 249–250.
"Einleitung" [introduction tot he thematic issues 6/1984 and 1/1985: Geschichte des Mathematikunterrichts], *Zentralblatt für Didaktik der Mathematik*, 1984, *16*, 185–186.
"L'apport des recherches en histoire de l'enseignement des mathématiques à la didactique des mathématiques," *Séminaire de Didactique des Mathématiques et de l'Informatique, année 1984–1985*, éd. Institut IMAG, Grénoble, 3–25.
"Die Erinnerungen von Karl Emil Gruhl (1833–1917) an sein Studium der Mathematik und Physik in Berlin (1853–1856)", *Jahrbuch Überblicke Mathematik*, 1985, *18*, 143–173.
"Die Entwicklung des mathematischen Seminars der Universität Bonn, 1864–1929", *Jahresberichte der Deutschen Mathematiker-Vereinigung*, 1985, *87*, 139–163.
"Das mathematische Seminar der Universität Münster, 1831/1875 bis 1951", *Sudhoffs Archiv*, 1985, *69*, 154–191.
"Die Geschichte des Mathematiklehrerberufs in mathematikdidaktischer Perspektive", *Zentralblatt für Didaktik der Mathematik*, 1985, *17*, 20–26.
"Mathematics Education, History of", *The International Encyclopedia of Education*, eds. Torsten Husen, T. Neville Postlethwaite (Oxford: Pergamon Press 1985), *vol. 6*, 3236–3239. (zus. mit C. Keitel u. R. J. K. Stowasser).
"Histoire des nombres négatifs—Ruptures dans leur statut mathématique", *Actes du Colloque INTER-IREM: Histoire et Epistémologie des Mathématiques* (Montpellier 1985), 65–73.
L'histoire de l'enseignement des mathématiques comme sujet de recherche en didactique des mathématiques, Cahiers de Didactique des Mathématiques, Ed. IREM Université Paris VII (Paris 1986), no.26 (33 S.)
Greek translation: "H istoría tis Mathematikís Ekpaídeusis os théma éreunas sti Didaktikí ton Mathematikón", *Stoicheía Istorías tis Mathematikís Ekpaídeusis* Tmima Mathematikón—Aristoteles Panepistímion Thessaloníki, Hrsg. Athanasios Gagatsis (Thessaloniki 1993), 23–73.
"Wilhelm Lorey (1873–1955) und die Methoden mathematikgeschichtlicher Forschung", *mathematica didactica*, 1986, *9*, 75–87.
"The three parts of the Dirichlet *Nachlaß*", *Historia Mathematica*, 13 (1986), *13*, 52–56
"Ruptures dans le statut mathématique des nombres négatifs", *petit x*, no.12, 1986, 5–32.
Portuguese translation: "Rupturas no estatuto matemático dos números negativos", *Boletim GEPEM*, I: no. 37, 2000, 51–64; II: no. 38, 2001, 73–93.

Probleme vergleichender Analyse historischer Lehrbücher, Occasional paper Nr. 79, IDM, Universität Bielefeld, 1986. (28 S.)
"Recherche et Documentation en didactique des Mathématiques", *L'informatique et l'enseignement des mathématiques dans les pays en voie de développement. Proceedings du premier symposium ICOMIDC*, Eds. M. Amara, M.[N.] Boudriga, & K. Harzallah (Tunis, Tunisie: UNESCO, ICOMIDC & Faculté des Sciences de Tunis 1986), 249–252.
"The *Nachlass* of Wilhelm Lorey", *Historia Mathematica*, 1987, *14*, 55–57.
The German Mathematical Community in the Nineteenth Century, The Open University: Topics in the History of Mathematics, MA 290 (BBC Production: June Harben, BBC Research: Gale Jagger). Milton Keynes 1987.
The Cross-Cultural 'Transmission' of Concepts—the first international mathematics curricular reform around 1900, with an Appendix on the Biography of F. Klein. Occasional paper Nr. 92, IDM (Universität Bielefeld), 1987. 39 S.
"On the methodology of analysing historical textbooks: Lacroix as textbook author", *for the learning of mathematics*, 1987, *7: 3*, 41–51. ("Errata", ibid., 1988, *8: 2*, 51).
Spanisch trtanslation by Alejandro Garciadiego: "Sobre la metodologia de análisis de libros de texto históricos: Lacroix como autor de libros de texto", *Mathesis. filosofia y historia de las matemáticas*, 1992, *8*, 273–298. ("Errata", ibid., 1994, *10*, 127).
"Der Lehrer: "ein Organ seines Lehrbuchs"? Staatliche Vorschrift kontra methodische Autonomie (1829)", *Der Mathematikunterricht*, 1988, *34:1*, 4–29.
"Ein früher "Aufruf: Rettet die mathematisch—naturwissenschaftliche Bildung!" -Die Denkschrift Schellbachs von 1860", *Der Mathematikunterricht*, 1988, *34:1*, 30–72.
"Historische Begriffsentwicklung und Lernprozeß aus der Sicht neuerer mathematikdidaktischer Konzeptionen (Fehler, "Obstacles", Transposition)", *Zentralblatt für Didaktik der Mathematik*, 1988, *20*, 138–148.
Greek translation: "I istorikí exélixi ton mathimatikón ennoión kai i didaktikasía tis mathisis apó ti skopiá ton neóteron antilípseon tis didaktikís ton mathimatikón (lathi, "empódia", metaphorá)" (T. Tokmakidis und A. Magarisioti), *Omilos gia tin Istoría ton Mathimatikón, Enimerotiko Philladio*, Teuchos 13, Maios 1990, 10–34.
Portuguese translation: "Desenvolvimento histórico do conceito e do processo de aprendizagem, a partir de recentes concepções matemático-didáticas (erro, obstáculos, transposição)", *Zetetiké*, 1998, *6*, no. 10, 9–34.
"Discussions épistémologiques sur le statut des nombres négatifs et leur représentation dans les manuels allemands et français de mathématiques entre 1795 et 1845", *Actes du Premier Colloque Franco-Allemande de Didactique des Mathématiques et de l'Informatique*. Ed. C. Laborde (Grenoble, éditions sauvage 1988), 137–145.
"Epistemologische Debatten über den Status negativer Zahlen und die Darstellung negativer Zahlen in deutschen und französischen Lehrbüchern 1795–1845", *Mathematische Semesterberichte*, 1988, *35:1*, 183–196.

Theoretical Categories for Investigations in the Social History of Mathematics Education and some Characteristic Patterns, Occasional Paper Nr. 109, IDM, Universität Bielefeld, 1988. (12 S.)
Greek translation: "Theoretikés kategoríes ereunón gia tin koinonikí istoría tis mathimatikís ekpaídeusis kai meriká charakteristiká ypodeígmata", *Omilos gia tin Istoría ton Mathimatikón*, No. 23, Aprilios 1989, 52–70.
Revised version published in: *A Multidimensional Approach to Learning in Mathematics and Sciences*, ed. Athanasios Gagatsis Nicosia/Cyprus (Intercollege Press, 1999), 217–224.
"Warum Karl Weierstraß beinahe in der Lehrerprüfung gescheitert wäre", *Der Mathematikunterricht*, 1989, *35: 1*, 13–29.
"Letter to the Editor: Dirichlet", *The Mathematical Intelligencer*, 1990, *12:1*, 5–6.
"Les échanges entre les mathématiciens Français et Allemands sur la rigueur dans les concepts d'Arithmétique et d'Analyse", *Echanges d'Influences Scientifiques et Techniques entre Pays Européens de 1780 à 1830*, Colloques du C. T. H. S., No. 5: Actes du 114e Congrès National des Sociétés Savantes (Paris 3–9 avril 1989), Section histoire des sciences et des techniques (Paris: Editions du CTHS 1990), 89–104.
"Der Süvernsche Lehrplan—"Ideales Muster" oder staatlicher Zugriff?", *Zeitschrift für Pädagogik*, 1990, *36*, 393–418.
"'Durchschnittsmenschen, ... nicht Genies.' Zu den Widerständen gegen die neuhumanistische Bildungsreform", *Zeitschrift für Pädagogik*, 1991, *37*, 347–370
"Bernard Bolzano—Not as Unknown to His Contemporaries as is Commonly Believed?", *Historia Mathematica,* 1993, *20,* 45–53.
"Continuités et fonctions variables dans le rôle social et éducatif des mathématiques", *Les mythes historiques, sociaux et culturels des mathématiques: leur impact sur l'éducation,* éd. Catherine Goldstein. Cahiers de Didactique des Mathématiques, Ed. IREM Université Paris VII (Paris 1993), no. 82, 13–19.
"Evolution du concept d'infiniment petit aux 18ème et 19ème siècles", *Histoire d'Infini*, Actes du 9e Colloque Inter-IREM 'Epistémologie et Histoire des mathématiques', éd. IREM de Brest (Brest: IREM de Brest 1994), 317–326.
"Der Übergang von der Sekundarstufe I zur Sekundarstufe II in den Gymnasiallehrplänen der Bundesländer", *Der Mathematikunterricht*, 1994, *40:2*, 6–14.
"Les enjeux épistémologiques des nombres négatifs", *Histoire et Épistémologie dans l'Éducation Mathématique. Actes de la Première Université d'Été Européenne*, Montpellier 1993 (Montpellier: IREM de Montpellier [1995]), 443–449.
"Bericht über die erste internationale Graßmann-Tagung", *Jahrbuch Überblicke Mathematik 1995*, Hrsg. Albrecht Beutelspacher u. a. (Braunschweig: vieweg 1995), 185–187.

"Differences in the Involvement of Mathematicians in the Political Life in France and in Germany", *Bollettino di Storia delle Scienze Matematiche*, 1995, *15*: 61–83. Errata: ibid., 1996: *16*:
"Tendencias actuales en la investigación sobre la historia institucional de la ciencia y su aplicación a la cultura islámica", *LLULL*, 1996, *19* (no. 36): 177–193.
"Die Verschiebung der "Grenzen"—Konflikte zwischen Schulen und Hochschulen", Übergänge III: Von der Sekundarstufe II zur Hochschule, *Der Mathematikunterricht*, 1996, *42: 4/5*, 7–19.
"Relações entre a História e o Ensino da Matemática", *II Encontro Luso-Brasileiro de História da Matemática e II Seminério Nacional de História da Matemática, Águas de São Pedro—São Paulo 1997. Anais -Actas*, Sergio Nobre, ed. (Rio Claro, SP, 1997).
"Die "Beschleunigung" in den Lehrzielen—zum sozialen Defizit der Mathematik-Didaktik", *Beiträge zum Mathematikunterricht. Vorträge auf der 32. Tagung für Didaktik der Mathematik vom 2. bis 6. 3. 1998 in München* (Hildesheim: franzbecker, 1998), 549–552.
"Le retour du refoulé. Les débats sur la supériorité de la méthode analytique ou de la méthode synthétique autour de 1800", *Histoires de Géométries. Textes du Séminaire de l'année 1996.* Èquipe Formalismes, Formes et Données sensibles: recherches historiques, philosophiques et mathématiques, éd. Dominique Flament (Paris: Fondation Maison des Sciences de l'Homme, [1998]), 31–37.
"Mathematische Wörterbücher im 18. Jahrhundert", *Das Achtzehnte Jahrhundert*, 1998, *22*: 114–128 (Heft 1: Enzyklopädien, Lexika und Wörterbücher im 18. Jahrhundert).
"An unknown part of Weierstraß's *Nachlaß*", *Historia Mathematica*, 1998, *25*: 423–430.
"Tendances actuelles des recherches sur l'histoire institutionelle des sciences et leur application à la culture islamique", *Vème Colloque Maghrébin sur l'Histoire des Mathématiques Arabes, Hammamet 1–3. Décembre 1994, Actes du Colloque*, éd. Association Tunisienne des Sciences Mathématiques (Tunis, 1998), 273–283.
"A pesquisa em história da matemática: questões metodológicas", *Anais III Seminário Nacional de História da Matemática, Vitória Março 1999*, ed. Circe M. Silva da Silva (Vitória, E.S.: Universidade Federal do Espírito Santo, 1999), 192–203.
"O primeiro movimento internacional de reforma curricular em matemática e o papel da Alemanha: um estudo de caso na transmissão de conceitos", *Zetetiké*, 1999, *7*, no. 11, 29–50.
English translation: "The first international curricular reform movement in mathematics and the role of Germany—a case study in the transmission of concepts". In: *Learning and assessment in mathematics and science,* A. Gagatsis et al. (eds.) (Department of Education, University of Cyprus: Nicosia, 2000), 265–287.

"Felix Kleins Gutachten zur Schulkonferenz 1900: Initiativen für den Systemzusammenhang von Schule und Hochschule, von Curriculum und Studium", Felix Klein und die Berliner Schulkonferenz des Jahres 1900, *Der Mathematikunterricht*, 2000, *46: 3*, 62–76.

"Kabinett—Seminar—Institut: Raum und Rahmen des forschenden Lernens", *Berichte zur Wissenschaftsgeschichte*, 2000, *23*: 269–285.

"Novas Fontes e Abordagens na História dos Números Negativos: Uma Análise de um Manuscrito do Monteiro da Rocha", *IV Seminário Nacional de História de Matemática. 8 a 11 de Avril de 2001. Anais*, John A. Fossa (ed.) (Editora da SBHMat: Rio Claro/SP, 2001), 95–108.

"Obstáculos epistemológicos: meios de ligar a história e o ensino matemático—O caso da multiplicação", *Cadernos de Pesquisa em Educação PPGE-UFES* [Vitória/ES], 2001, vol. *7*, n. 14 Julho/Dezembro, 231–259.

"Production Mathématique, Enseignement et Communication. Remarques sur la note de Bruno Belhoste, "Pour une réévaluation du rôle de l'enseignement dans l'histoire des mathématiques" parue dans la RHM 4(1998), pp. 289–304", *Revue d'histoire des mathématiques*, 2001, *7*: 295–305.

Reprint in: *Educação, Matemática, Pesquisa* (PUC de São Paulo), 2002, vol. 4, no.: 29–39.

Mathematics between Propaedeutics and Professional use: a Comparison of Institutional Developments. IDM—Universität Bielefeld. Occasional Paper Nr. 183, 2002.

"A Framework for Comparing Transmission Processes of Mathematics to the Americas", *Revista Brasileira de História da Matemática*, 2002, vol. *2*, no. 3, 45–63.

"A Noção de Multiplicação: Um "obstáculo" desconhecido na História da Matemática", *Bolema* (Rio Claro/SP), 2002, ano *15*, no. 18, 26–52.

"Mathematik-Geschichte im Mathematikunterricht und in der Mathematiklehrer-Ausbildung", *Didaktik der Mathematik in den alten Bundesländern—Methodik des Mathematikunterrichts in der DDR.* Bericht über eine Doppeltagung zur gemeinsamen Aufarbeitung einer getrennten Geschichte, Hrsg. H. Henning, P. Bender (Magdeburg, 2003), 101–104.

"*L'Enseignement Mathématique* and the First International Commission (IMUK): The Emergence of International Communication and Cooperation", *One Hundred Years of L'Enseignement Mathématique. Moments of Mathematics Education in the Twentieth Century.* Proceedings of the EM-ICMI Symposium Geneva, 20–22 October 2000, eds. Daniel Coray, F. Furinghetti, H. Gispert, B.H. Hodgson, G. Schubring (Geneva: L'Enseignement Mathématique, 2003), 47–65.

"E.H. Dirksens Beiträge zu den Grundlagen der Analysis", *N.T.M.*, 2003, *11*: 90–99.

"A algebrização do conceito do limite no século XVIII", *V Seminário Nacional de História da Matemática 13–16.4 2003. Anais*, eds. M. Teixeira, S. Nobre (Rio Claro/SP: UNESP, 2003), 93–113.

"Recent Developments in Research on the Institutional History of Mathematics", *Llull*, 2003, vol. *26*, no. 57, 1045–1059.

"Relações Culturais entre Alemanha e Brasil: 'Imperialismo Cultural' versus 'Nacionalização'", *Zetetiké*, 2003, vol. *11*, no. 20: 56–109.

"Deutsch-brasilianische Kulturbeziehungen—'Cultural Imperialism' versus 'Nationalisierung'", *Jahrbuch für Historische Bildungsforschung*, 2004, *10*: 161–192.

"A Case Study in Generalisation: The Notion of Multiplication", *Activity and Sign—Grounding Mathematics Education. Festschrift for Michael Otte*, eds. M. Hoffmann, J. Lenhard, F. Seeger (New York: Springer, 2005), 275–285.

"Pesquisar sobre a história do ensino da matemática: metodologia, abordagens e perspectivas", *História do Ensino da Matemática em Portugal*, organ. Darlinda Moreira, José Manuel Matos (Lisboa: Sociedade Portuguesa de Ciências da Educação, 2005), 5–20.

'Genikeyontas tin ennoia tou pollaplasiasmou—Epistemologikes synepeies tis schesis metaxy posotitas kai arithmou", M. Korkoulos, G. Troulis, C. Tzanakis (eds.), *Proceedings 4th International Colloquium on the Didactics of Mathematics* (Rethymnon: Department of Education, University of Crete, 2005), tomos I, 117–130. [in Greek]

English version "Generalizing the concept of multiplication—Epistemological implications of the relation between quantity and number" in: ibid., tomos II, 119–130.

"100 Jahre Meraner Reform. Ausstrahlung und Realität einer Reform des Mathematikunterrichts", *forum schule heute. Pädagogische Zeitschrift für die Schule in Südtirol*, 2005, *19/5*: 34–35.

"Les processus d'algébrisation comme déterminants de l'histoire des mathématiques" In: *Actes du Huitième Colloque Maghtrebin sur l'Histoire des Mathématiques Arabes*, Tunis 18–20 Décembre 2004, Mahdi Abdeljaouad (éd.) (Tunis: Association Tunisienne des Sciences Mathématiques, 2005), 265–274.

"A História da Profissão de Professor de matemática", *Anais do 1° seminário paulista de história e educação matemática. Possibilidades de diálogos. Out. 2005*, eds. A.C. Brolezzi, O.J. Abdounur (São Paulo: IME-USP, 2005), 23–32.

"Méthodes d'analyse des manuels historiques des mathématiques", *Actes du 7ème Colloque Maghrébin sur L'Histoire des Mathématiques Arabes.* Marrakech Mai-Juin 2002. Vol. 2 (École Normale Superieure: Marrakech, 2005), 191–207.

"Un programma di matematica all'inizio dell'Ottocento", *Archimede*, 2006, *58/2*: 78–79.

"Introduction. History of Teaching and Learning Mathematics", *Paedagogica Historica*, Special Issue: History of Teaching and Learning Mathematics, ed. by Gert Schubring, 2006, *XLII*: IV&V, 511–514.

"Synthesis. Researching into the History of Teaching and Learning Mathematics: The State of the Art", *Paedagogica Historica*, Special Issue: History of Teaching and Learning Mathematics, ed. by Gert Schubring, 2006, *XLII*: IV&V, 665–678.

"Stabilität und Wandel: Vergleich von Entwicklungs-Mustern für eine Geschichte mathematischer Lehrbücher seit der Antike", *Von der Tontafel zum Internet. Der Einfluß des Mediums auf die Entwicklung der Mathematik. VIII. Österreichisches Symposium zur Geschichte der Mathematik*, Hrsg. Christa Binder (Wien Technische Mathematik, 2006), 86–93.

"Editorial", *International Journal for the History of Mathematics Education*, 2006, *1*:*1*, 1–5.

"Sources for the History of Mathematics Education in Brazil", *International Journal for the History of Mathematics Education*, 1 (2006), *1*: 87–90.

"Il primo movimento internazionale di riforma dell'insegnamento della matematica: Felix Klein e il caso dei Stati Tedeschi", *Associazione Subalpina: Conferenze e Seminari, 2005–2006*, a cura di L. Giacardi et al. (Torino 2006), 63–76.

"Ontogeny and Phylogeny. Categories for cognitive development", *Proceedings of HPM 2004 & ESU 4* (ICME10 Satellite Meeting of the HPM Group & Fourth European Summer University 12–17 July 2004 Uppsala), eds. F. Furinghetti, S. Kaijser, C. Tzanakis (Iraklion: University of Crete, 2006), 329–339.

"Der Aufbruch zum 'funktionalen Denken': Geschichte des Mathematikunterrichts im Kaiserreich", *N.T. M.*, 2007, *15*: 1–17.

"Documents on the mathematical education of Edmund Külp (1800–1862), the mathematics teacher of Georg Cantor", *ZDM The International Journal for Mathematics Education*, 2007, *39*: 107–118.

"Felix Klein: Zur Gründung der IMUK (ICMI) vor hundert Jahren", *Mitteilungen der DMV*, 2007, 15: 2, 122–127.

"Berufsfeldanalysen für die Mathematik", *Mitteilungen der DMV*, 2007, 15: 3, 192–195.

"Um outro caso de obstáculos epistemológicos—o princípio de permanência", *BOLEMA*, 2007, 20, no. 28, 1–20.—"Errata", *BOLEMA*, 2008, *21*, no. 29, 245.

"La diffusion internationale de la géométrie de Legendre: différentes visions des mathématiques", *Raisons—Comparaisons—Éducations. La Revue française d'éducation comparée*, 2007, 2: 31–54.

"Hüseyin Tevfik Pasha—the Inventor of 'Linear Algebra'"/"Hüesyin Tefvik Paşa: 'Lineer Cebir'in Mucidi", *Studies in Ottoman Science (Osmanli Bîlîmî Araştirmalari)*, 2007, VIII: Nr. 2, 43–48/49–54.

Arlete de Jesus Brito & Gert Schubring, "La práctica de la matemática ,normal' en el siglo XVII: el caso de la geografía matemática de Varenius", *Llull*, 2008, *31*: 67, 41–60.

"The Origins and the Early History of ICMI", *International Journal for the History of Mathematics Education*, 2008, *3: 2*, 3–33.

Gert Schubring & Yasuhiro Sekiguchi, "Report on TSG 29: The history of the teaching and the learning of mathematics", *Proceedings of ICME 10 2004* (10th International Congress on Mathematical Education), ed. Mogens Niss (IMFUFA, Roskilde University: Roskilde, 2008), 422–425.

Hélène Gispert & Gert Schubring, "The History of Mathematics Education and its Contexts in 20th Century France and England. Plenary Lecture", *History and Epistemology in Mathematics Education. Proceedings of the 5th European Summer University (ESU 5)*, eds. E. Barbin, N. Stehliková, C. Tzanakis (Plzen: Vydavatelsky servis, 2008), 707–718.

Gert Schubring & Hélène Gispert & Nikos Kastanis & Livia Giacardi, "The Emergence of Mathematics as a Major Teaching Subject in Secondary Schools. Panel Discusssion", *History and Epistemology in Mathematics Education. Proceedings of the 5th European Summer University (ESU 5)*, eds. E. Barbin, N. Stehliková, C. Tzanakis (Plzen: Vydavatelsky servis, 2008), 719–730.

"The debate on a ‚geometric algebra' and methodological implications", *Proceedings HPM Mexico 2008* (Mexico City, 2008), on CD-ROM.

"Discussioni epistemologiche sullo statuto dei numeri negativi e delle loro rappresentazioni nei manuali di matematica tedeschi e francesi tra il 1795 e il 1845". In: *Atti del Convegno di didattici della matematica 2008*, Quaderni Alta Scuola Pedagogica. Centro didattico cantorale (Locarno, 2008), 49–54.

"Gauss e a Tábua dos Logaritmos", *Relime*, 2008, *11: 3*, 383–412.

"Reforma e Contra-Reforma na Matemática—o Papel dos Jesuitas", *Perspectivas na Educação Matemática* (Campo Grande/MS), 2008, *1: 2*, 23–38.

Arlete de Jesus Brito & Gert Schubring, "Varenius e o conhecimento matemático do século XVII", *Ciência & Educação*, 2009, *15*: 1, 139–153.

"Hermann Günther Graßmann (1809–1877)—Ein vielseitiger Innovator", DMV-*Mitteilungen der Deutschen Mathematiker Vereinigung*, 2009, *17: 3*, 181–189.

"How to relate regional history to general patterns of history?—The case of mathematics teaching in Westphalia", *"Dig where you stand". Proceedings of the conference "On-going research in the History of Mathematics Education"*. Bjarnadóttir, K., Furinghetti, F., & Schubring, G. (eds.) (Reykjavik: University of Iceland—School of Education, 2009), 181–195.

"Fundstück: Hermann Graßmann—zwei differente Lebensläufe", *N.T.M.—Zeitschrift für Geschichte der Wissenschaften, Technik und Medizin* [NTM Journal of the History of Science, Technology and Medicine], 2010, *18:2*, 197–230.

"Como relacionar Histórias Regionais a Padrões Gerais de História? O caso do ensino da matemática na Westphalia—How to relate regional history to general patterns of history?—The case of mathematics teaching in Westphalia", *bolema*, 2010, vol. 12, no. 35A, 101–122.

"120 Jahre Deutsche Mathematiker-Vereinigung: Neue Ergebnisse zu ihrer Geschichte", DMV-*Mitteilungen der Deutschen Mathematiker Vereinigung*, 2010, *18: 2*, 103–108.

"Historical comments on the use of technology and devices in ICMEs and ICMI", *ZDM—The International Journal on Mathematics Education*, 2010, *42:1*, 5–9.

"Recherches recentes sur l'histoire de l'enseignement des mathématiques", *Actes du 10ème Colloque Maghrébin sur l'Histoire des Mathématiques Árabes* (Association Tunisienne des Sciences Mathématiques: Tunis, 2010), 270–278.

Marcel Almeida & Gert Schubring, "O Tratado de Álgebra de John Wallis e a defesa do trabalho de Thomas Harriot". In: *Anais do V Colóquio de História e Tecnologia no Ensino da Matemática.* Recife/Brasil: Universidade Federal de Pernambuco, 25 a 30 de julho 2010. CD-ROM.

Hélène Gispert & Gert Schubring, "Societal, structural and conceptual changes in mathematics teaching: reform processes in France and Germany over the twentieth century and the international dynamics", *Science in Context*, 2011, 24:1, 73–106.

"A matematização do calor", *Ciência hoje*, 2011, *47*, no. 279, Março, 68–70.

"Conceptions for Relating the Evolution of Mathematical concepts to Mathematics Learning—Epistemology, History, and Semiotics Interacting", *Educational Studies in Mathematics*, 2011, *77: 1*, 79–104.

Luis Radford, Gert Schubring & Falk Seeger, "Signifying and meaning-making in mathematical thinking, teaching, and learning", *Educational Studies in Mathematics*, 2011, *77: 2–3*, 149–156.

Luiz Carlos Guimarães, João Bosco Pitombeira & Gert Schubring, "Manuel de Campos, um Precursor", *Llull*, 2011, vol. *34*, no. 74, 315–340.

"On how *n* and *i* turned out to become indices in mathematical sequences and formulae", *Educational Studies in Mathematics*, 2011, *77: 2–3*, 331–350.

"Um primeiro livro didático sobre geometria não-euclidiana, em 1874—o papel do ensino na história da matemática". In: *Anais da XIII Conferência Interamericana de Educação Matemática (26 a 30 de junho 2011)*, Refice/Brasil: Universidade Federal de Pernambuco. CD-ROM.

"Searching the roots of PME—the case of ‚experimental pedagogy' in Germany", *Proceedings of the 35th Conference of the International Group for the Psychology of Mathematics Education*, Behiye Ubuz (ed.), vol. 4 (Ankara 2011), 121–128.

Andrea Verdugo Rohrer & Gert Schubring, "Ethnomathematics in the 1930s—The contribution of Ewald Fettweis to the history of Ethnomathematics", *for the Learning of Mathematics*, 2011, *31: 2*, 35–39.

"Antagonisms between German states regarding the status of mathematics teaching during the 19th century: processes of reconciling them", *ZDM—The International Journal on Mathematics Education*, 2012, *44: 4*, 525–535.

Gert Schubring, Fulvia Furinghetti & Man-Keung Siu, "Introduction: the history of mathematics teaching. Indicators for modernization processes in societies", *ZDM—The international journal on Mathematics Education*, 2012, *44: 4*, 457–459.

Marcel Almeida & Gert Schubring, "O Tratado de Álgebra de John Wallis e a defesa do trabalho de Thomas Harriot", *Revista Brasileira de História da Matemática*, 2012, vol. 12, no. 25, 51–63.

"The road not taken—empirical research at the beginning of New Math", *Proceedings of the 36th Conference of the International Group for the Psychology of Mathematics Education*, ed. Tai-Yih Tso (Taipei/Taiwan: PME, 2012, vol. 4, 320.

"'Experimental pedagogy' in Germany—a case study in searching for the roots of PME", *Research in Mathematics Education*, 2012, Vol. 14, No. 3, December, 221–235.

"A biografia de Argand—a historiografia e as fontes", *Anais do 13° Seminário Nacional de História da Ciência e da Tecnologia*, org.s Márcia Regina Barros da Silva, Thomás A. S. Haddad (São Paulo: EACH/USP, 2012). [Documento eletrônico. Modo de acesso ao texto: http://www.sbhc.org.br/], 1–9.

"Exchanges between German and Italian Mathematicians: their first culmination during the 19th century", *Matematica e Risorgimento Italiano*. Studi vol. 18, a cura di Luigi Pepe (Bologna: CLUEB, 2012), 27–35.

"From the few to the many: Historical perspectives on who should learn mathematics". In *"Dig where you stand." 2. Proceedings of the Second International Conference on the History of Mathematics Education*, Lisbon, October 2–6, 2011, eds. Krstín Bjarnadottir, Fulvia Furinghetti, José Matos, Gert Schubring (Caparica: UIED, 2012), 443–462.

"El concepte de límit de Newton a Cauchy: entre la geometria i l'àlgebra i el paper dels signes [Primera Part]", *noubiaix*, 2013, *32* (abril), 14–27; [Segona Part], *noubiaix*, 2013, 33 (octubre), 6–21.

"From 'Armchair Pedagogy' To Experimental Research And To Case Studies". In: *Proceedings of the 37th Conference of the International Group for the Psychology of Mathematics Education*, Lindmeier, A. M. & Heinze, A. (Eds.), vol. 4 (Kiel 2013), 169–176.

Andrea Rohrer & Gert Schubring, "The interdisciplinarity of ethnomathematics: challenges of ethnomathematics to mathematics and its education", *Revista Latinoamericana de Etnomatemática*, 2013, *6*: 3, 78–87.

Bruder, Regina, Bärbel Barzel, Michael Neubrand, Silke Ruwisch, Gert Schubring, Hans-Dieter Sill, and Rudolf Sträßer. On German research into the didactics of mathematics across the life span—National Presentation at PME 37. In: *Proceedings of the 37th conference of the international group for the psychology of mathematics education*, eds. Anke M. Lindmeier and Aiso Heinze, 233–276 (Kiel: IPN, 2013).

Gert Schubring & Tatiana Roque, "O papel da régua e do compasso nos Elementos de Euclides: uma prática interpretada como regra", *História Unisinos*, 2014, *18*: 1, 91–103.

"The road not taken—The Failure of Experimental Pedagogy at the Royaumont Seminar 1959", *Journal für Mathematik-Didaktik*, 2014, *35*: 1, 159–171.

"A correspondência de Karl Weierstraß—resultados de pesquisa em andamento". In: Sergio Nobre et al. (eds.), Anais/Actas do 6º Encontro Luso-Brasileiro de História da Matemática (Natal: Sociedade Brasileira de História da Matemática, 2014), 225–240.

"Conceptions for Relating the evolution of Mathematical concepts to Mathematics Learning—epistemology, History, and semiotics interacting", in: ICMI (ed.), *Materials from ICME 11, Mexico 2008: Regular Lectures*, 129–155. [n.d.] url: http://www.mathunion.org/icmi/publications/icme-proceedings/materials-from-icme-11-mexico/regular-lectures/ [2014]

"The original conclusions of the Royaumont Seminar 1959", edited and commented by Gert Schubring, *International Journal for the History of Mathematics Education*, 2014, *9*: 1, 89–101.

"Classical Papers in Mathematics Education. Beke's Report: Results obtained in the Introduction of Differential and Integral Calculus", edited and commented by Gert Schubring, *International Journal for the History of Mathematics Education*, 2014, *9*: 1, 33–70.

"Problems of transmission and translation—The case of MacLaurin on negative numbers", *Recherches en didactique des mathématiques*, 2014, *34: 2–3*, 285–296.

"Paulus Gerdes †—Instead of an Interview: an Obituary", *International Journal for the History of Mathematics Education*, 2014, 9:2, 71–76.

"The emergence of the profession of mathematics teachers—an international analysis of characteristic patterns". In: *"Dig where you stand." 3. Proceedings of the Third International Conference on the History of Mathematics Education*, Uppsala, September 25–28, 2013, eds. Kristín Bjarnadottir, Fulvia Furinghetti, Johan Prytz, Gert Schubring (Uppsala: Uppsala University—Department of Education, 2015), 389–403.

Marcello Santos Amadeo & Gert Schubring, "A *École Polytechnique* de Paris: mitos, fontes e fatos", *BOLEMA* (Boletim de Educação Matemática), 2015, vol. 29, no. 52, 435–451.

"Tentativas de estabelecer o conceito de número como fundamento para cursos de análise", *Caminhos da Educação Matemática em Revista*, vol. 3, pp. 111–125, 2015.

"New approaches and results in the history of teaching and learning mathematics". In: *History and Epistemology in Mathematics Education. Proceedings of the Seventh European Summer University,* Barbin, Evelyne, Uffe Th. Jankvist & Tinne Hoff Kjeldsen (eds.) (Copenhagen: Danish Scool of Edcation, 2015), 659–678.

"From the Few to the Many: On the Emergence of *Mathematics for All*", *Recherches en didactique des mathématiques*, 2015, *35: 2*, 222–260.

"Die Entwicklung der Mathematikdidaktik in Deutschland", *Mathematische Semesterberichte*, 2016, *63:1*, 3–18.

"Comments on a Paper on Alleged Misconceptions Regarding the History of Analysis: Who Has Misconceptions?", *Foundations of Science*, 2016, vol. 21:3, 527–532.

Everaldo Paulo da Silva & Gert Schubring, "Cálculo em Matemática: Um Assunto para o Ensino em Geral ou Específico para o Ensino Técnico", *História da Educaçao (online),* Porto Alegre, 2016, vol. 20, no. 49, 65–80.

"L'evoluzione della professione di insegnante di matemática in varie culture", *Associazione Subalpina Mathesis: Conferenze e Seminari, 2015–2016*, a cura di F. Ferrara et al. (Torino 2016), 219–234.

"Novas pesquisas sobre a história da convergência uniforme e dos seus aspetos da didática", *Cuadernos de Investigación y Formación en Educación Matemática*, 2016, ano 11, número 11, 151–158. [Costa Rica].

"Searches for the origins of the epistemological concept of model in mathematics", *Archive for History of Exact Sciences*, 2017, *71: 3*, 245–278.

"Mathematics teaching in the process of decolonisation". In: *"DIG WHERE YOU STAND" 4. Proceedings of the Fourth International Conference on the History of Mathematics Education, September 23–26, 2015, at University of Turin, Italy,* Bjarnadóttir, Kristín & Fulvia Furinghetti, Marta Menghini, Johan Prytz, Gert Schubring (eds.) (Roma: Edizioni Nuova Cultura, 2017), 349–368.

Hans-Georg Weigand, William McCallum, Marta Menghini, Michael Neubrand, Gert Schubring & Renate Tobies, "What Is and What Might Be the Legacy of Felix Klein?", in: Gabriele Kaiser (ed.), *Proceedings of the 13th International Congress on Mathematical Education*, ICME-13 Monographs (New York: SpringerOpen, 2017), 321–334.

"Au delà de l'approche synthétique grecque—L'essai de généraliser les théorèmes d'Euclide dans Qutb al-Din al-Shīrāzī's Appendice au Livre I". In: *Actes du XIIè Colloque Maghrébin sur l'Histoire des Mathématiques Arabes, Marrakech 26 au 28 mai 2016,* Ezzaim Laabid (éd.) (École Normale Supérieure: Marrakech, 2018), 266–277.

"A biografia de Argand: as lendas e as fontes", *revista do professor da matemática*, 2018, ano *36*, no. 95, 2–8.

Ricardo Motta Telo & Gert Schubring, "A Comissão Nacional do Livro Didático e a Avaliação dos Livros de Matemática entre 1938 e 1969", *Revista Brasileira de História da Educação*, 2018, *18*, 1–27.

Márcia Maria Fusaro Pinto & Gert Schubring, "The conception and development of textbooks for distance learning courses: a case study of a teacher education course", *ZDM—Mathematics Education*, 2018, *50: 5*, 893–906.

Gert Schubring & Lianghuo Fan, "Recent advances in mathematics textbook research and development: an overview", *ZDM—Mathematics Education*, 2018, *50: 5*, 765–771.

Alejandre J. F. de Sousa & Gert Schubring, ""*Exame de Bombeiros*": Como Fazer os Bombeiros Brasileiros no Século XVIII Servir Melhor a Coroa Portuguesa Através da Matemática". In: *Actas/Anais do 7.° Encontro Luso-Brasileiro de História da Matemática* (Óbido, Portugal, 15 a 19 de outubro de 2014), volume II, António Costa Canas, Joao Caramalho Domingues & Luis Saraiva (eds.) (Lisboa: Sociedade Portuguesa de Matemática, 2018), 405–422.

Vinicius Mendes Pereira & Gert Schubring, "A Matemática Desconhecida da Escola de Minas de Ouro Preto", In: *Actas/Anais do 7.° Encontro Luso-Brasileiro de História da Matemática,* António Costa Canas, Joao Caramalho Domingues & Luis Saraiva (eds.) (Óbido, Portugal, 15 a 19 de outubro de 2014), volume II (Lisboa: Sociedade Portuguesa de Matemática, 2018), 513–531.

"Conflitos entre matemáticos profissionais e amadores: tentando resolver a quadratura do círculo e o ultimo teorema de Fermat", *Cuadernos de Investigación y Formación en Educación Matemática*, 2018, ano 13, número 17, 105–119. [Costa Rica]

"O Que é e ao Que Serve a Hermenêutica?" *Jornal Internacional de Estudos em Educaçao Matemática*, 2018, vol. 11, n. 2, 194–200.

"Sur le concept et les origines de l'algèbre cossiste." In : *Actes du XIIIème Colloque Maghrébin sur l'histoire des mathématiques Arabes, Tunis 2018.* Mahdi Abdeljaouad & Hmida Hedfi (éd.s) Tunis: Graphika Imprimérie, 2018, 321–336.

"Letters to Weierstraß by Italian Mathematicians", *Bollettino di Storia delle Scienze Matematiche*, 2019, *39*: 287–333.

Oliveira, Carlos & Gert Schubring, "Enseñanza de Matemáticas en Colombia y Venezuela en el Siglo XIX: Un Estudio de Caso del Concepto de Transmisión de Idea". In: *V Congresso Iberoamericano de Historia de la Educación Matemática. Memorias,* Gert Schubring, Jhon Helver Bello Chávez & Harold Vacca González (eds.) (Universidad Distrital Francisco José de Caldas, Bogotá, 2019), 1–19.

"The development of forms to study mathematics", in ***Dig where you stand. 6.*** *Proceedings of the Sixth International Conference on the History of Mathematics Education, September 16–20, 2019, at the CIRM (Luminy) France*, eds. Évelyne Barbin, Kristín Bjarnadóttir, Fulvia Furinghetti, Alexander Karp, Guillaume Moussard, Johan Prytz, Gert Schubring & Harm Jan Smid (Münster: WTM Verlag, 2020), 289–302.

Azelman de Fabia Andreu, Livia, Ana Mary Fonseca Barreto de Almeida, Magno Luis Ferreira, Carlos Antonio Assis de Oliveira & Gert Schubring, "A história da matemática nos livros-texto de Cajori, Eves, Boyer e Struik", *Revista Brasileira de História da Ciência*, 2020, vol. 13, no.2, 280–297.

Michel Santos Salazar & Gert Schubring, "A Justificação das Operações algébricas na Inglaterra do Século XVII a XIX—O Caso das Quantidades Negativas", *Revista Brasileira de Hisória da Matemática*, 2020, Vol. 20, n° 40, pp. 46–79.s

Michel Pacheco & Gert Schubring, "As contribuições de Francisco de Borja Garção Stockler (1759–1829) aos fundamentos da Análise", *Boletim GEPEM*, 2020, *77*: xx–xx [1–21].

"Interactions between epistemologies of mathematics and educational systems—the emergence of mathematical communities according to cultures and states in 19th century Europe". *Matemáticas, Educación y Sociedad,* 2021, *4*(1), 1–16.

"On processes of coloniality and decoloniality of knowledge: notions for analysing the international history of mathematics teaching". *ZDM Mathematics Education*, 2021-7: 1455–1469. https://doi.org/10.1007/s11858-021-01261-2

"Álgebra Cossista Alemã—Significado e Impactos". In: *Anais/Actas do 8° Encontro Luso-Brasileiro de História da Matemática* (agosto de 2018), Marcos Lübeck & Sergio Nobre (org.s) Foz do Iguaçu: Universidade Estadual do Oeste do Paraná, 2021, 101–114.

Almeida, Ana Mary Fonseca Barreto de & Gert Schubring, "As Transformações da Geometria Descritiva: Do Coupe de Las Pierres à Disciplina Principal no Ensino Superior e Técnico". In: *XIV Seminário Nacional de História da*

Matemática. Anais... Uberaba (MG) Universidade Federal do Triângulo Mineiro (UFTM), 2021.
https//www.even3.com.br/anais/xivsnhm/324238-AS-TRANSFORMACOES-DA-GEOMETRIA-DESCRITIVA-DO-COUPE-DE-LAS-PIERRES-A-DISCIPLINA-PRINCIPAL-NO-ENSINO-SUPERIOR-E-TE.

Oliveira, Carlos Antonio Assis de & Gert Schubring, "A Transmissão da Geometria de Legendre: O Casos da Colômbia e da Venezuela". In: XIV Seminário Nacional de História da Matemática. Anais...Uberaba (MG) Universidade Federal do Triângulo Mineiro (UFTM), 2021.
https//www.even3.com.br/anais/xivsnhm/326416-A-TRANSMISSAO-DA-GEOMETRIA-DE-LEGENDRE-O-CASOS-DA-COLOMBIA-E-DA-VENEZUELA.

Oliveira, Carlos & Gert Schubring, "As transmissões duplas dos livros-texto de Lacroix e Legendre no século XIX: o caso da Colômbia e da Venezuela." *Matemáticas, Educación y Sociedad*, 2021, 4(2), 1–20

Salazar, Michel & Gert Schubring, "A Justificação das Operações Algébricas na Inglaterra do Século XVII a XIX—O Caso das Quantidades Negativas". *Revista Brasileira de História da Matemática*, 2021, vol. 20, no. 40, 46–79.

Renata Cardoso & Gert Schubring, "Os matemáticos Martim Machado e Garção Stockler: Uma breve biografia e a recepção ao plano de Condorcet para ensino de matemática", *Boletim GEPEM*, 2021, 79: 22–36.

Linder, Vinicius & Gert Schubring, "Relações entre matemática e filosofia na emergência da matemática pura: a matemática como fundamento da pensabilidade". *Educação, Matemática, Pesquisa*, 2022, *24*(2): 108–146.

Débora Ferreira & Gert Schubring, "Complex numbers" and the problem of multiplication between quantities", *Historia Mathematica*, 2022, 59: 119–145.

Reviews

"Probleme der Geschichtsschreibung des Mathematikunterrichts". Rezension des Buches: "Die Reform des gymnasialen Mathematikunterrichts zwischen 1890 und 1914" von Heide Inhetveen, *Zentralblatt für Didaktik der Mathematik*, 1978, *10*, 27–30.

Terry Shinn, *L'École Polytechnique 1794–1914. Savoir Scientifique et Pouvoir Social.* Paris: Presses de la fondation nationale des sciences politiques, 1980, in: *Historische Zeitschrift*, 1982, 234:3, 709–710.

Klaus Volkert, *Die Krise der Anschauung.* Göttingen: Vandenhoeck und Ruprecht, 1986, in: *Historia Mathematica*, 1990, *17*: 178–180.

Ivor Grattan-Guinness, *Convolutions in French Mathematics, 1800–1840. From the Calculus and Mechanics to Mathematical Analysis and Mathematical Physics* (Basel: Birkhäuser, 1990), *NTM*, Neue Serie, 1993, 2:121–123.

Siegbert Schmidt, *Rechenunterricht und Rechendidaktik an den rheinischen Lehrerseminaren im 19. Jahrhundert* (Köln: Böhlau, 1991), *Zentralblatt für Didaktik der Mathematik*, 1993, *25*: 146–150.

Yves Cohen, Klaus Manfrass (eds.), *Frankreich und Deutschland: Forschung, Technologie und industrielle Entwicklung im 19. und 20. Jahrhundert* (München: C.H. Beck 1990), British Journal for the History of Science, 1993, 26: 119–121.
Michael Price, *Mathematics for the Multitude? A History of the Mathematical Association.* Leicester: Mathematical Association, 1994. *NTM*, N.S., 1996, *4*: 189–190.
Condorcet: Arithmétique politique. Textes rares ou inédits (1767–1789). Édition critique commentée par Bernard Bru et Pierre Crépel. Paris: Institut National d'Études Démographiques/Presses Universitaires de France, in: *LLULL*, 1997, *20*, no. 38, 377–378. [in Englisch]
Condorcet: Arithmétique politique. Textes rares ou inédits (1767–1789). Édition critique commentée par Bernard Bru et Pierre Crépel. Paris: Institut National d'Études Démographiques/Presses Universitaires de France, in: *NTM*, N. S., 1998, *6*: 176–177. [in deutsch]
Christoph Schöner, *Mathematik und Astronomie an der Universität Ingolstadt im 15. und 16. Jahrhundert.* Berlin: Duncker und Humblot, 1994, in: *LLULL*, 1998, *21*, no. 42, 850–851.
Hans Niels Jahnke, Norbert Knoche & Michael Otte (eds.), *History of Mathematics and Education: Ideas and Experiences* (Göttingen: Vandenhoeck und Ruprecht, 1996), in: *Paedagogica Historica*, 1998, *34*:598–600.
Michael Toepell, *Mathematiker und Mathematik an der Universität München. 500 Jahre Lehre und Forschung* (München: Inst. für Geschichte der Naturwissenschaften, 1996), in: *Paedagogica Historica*, 1998, *34*:600–604.
Mario Otero, *Joseph-Diez Gergonne (1771–1859): Histoire [Logique] et philosophie des sciences.* Sciences et Techniques en Perspective, No. 37 (Nantes: Centre François Viète, 1997), in: *ISIS*, 1999, *90*, 603–604.
The History of Mathematics from the Antiquity to the Present: A Selective Annotated Bibliography. Edited by Joseph W. Dauben; Revised Edition in CD-ROM edited by Albert C. Lewis. American Mathematical Society, 2000, in: *História & Educação Matemática* (Rio Claro, SP), 2001, *1*: 65–67.
Hermann Graßmann, *Extension Theory*, translated by LLoyd C. Kannenberg (Providence, R.I.: American Mathematical Society, 2000), in: *ISIS*, 2003, 94, *2*: 386–387.
Olaf Breidbach, Paul Ziche (eds.), *Naturwissenschaften um 1800: Wissenschaftskultur in Jena-Weimar* (Weimar: Böhlau, 2001), in: *ISIS*, 2004, *95*: 305–306
Antonella Romano, *La contre-réforme mathématique: constitution et diffusion d'une culture mathématique jésuite à la Renaissance; (1540–1640)* (Roma: École française de Rome, 1999), in: *Llull*, 2003, vol. 26, no. 57: 1069–1076.
Helena Pycior, *Symbols, impossible numbers, and geometric entanglements: British algebra through the commentaries on Newton's Universal Arithmetick* (Cambridge: Cambridge Univ. Press, 1997), in: *Historia Mathematica*, 2005, 32: 363–366.
Reinhard Siegmund-Schultze, *Rockefeller and the internationalization of mathematics between the two world wars: documents and studies for the social history of*

mathematics in the 20th century (Basel: Birkhäuser, 2001), in: *Annals of Science*, 2005, *62*: 549–550.
Leung, Frederick K.S. et al. (eds.), *Mathematics Education in Different Cultural Traditions. A Comparative Study of East Asia and the West* (New York: Springer, 2006), in: *International Journal for the History of Mathematics Education*, 2007, *2: 1*, 62–65.
Livia Giacardi (ed.), *Da Casati a Gentile: momenti di storia dell'insegnamento secondario della matematica in Italia,* in: *International Journal for the History of Mathematics Education*, 2007, *2: 1*, 66–67
Hans-Joachim Petsche, *Graßmann (Vita Mathematica,)* (Basel: Birkhäuser, 2006), in: *The Mathematical Intelligencer*, 2008, *30*: 4, 62–64.
Arlete de Jesus Brito, *A geometria de Euclides a Lobatschewski: um estudo histórico-pedagógico*, in: *Zetetiké*, 2008, vol. *16*, no. *30*, 255–258.
Peggy Aldrich Kidwell, Amy Ackerberg-Hastings, David Lindsay Roberts, *Tools of American mathematics teaching, 1800–2000*, in: *International Journal for the History of Mathematics Education*, 2009, *4: 1*, 107–109
Karl-Heinz Böttcher & Bertram Maurer, *Stuttgarter Mathematiker. Geschichte der Mathematik an der Universität Stuttgart von 1829 bis 1945 in Biographien*, in: Historia Mathematica, 2009, 36: 3, 285–287
Hans-Joachim Petsche, *Hermann Graßmann. Biography.* Translated by Mark Minnes. Birkhäuser, 2009; Hans-Joachim Petsche, Lloyd Kannenberg, Gottfried Kessler & Jolanta Liskowacka (eds.), *Hermann Graßmann. Roots and Traces.* Birkhäuser 2009, in: Mathematische Semesterberichte, 2011, 58: 1, 114–117.
Laurent Kuzniak & Rossana Tazzioli, *Mathematicians at War. Volterra and his French colleagues in World War I.* Dordrecht, Heidelberg, London, New York (Springer), 2009, in: *Historia Mathematica*, 39 (2012), pp. 347–350.
Luigi Maierú, *Scienza Geometria, Geometrie [Science, Geometry, Geometries. A historical-educational itinerary].* Soveria Mannelli (Rubbettino Editore), 2008, in: *Historia Mathematica,* 40 (2013), pp. 206–207.
Kerrin Klinger: *Zwischen Gelehrtenwissen und handwerklicher Praxis. Zum mathematischen Unterricht in Weimar um 1800* [Between scholarly knowledge and practice of craftsmanship. On mathematics teaching in Weimar around 1800], in: *International Journal for the History of Mathematics Education*, 2014, *9: 2*, 67–70.
Pieter Josephson, Thomas Karlsohn & Johan Östling. *The Humboldtian Tradition. Origins and Legacies*, in: *ISIS*, 2016, 107: 2, 414–416.
Mechthild Koreuber: "Emmy Noether,
die Noether-Schule und die moderne Algebra" Springer, 2015, in: *Jahresbericht der Deutschen Mathematiker-Vereinigung*, 2019, *121: 1*, 41–48.
Wolfgang König & Jürgen Sprekels (eds.), Karl Weierstraß (1815–1897). Aspekte seines Lebens und Wirkens—Aspects of his Life and Work. Wiesbaden: Springer Fachmedien. 2016, in: *Historia Mathematica*, 2019, *48*. 96–100.
Uta Merzbach: Dirichlet. A Biography. *British Journal for the History of Mathematics*, 2021, 36: 1, 55–58

Egbert Brieskorn & Walter Purkert: Biographie. Felix Hausdorff, Gesammelte Werke. Band IB. Berlin: Springer, 2018. *Mathematical Intelligencer*, 2021, *43*: 94–98.

Index

A. Verdugo Rohrer and J. Zender (eds.), *History of Mathematics and Its Contexts*, Trends in the History of Science, https://doi.org/10.1007/978-3-031-60828-5